Programming in Siemens Step 7 (TIA Portal), a Practical and Understandable Approach

Jon Stenerson

David Deeg

ISBN-10: 151503657X
ISBN-13: 978-1515036579

DEDICATION

Dedicated to the great teachers that have had a tremendous impact on my life.

Jon Stenerson

Dedicated to Becky Deeg.
With her love and support I have been blessed with much happiness and success.

David Deeg

CONTENTS

ACKNOWLEDGMENTS

A tremendous thank you to the Siemens Cooperates with Education (SCE) Program for their help in developing this book.

Chapter 1

Introduction to PLCs

Objectives

Upon completion of this chapter, the reader will be able to:

Describe the components of a typical PLC system.

Define terminology such as: chassis, rack, module, backplane, CPU, memory, discrete, analog, and so on.

Describe the IEC 61131standard.

Overview

Programmable Logic Controllers (PLCs) are the brain and backbone of industrial automation. PLCs were designed to be easy for electricians and maintenance technicians to work with. PLCs have been widely used in industry since their introduction in the early 1980s. Their capabilities have expanded tremendously. They have also become much easier to program and integrate.

History of PLC's

In the old days (only a few decades ago), automation was done with hard-wired relays. Relays, being mechanical, also were prone to failure, which would shut down the line. Reliability was low and troubleshooting was cumbersome and time consuming. Control cabinets could contain hundreds of electromechanical relays, timers and counters. To change a process production was halted, the manufacturing line was shut down and wiring and relays were modified. Change was avoided whenever possible.

General Motors Corporation (GM) was one of the first to see the need for a programmable computer that could replace hardwired relay logic and complex control cabinets. GM thought that the program in a programmable control device could be changed to change the way a system operated. GM also needed a programmable device that electricians could program, wire devices to, troubleshoot and use without learning a computer language.

The first PLCs were simple devices that were designed to replace hardwired relays. The language that was developed for PLCs was called ladder logic. It was a very graphical language and looked a lot like the electrical diagrams that electricians were already very familiar with. PLC input and output modules were designed so that electricians could easily connect inputs and outputs. Imagine an automated automobile line before PLCs. There were many huge electrical control cabinets each filled miles of wire and relays, motor controls, etc. The logic to run the system was dependent on how all of the hardware relays and devices were wired.

If the assembly process had to be changed the whole system would be shut down and wiring would have to be changed. This was tremendously costly and made changes expensive and undesirable. If something went wrong the system would have to be turned off and technicians would have to troubleshoot all of the hardwired logic and devices. This is very costly and time consuming. PLCs made it possible to wire devices one time and just change the program to change how the system operated. It also eliminated the need for hardware such as logic relays, hardware timing and counting relays, etc. The PLC program is used to create the logic and timing, counting, etc. Today, one PLC can take the place of thousands of hard-wired relays. Remember also that with the PLC, no wiring has to be changed to change the operation of the system. The programmer simply changes the program.

PLCs enable companies to automate processes rapidly and at low cost. Automating a process improves productivity, quality, and dramatically reduces scrap and rework. PLCs can enable a manufacturing process to be flexible so processes can produce products to the individual specifications that each customer has ordered. PLCs also enable very rapid, even instantaneous product changeover though program logic. Dramatic advances have been made in PLC capabilities. Recently alternative languages have become more widely accepted. In addition to ladder logic, languages such as instruction list, sequential function chart, and function block have been increasing in use.

There are many similarities between a personal computer and a PLC. Both have inputs. A computer has a keyboard and a mouse for inputs. The computer can also access the hard drive, CD/DVD, and internet to get input. A PLC can get inputs from sensors and other devices such as robots, other controllers, etc. These inputs come into the PLC via input modules. Figure 1-1 shows a block diagram of a simple PLC system.

Computers and PLC have a central processing unit (CPU) and memory. The CPU runs the user program, evaluates the inputs and generates outputs. A computer and a PLC both have outputs. A computer can output to a printer, or send email out over the internet, or store a file on a drive. A PLC has outputs such as motors, drives, lights, etc. PLCs output their signals through output modules. A programming device is used to enter the program into a CPU and also for troubleshooting and monitoring the program. A computer is normally used for programming. A touch screen can provide an operator with the ability to monitor the process and to enter and retrieve data from the PLC.

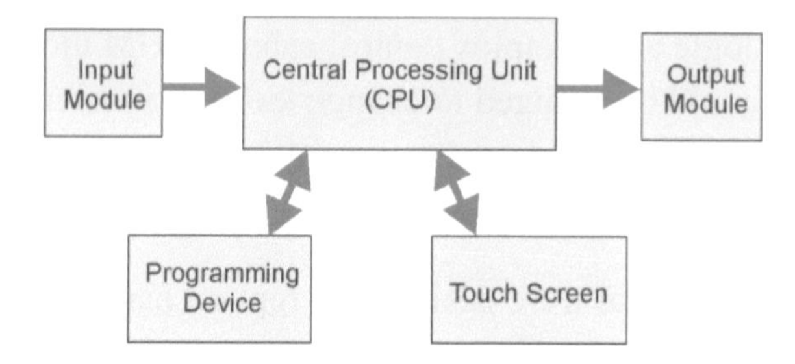

Figure 1-1. Block diagram of a PLC system.

A central processor unit (CPU) is a microprocessor-based system. The CPU has system memory and is the PLC's brain. The CPU monitors inputs, outputs, and program variables and makes decisions based on the program in its memory.

Figure 1-2 shows a functional illustration of a PLC system. Note the various inputs shown such as: switches, sensors, and touch panel. Note also that the outputs are used to control an application's devices such as motor starters, indicators, etc.

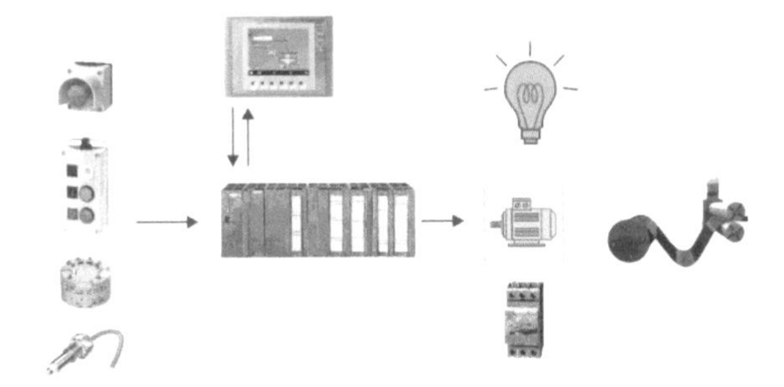

Figure 1-2. Functional diagram of a PLC system. Courtesy Siemens.

Figure 1-3 shows a simplified overview of a PLC. Switches S1 and S2 provide true or false input signals to the input module of the PLC. The CPU evaluates the program logic based on the states of the I/O and then changes the states of outputs. The output module turns on outputs based on what the CPU has written to output memory.

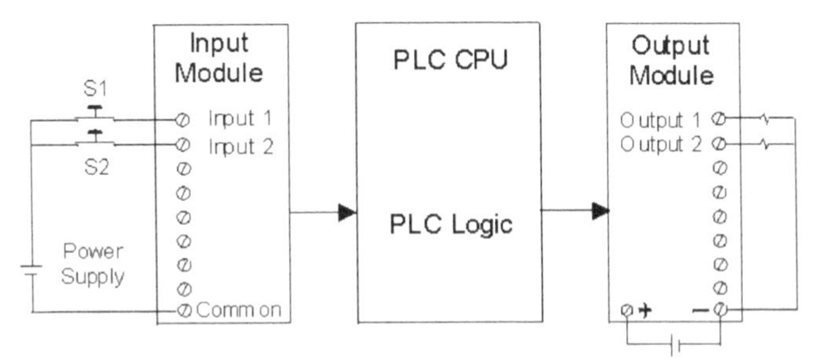

Figure 1-3. Simplified view of a PLC.

Modular PLCs

Most PLCs are modular. They allow the user to purchase and install modules to accomplish a task. This enables a user to choose the input and output modules needed for the particular application. A modular PLC often begins with a rack.

Figure 1-4 shows a Siemens S7-400 PLC. The main components have been identified. The power supply is located on the left side of the chassis. The rack is chosen to hold the number of required modules and the CPU.

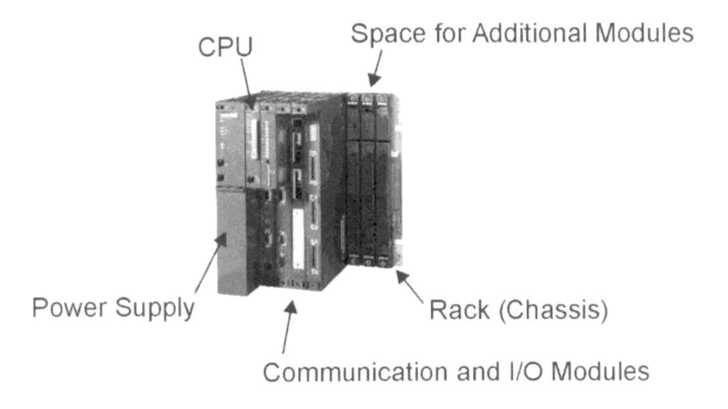

Figure 1-4. Components of a typical PLC. Courtesy Siemens.

PLCs are available with various memory sizes. Some PLC CPU modules allow the user to add additional non-volatile memory to the CPU module.

Central Processing Unit

The central processing unit (CPU) is the brain of the PLC. The CPU takes input information, examines the logic in the program in the CPU and then controls the states of outputs.

The CPU is really just a microcomputer. It has a microprocessor just like a personal computer. The main difference between the PLC CPU and a personal computer is the program. Until recently the language that PLCs were programmed in was a graphical language called ladder logic. Ladder logic was designed to look like a normal industrial electrical print and be easy for electricians and technicians to understand and troubleshoot. There are additional languages that can be used in many PLCs today.

S7-1200 CPU Modules

One of the newer models of PLC available from Siemens is the S7-1200 (see Figure 1-5). They are available in three voltage combinations (see Figure 1-6). The inputs/outputs refer to the I/O that is integrated into the CPU. Modules can be added to the CPUs to utilize AC/DC/ or relay. The integrated relay outputs can handle up to 30 VDC or up to 250 VAC.

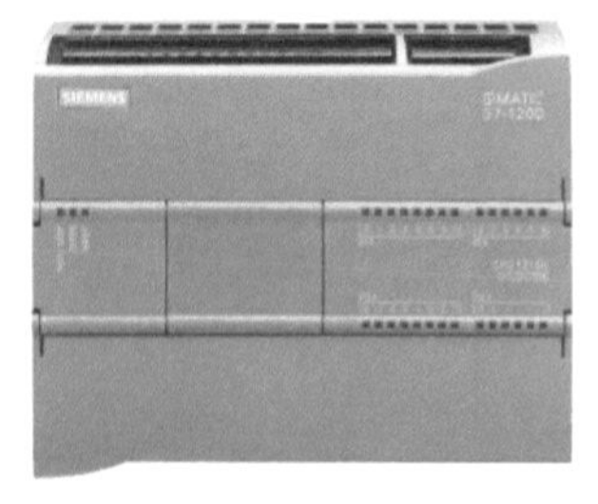

Figure 1-5. A Siemens 1215 CPU.

The available 1200 CPUs vary in supply voltage, the number of integrated inputs and outputs, memory size, and the expansion capabilities with signal modules (SMs).

CPU Power	Input Type	Output Type
DC	DC	DC
AC	DC	Relay
DC	DC	Relay

Figure 1-6. Types of 1200 CPU voltages that are available.

In addition to digital inputs and outputs, each CPU has two analog input channels for 0 – 10 VDC or 4-20 mA. The inputs have 10-bit resolution. Analog resolution will be covered in Chapter 5.

CPU Scan Cycle

The PLC program is repeatedly executed. This repetitive execution is called a scan. A PLC scan starts with the CPU reading the status of inputs (see Figure 1-7). Note that these states are essentially written down and frozen at this stage before the next step. Next, the program logic is evaluated based on the I/O states from the previous steps. Next the CPU performs housekeeping tasks such as internal diagnostics and communication tasks. Finally, the CPU updates the status of outputs based on the result of the logic evaluation. The scan then begins again with the PLC reading the states of inputs. Scanning continues as long as the CPU remains in the run mode. Scan time depends on the CPU model, the size and complexity of the program, the number of I/O, and other factors. Scan time is generally in milliseconds or less.

Figure 1-7. A scan cycle. Courtesy Siemens.

Steps in the Scan Cycle

Every scan cycle includes writing the outputs, reading the inputs, executing the user program, and performing system maintenance or background processing. Under default conditions, internal memory, called the process image, is used to update all digital and analog I/O points synchronously. The process image contains a snapshot of the physical inputs. Individual modules can be excluded from automatic updating of the process image. This can be setup in module configuration. The signal states of modules set up in this manner are then accessed by the program using the direct access I/O area (P).

Figure 1-8 shows a block diagram of the startup and runtime mode scanning.

Startup Mode

A - The input memory area (I) of the process image is cleared.

B - The outputs are initialized with the last value.

C - Any start-up logic (contained in special code blocks) is executed.

D - The state of the physical inputs is copied to I memory.

E - Any interrupt events are queued for processing during RUN mode.

F - The writing of the output area (Q memory) of the process image to the physical outputs is enabled.

Run Mode

1 – Output (Q) memory is written to the physical outputs.

2 - The states of the physical inputs are copied to I memory.

3 - The logic of the user program is executed.

4 - The self-test diagnostics are performed.

5 - Interrupts and communications are processed during any part of the scan cycle.

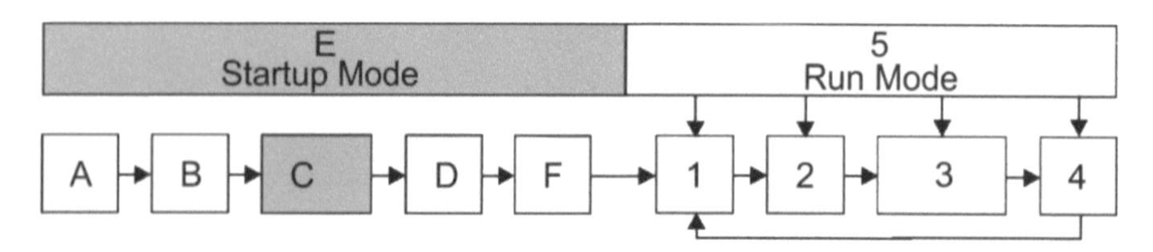

Figure 1-8. Scanning during startup and run mode. Courtesy Siemens.

Cycle Time

Cycle processing time consists of:

- The total time of the main program and all user organizational blocks assigned to the main program.
- Processing times for the higher priority classes that interrupt the main program.
- Time required to update the process image.
- Time used for the operating system communication processes such as communicating with a programming computer for program status.

The cycle time can be accessed when online using the Online and Diagnostics. The cycle time display will show the shortest, current, and longest cycle time in ms.

CPU Operating Modes

The CPU has three operating modes: STOP mode, STARTUP mode, and RUN mode.

There are status LEDs on the front of the CPU which indicate operating mode (see Figure 1-9).

The color of the RUN/STOP LED indicates the operating mode of the CPU:

- Yellow indicates STOP mode.
- Green indicates RUN mode.
- Flashing indicates STARTUP mode.

The ERROR LED flashes red if an error has been detected.

The maintenance (MAINT) LED is a continuous yellow light to indicate that a previously configured maintenance request is now present. Note that all LEDs will flash if the CPU firmware is faulty.

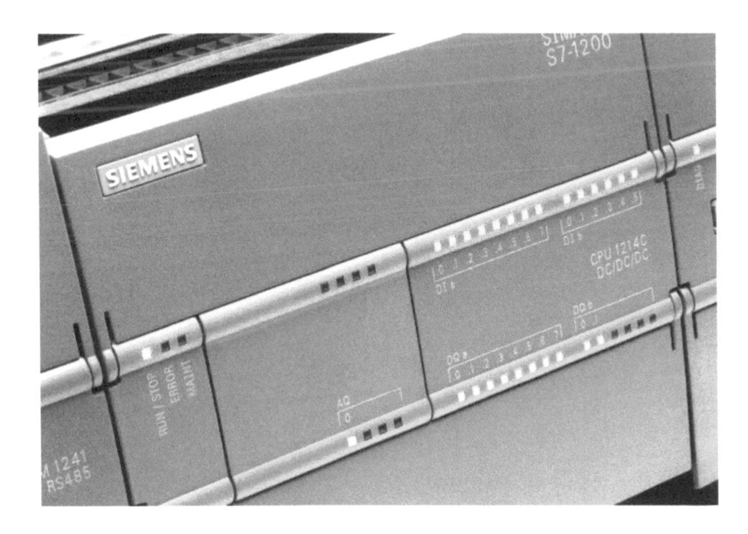

Figure 1-9. CPU status LEDs.

- In stop mode, the CPU is not executing the program, and a project can be downloaded.
- In startup mode, the CPU executes any startup logic (if present). Interrupt events are not processed in startup mode.
- In run mode, the scan cycle is repeatedly executed. Interrupt events can occur and be processed at any point within the program cycle phase.

The CPU does not have a physical switch for changing the operating modes (STOP or RUN). When the CPU is configured in the device configuration, you can configure the start-up behavior in the properties of the CPU.

CPU Memory

The user memory of a PLC holds the user program and data. The amount of memory depends on the CPU type and model. The user program is developed using a programming device, such as a computer, and then loaded into the program memory of the CPU. Figure 1-10 shows a graphic of the CPU memory areas.

The peripheral inputs are connected to the input modules. The CPU automatically copies their states into the input process image prior to each scan cycle of the program. The input process image is located in CPU system memory. The input process image input states are used during logic execution. This means that the actual input modules are not directly scanned during logic execution (evaluation); the input process image states are scanned.

Peripheral outputs contain the output states and are connected to the output devices. The CPU automatically transfers the signal states from the output process image to the peripheral outputs prior to each scan cycle of the program. The output modules are not directly written to normally. The CPU writes the states to the output process image.

Bit memory is internal memory used to store binary signal states. They can be used like outputs, but they are not connected to the outside of the PLC. Bit memory can be read from or written to. Some bit memories can be set to be retentive. Then the bits will retain their states even when de-energized. Retentive memory always starts at memory byte 0 and ends at the upper limit that you set. Retentivity can be set when you configure the CPU parameters.

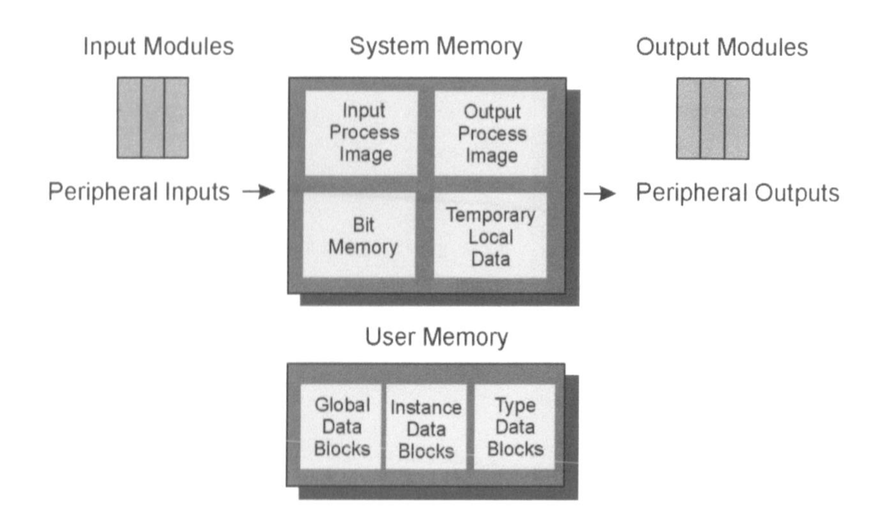

Figure 1-10. CPU Memory.

Work memory is provided with a floating boundary between the user program and user data. Work memory is kept in RAM memory (see Figure 1-11). Random Access Memory (RAM) is used as a temporary memory storage area. RAM enables data to be written to or read from any address in the memory chip. RAM memory is volatile. This means that data stored in RAM will be lost if power is lost. Battery backup is used to avoid losing data. A capacitor is used to maintain a charge for a period of time.

The capacitor protects data stored in RAM in the event of a power loss. The capacitor provides protection for 40 – 100 hours. It also protects RAM memory while that battery is being changed.

Read Only Memory (ROM) is nonvolatile memory. It is not lost if there is no power. The original data stored in ROM can be read, but not changed. ROM is normally used to store the PLCs operating system.

EEPROM (Electrically Erasable Programmable Read-Only Memory) is non-volatile memory used to store programs or data that must be saved when power is removed. EEPROM is ROM that can be erased and reprogrammed repeatedly by the user.

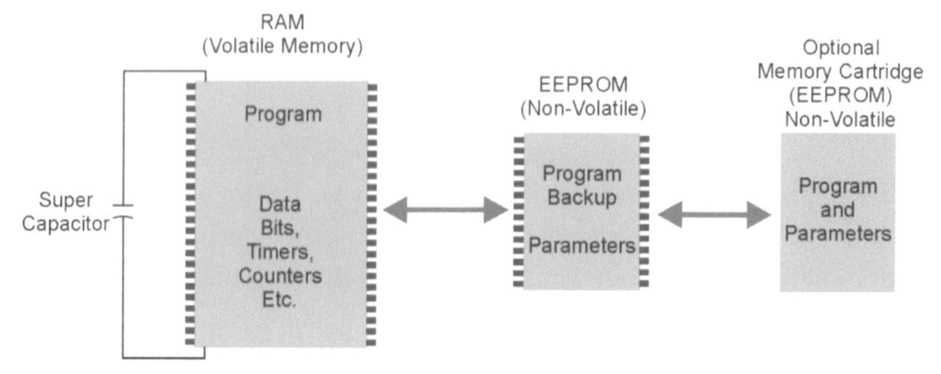

Figure 1-11. PLC memory.

Data Management in a Siemens System

Figure 1-12 shows how data is managed in a Siemens PLC system. The programming device (computer) is used to create the project. The project is stored on the computer's hard drive. Programming can be done offline or online. Once a project is developed it is compiled and downloaded to the CPU. The project can be downloaded directly from the computer to the CPU or the computer can save the project to a Siemens memory card and the card can be loaded into the CPU and the project loaded to the CPU from the memory card.

Memory in the CPU is divided into three areas: load memory; work memory; and retentive memory. The load memory contains the complete control program and the configuration data. Work memory contains the executable control program and the current data. Retentive memory contains the tags whose current values are protected from power loss.
The Siemens memory card can be used as a way to transfer a program from the computer to the CPU or as a program card to expand the CPU's internal load memory.

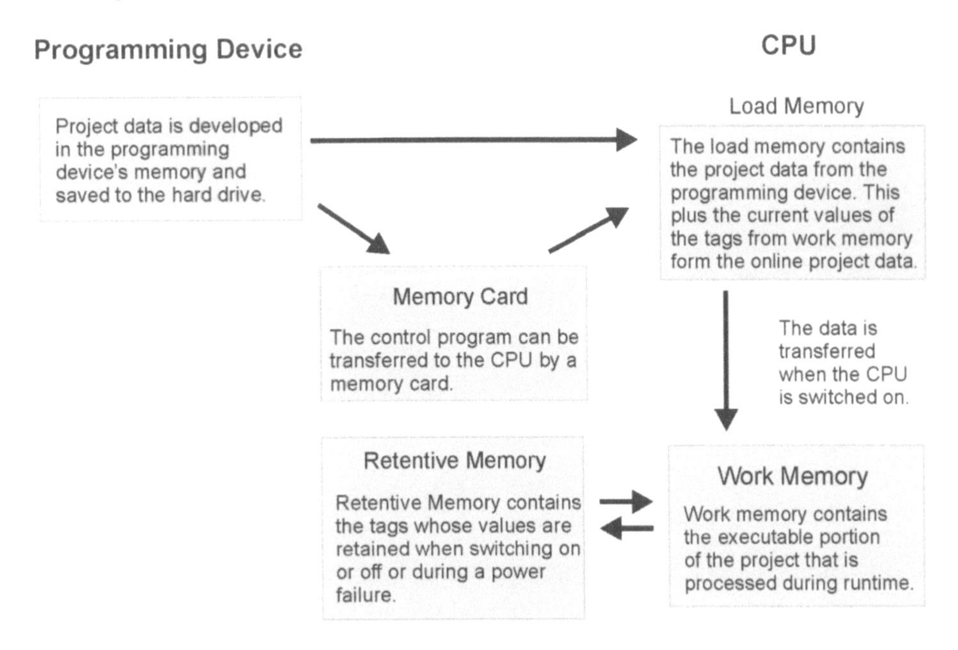

Figure 1-12. Siemens memory data management.

Micro Memory Card

A memory card can be used to store files or to update the firmware of the system (see Figure 1-13). The memory card can also be used to transfer programs to multiple CPUs.

Figure 1-13. A SIMATIC memory card. Courtesy Siemens.

Racks

Racks are also called chassis. An I/O rack or chassis is a housing in which modules are installed (see Figure 1-14). Racks are available in various sizes. The user decides which modules will be required for inputs and outputs, communications and other special purposes such as motion control. The user chooses the size chassis that is needed to hold the number of modules required for the application. Racks have slots which locate and power the modules. The slots connect the module to the backplane. The backplane passes power to operate modules and also provides enables the modules to communicate with the CPU and other modules.

Some applications require more modules than will fit in one rack. Some applications may have some I/O that is located a long distance from the PLC. Remote chassis can be used in these applications.

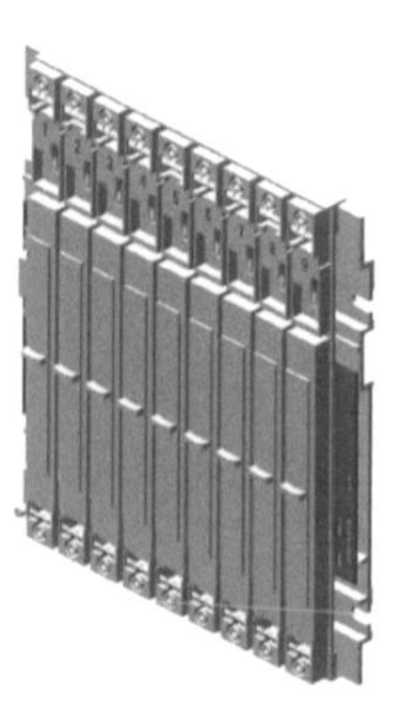

Figure 1-14. A 9-slot rack. Courtesy Siemens.

Not all PLCs require a rack or chassis. PLCs such as the S7-1200 have an internally designed bus connector system. Others such as the S7-300 use a u-shaped connector to connect CPU with I/O and communications modules.

Input to a PLC

Input modules provide the link between the outside world and the PLC's CPU. The main function of PLC input modules is to take information from the real world and convert them to signals that the PLC CPU can work with. Input modules also protect the CPU from the outside world.

Figure 1-15 shows an example of a sensor connected to an input on an input module. Note the power supply in the circuit. If the sensor is True (ON), 24 volts would be seen at input 1. The input module will convert the 24 volts to a five-volt signal that the CPU will see as a 1 (True). Many input modules run diagnostics to detect broken input wiring. Note also that the input module does not supply the power for the devices. The sensor is powered by an external power supply.

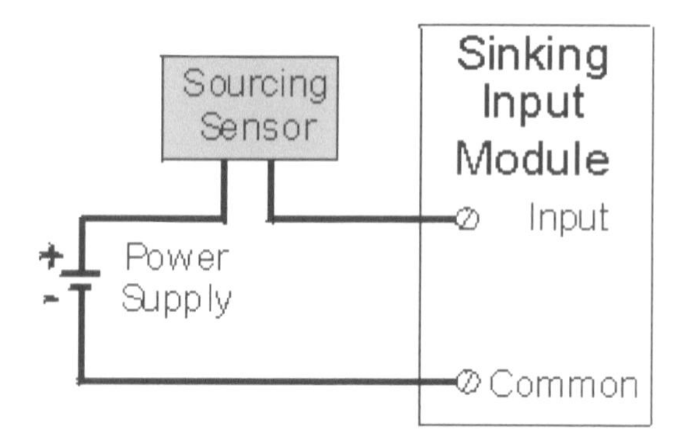

Figure 1-15. Sensor connected to an input.

Figure 1-16 shows how the electrical signal is converted to its binary equivalent in memory. The input module converts the electrical signal to a binary 1 or 0 and stores it in the memory bit that represents that input's state in memory.

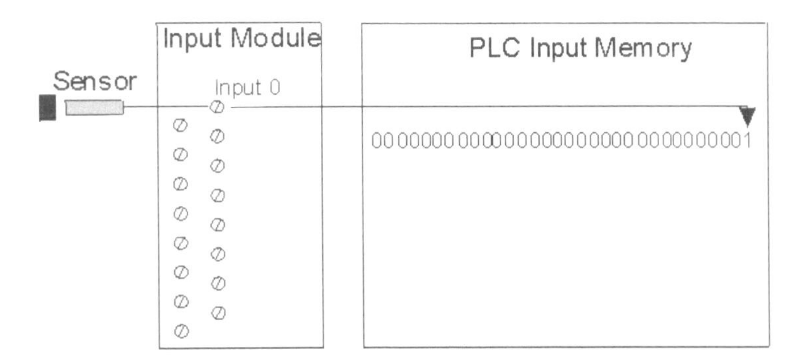

Figure 1-16. How a real-world input's state is stored in memory. The sensor is sensing an object so this sensor's output is true.

Discrete Input Modules

Discrete means that the module only accepts on/off type signals. The inputs have only two states: true or false. This is also called a digital signal. Pushbuttons, toggle switches, limit switches, proximity switches, and relay contacts, presence/absence sensors are examples of inputs that could be connected to a discrete input module.

The state of the input to the module is represented by the presence or absence of a voltage. Assume a 24 VDC input module. If 24 VDC is present at the input pin for input 0 (see Figure 1-17), the module puts a 1 in the input memory bit for input 0. 1 represents a 1 or a true input. If there is no voltage present at input 0, the module puts a 0 in input memory for input 0. Input modules often have LEDs that represent the state of each input.

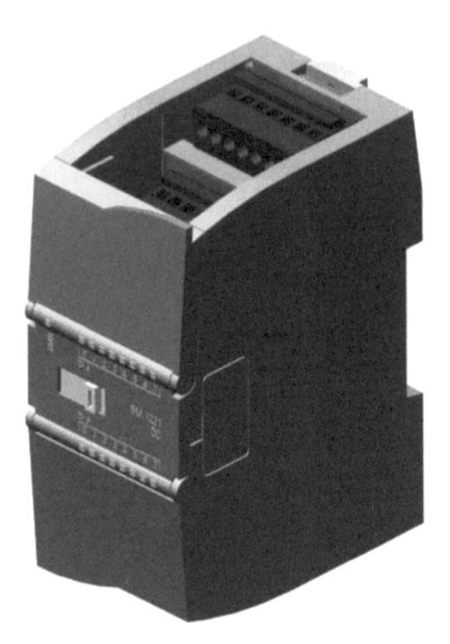

Figure 1-17. A S7-1200 digital input module. Note the status LEDS on the front of the module. Courtesy Siemens.

Figure 1-18 shows the input connections on a S7-1200 PLC. Note the 24 VDC digital inputs and the analog inputs. You can also see the memory card (MC) in this figure.

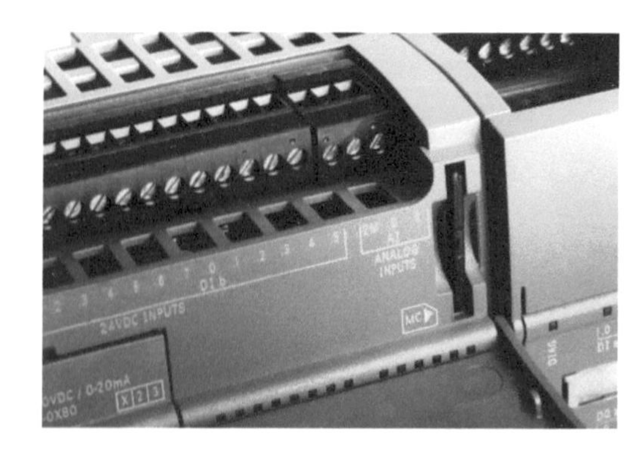

Figure 1-18. S7-1200 digital and analog connections. Courtesy Siemens.

Discrete input modules are available for various ranges of AC and DC voltages. Discrete modules are available with various numbers of inputs also. Modules are commonly available with 8, 16 and 32 inputs.

Analog Input Modules

Before analog values can be processed by the CPU they must be converted from an analog value to a digital value. Modules are available that perform this conversion from an analog value to a digital equivalent value that the CPU can work with. Modules are commonly available to take input signals of 0 – 10 VDC, -10 VDC to + 10 VDC, and 4-20 mA. These are useful for taking the input from analog input devices. These input devices are often called transducers. Transducers convert changes in things such as temperature, velocity, level, and so on, into standard analog signals such as plus or minus 10 VDC, plus or minus 20 mA, or 4 to 20 mA. A temperature measuring device such as a thermocouple with a converter is one example. For example, low voltage would represent a low temperature. 10 VDC might represent the maximum temperature.

Many industrial measurement devices produce a current signal such as 4-20 mA. A flow measurement device for example might have a 4-20 mA output. For this example, 4 mA might be no flow and 20 mA would be maximum flow. Any value between 4 and 20 would represent a different flow rate.

A transducer is used to convert the real-world analog value, such as temperature, to a voltage or current signal to the analog module. An analog module can be configured to measure voltage or current.

PLC Outputs

Discrete Output Modules

Discrete means that the output module only outputs on or off type signals. This is also sometimes called a digital signal. A digital output could be used to turn a valve or a motor on or off. It could be used to turn lights on or off or send an on/off signal to a robot or other equipment.

The output for the module is represented by a 1 or a zero in memory. Figure 1-19 shows an output connected to output 0 of an output module. The output module converts a binary 1 or 0 that represents the output's state in memory to an electrical signal that controls the actual state of the real-world output.

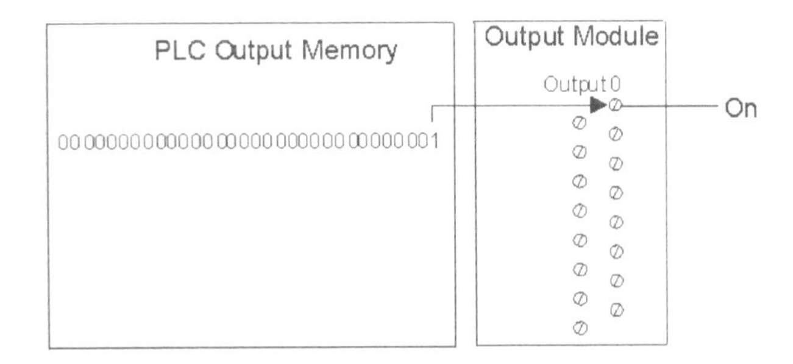

Figure 1-19. How an output state in memory is converted to an electrical signal to control an output.

Figure 1-20 shows a generic output module. The module is typically installed on a DIN rail or mounted directly to the control panel. The internal bus connector is used to power the module and also for communications with the CPU. Note the output status LEDs in Figure 1-20. Status LEDs are very useful for troubleshooting.

Figure 1-20. S7-1200 digital output module. Courtesy Siemens.

Typical Digital Outputs

Solenoids, relay and contactor coils, motor starters, indicator lamps are examples of devices often connected to the discrete outputs of a PLC. Figure 1-21 shows the removable terminal block. Output wiring is connected to the removable terminal block. This makes it possible to switch a defective module very quickly.

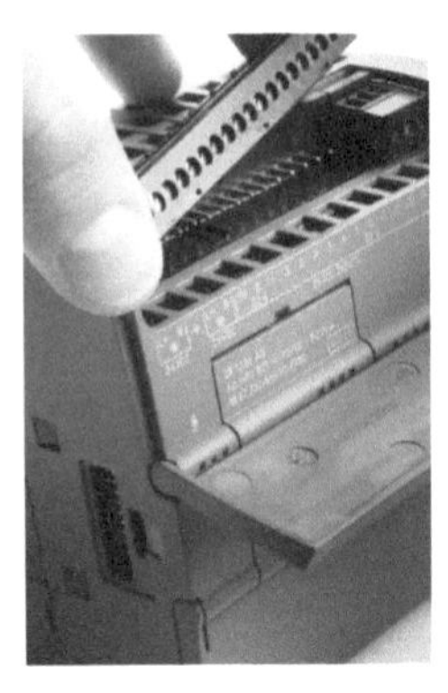

Figure 1-21. Removable terminal wiring block. Courtesy Siemens.

Output modules are available for AC and DC and various ranges of voltages. Modules are commonly available with 8, 16 and 32 inputs. Output modules are available with relay or solid-state outputs. Solid-state outputs are the most common.

Analog Output Modules

Modules are also available that can output an analog signal. Modules are commonly available to output 0 – 10 VDC, -10 VDC to + 10 VDC, and 4-20 mA. These are useful for controlling analog output devices. A motor drive is one example. A drive that is capable of clockwise and counter-clockwise rotation at various velocities might require a -10 VDC to + 10 VDC signal from a PLC to control direction and velocity. If the output is negative the drive may move in a counter-clockwise direction. If the output is positive the drive may move in a clockwise direction. The magnitude of the signal would control the velocity. For example, low voltage would be slow speed. 10 VDC would represent the maximum speed. Figure 1-22 shows the connection terminals for the 2 analog inputs on a S7-1200 CPU.

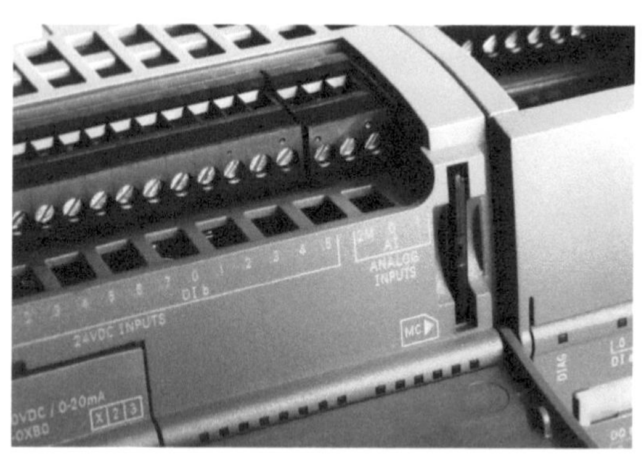

Figure 1-22. Analog input terminals. Courtesy Siemens.

Many process devices require a current signal such as 4-20 mA. A valve to control the rate of flow would be an example. For this example, 4 mA would be zero flow (valve completely shut) and 20 mA would make the valve wide open. Any value between 4 and 20 would set a different flow rate.

Programming

SIMATIC STEP 7 (TIA Portal) software is the programming software that is used for Siemens PLCs and WinnCC for human machine interfaces (HMIs). SIMATIC software with STEP 7 supports the hardware configuration of the system and setting up the parameters of the modules.

A variety of programming languages can be used. This includes basic programming languages such as ladder diagram, statement list (STL), and function block diagram (FBD), higher-level languages structured text and sequential function chart, and advanced engineering tools such as: S7 Structured Control Language, S7-Graph, S7-PLCSIM, and Continuous Function Chart.

The most commonly used PLC programming language is ladder logic. Figure 1-23 shows a simple example of ladder logic. It is not important to understand the logic at this point. There is a vertical line on the left and short vertical ones on the right of ladder logic. The vertical lines are sometimes called power rails. The rails represent power. If we connect the left rail to the right rail, power flows. The horizontal lines represent rungs of logic. The symbols on the left of rungs (contacts) represent input states or conditions. The symbol on the right of the rung (coil) represents an output. If the conditions on a rung are true, the output is turned on.

Ladder logic is still the most widely used PLC language. Recently other languages have been rapidly increasing in use. Programs can be written online or offline. Programming software also has error checking to make sure that the program addressing matches the available I/O as well as syntax errors. Programming software also provides many troubleshooting tools to help find and correct errors in program logic.

Figure 1-23 shows an example of ladder logic. The elements on the left are called contacts. Contacts represent input conditions. The elements on the right named Run and Pump are called coils. Coils represent outputs.

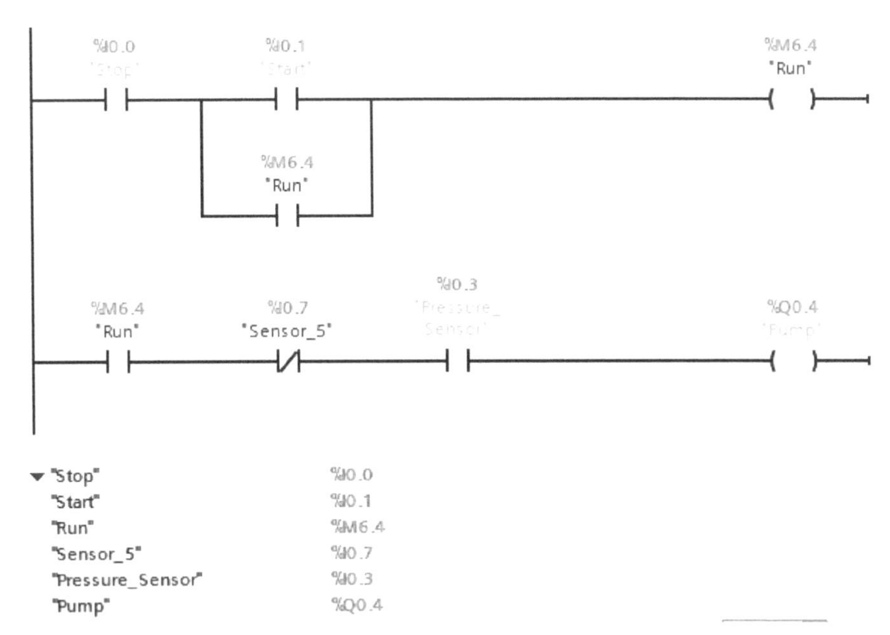

Figure 1-23. Simple example of a ladder logic program.

While ladder logic programs are still the most common, there are other ways to program PLCs. Note that not all PLC models can use all of the languages. Other common examples are statement list (STL) function block diagram (FBD), and structured control language (SCL). One of the great things about multiple languages is that the programmer can develop code blocks in various languages and then call them in a ladder diagram program. This enables the programmer to choose the best language for various parts of an application.

Statement List (STL) Programming

Statement list is a low-level computer language. This means that it is close to the assembly language that the microprocessor in the CPU actually uses to operate. High level languages are more user friendly. C+ and visual basic are examples of high-level languages.

Statement list (STL) instructions have an operation and an operand (see Figure 1-24). The operation to be executed is shown on the left of the statement list program.

The operand, the item to be operated on, is shown on the right of the statement list program. In the AND example the LD I0.5 tells the microprocessor to load the status of I0.0 (an input) into memory. The next line in the statement list program tells the microprocessor to perform an AND operation using the number just loaded in memory and I0.7 (another input).

In an AND instruction, if both are true, the result of the AND is true. If one or both are false the result of the AND is false. The result of the AND operation true (1) or false (0)) is put into Q0.4 (an output). The ladder logic on the right in the figure would be the equivalent operation. Note that STL cannot be used with the S7-1200.

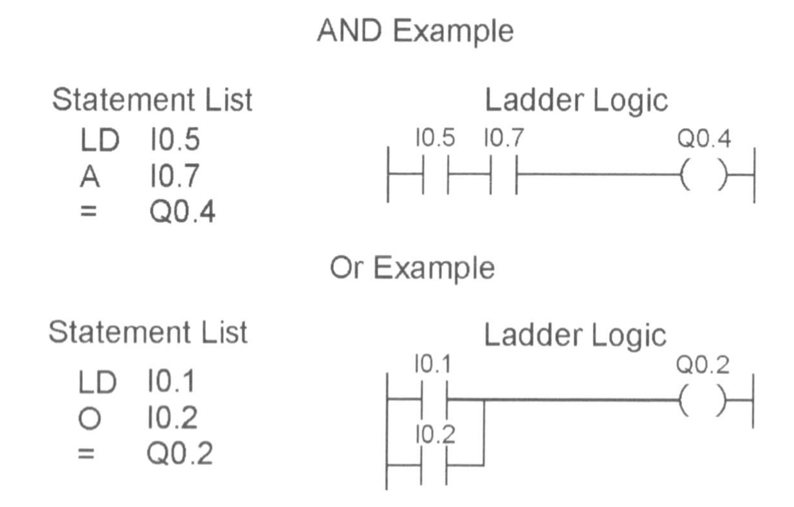

Figure 1-24. Simple example of Statement List (STL) versus ladder logic.

Function Block Diagram (FBD) Programming

Function block diagrams (FBDs) are block-shaped (rectangular) instructions (see Figure 1-25). Inputs are shown on the left side of the rectangle and outputs are shown on the right side.

In addition to LAD, STL, and FBD, other programming languages are also available. The choice of language is often determined by how comfortable the programmer is with each language. Different languages each have their own strengths and best areas of application. Many applications are best developed using multiple languages, each performing a portion of the logic.

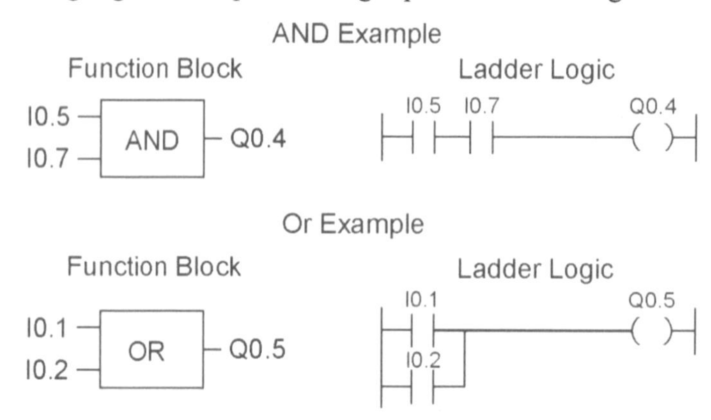

Figure 1-25. Simple example of Function Block versus ladder logic.

Structured Control Language (SCL)

SCL (Structured Control Language) is a high-level text-based programming language based on PASCAL. It is also based on the international standard for PLCs (EN-61131-3 (IEC 1131-3)). The standard specifies standardized requirements for programming languages for programmable controllers.

In addition to high-level language elements, SCL also includes language elements typical of PLCs such as inputs, outputs, timers, bit memory, block calls, etc. In other words, SCL complements and extends the STEP 7 programming software and its programming languages Ladder Logic and Statement List. An example of a function developed in SCL is shown below. Note that this code block could then be called from a ladder logic program.

```
FUNCTION SQUARE : INT

(*********************************************************

This function will return the square of the input value. If there is an overflow, it will return the maximum value that can be represented as an integer.

*********************************************************)

VAR_INPUT

value : INT;

END_VAR

BEGIN

IF value <= 181 THEN

SQUARE := value * value; //Calculation of function value

ELSE

SQUARE := 32_767; // If overflow, set maximum value

END_IF;

END_FUNCTION
```

International Standard IEC 61131

IEC 61131-3 is an international standard for PLCs. This standard is actually a collection of standards for PLCs and their associated peripherals. The standard consists of these parts:

Part 1: General information.

Part 2: Equipment requirements and tests.

Part 3: Programming languages.

Part 4: User guidelines.

Part 5: Communications.

Part 6: Reserved for future use.

Part 7: Fuzzy control programming.

Part 8: Guidelines for the application and implementation of programming languages.

Part 3 (IEC 61131-3) is the most important to the PLC programmer. Part three specifies the following languages: ladder diagram, instruction list, function block diagram, structured text, and sequential function chart. Instruction list is very similar to the language that is used to program the microprocessor. It is very detailed in nature and not friendly to people who have not studied microprocessor programming. The standard is intended to make the languages from different manufacturers more standard. It will never mean that programming software from one manufacturer can be used to program a PLC from another manufacturer. But the actual logic should be very similar. The standard essentially establishes base languages and elements for each language. Manufacturers must be compatible on these items to be IEC 61131-3 compliant. Manufacturers are free to add additional elements to the languages. Ladder logic, and function block diagram (FBD) programming will be covered in later chapters.

The Siemens Family of PLCs

Siemens refers to their line of industrial PLC controls as SIMATIC Industrial Automation Systems. Siemens has models from micro-PLCs up to PLCs that could control a whole manufacturing plant.

Modular SIMATIC Controllers

Siemens PLCs include: the LOGO! Logic Module, and S7-1200, S7-300, S7-1500 and S7-400 modular system PLCs. The micro-PLC LOGO! is programmed with the LOGO! Soft Comfort software. A LOGO! Micro PLC and a text screen are shown in Figure 1-26.

The rest of the PLC models begin with the S7 in their name. The S7 stands for Step 7 which is the Siemens PLC programming language. Models 1200, 300, 1500, and 400 are now programmed with S7 TIA Portal software. Figure 1-23 shows various Siemens PLCs. Siemens calls a PLC and all of its associated I/O modules a station.

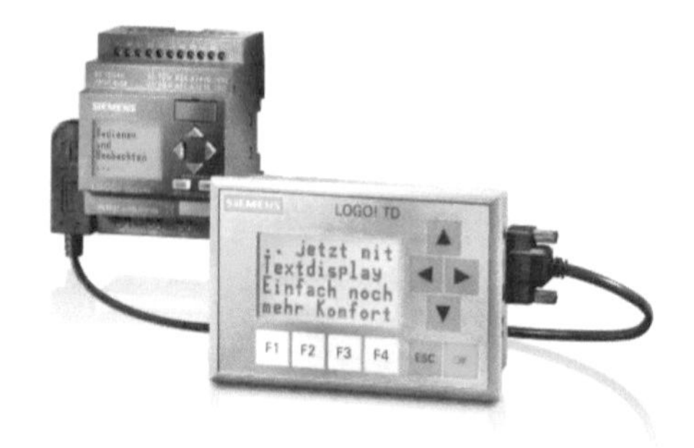

Figure 1-26. A LOGO! PLC and display. Courtesy Siemens.

SIMATIC S7-1200 PLC

One signal board can be added to any CPU to expand the digital or analog I/O (see Figure 1-27). Signal (I/O) modules can be connected to the right side of the CPU to Increase the number of digital or analog I/O. Some models of S7-1200 can have up to eight signal modules. S7-1200 CPUs can have with up to three communication modules on the left side of the CPU.

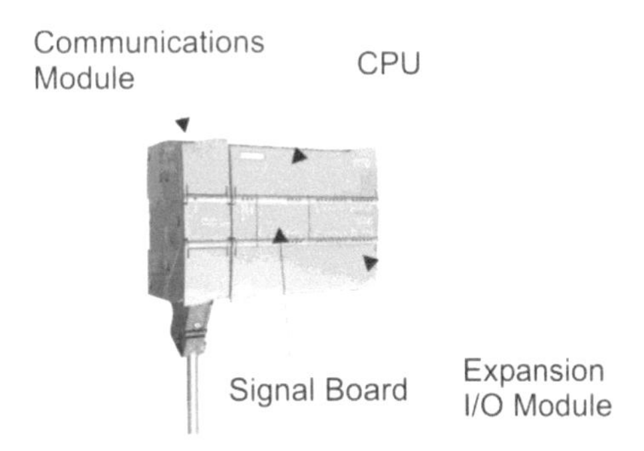

Figure 1-27. An example of a S7-1200 PLC system. Courtesy Siemens.

The S7-1200 can be used to control the speed and position of stepper motors and servo drives. The 1200 also has high-speed counters available for use. They can be used for monitoring of incremental encoders, frequency counting or high-speed counting of events. The S7-1200 has two high-speed outputs that can be used for either pulse train outputs (PTO) or pulse-width-modulated (PWM) outputs. PWM outputs could be used to control motor speed, valve position, a heating element, or many other analog output devices. The S7-1200 is also well suited to process control applications. The S7-1200 can control and perform auto-tuning on up to 16 PID process control loops.

For communication the S7-1200 has a noise-immune RJ45 connector that supports up to 16 Ethernet connections. Figure 1-28 shows a S7-1200 CPU and the connections for power, communications, I/O and also the status LEDs. Ethernet devices have unique fixed address by their manufacturer. It is called the MAC (Media Access Control) address. It is used as the unique identification in a local network. The device is then assigned an Ethernet IP address by the user. The assigned IP address must be in the same subnet as the IP address of the programming device being used to communicate with it. Subnet masks are used to divide networks into sub-networks.

STEP 7 TIA Portal software enables the programmer to develop the PLC program and the HMI touch panel application in the same integrated environment.

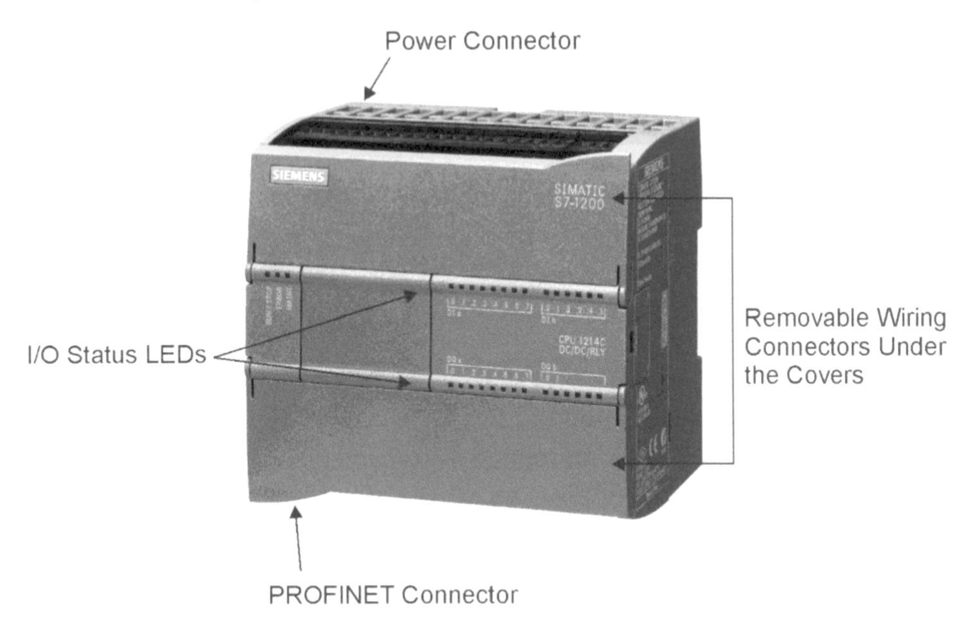

Figure 1-28. Connectors and status LEDs on an S7-1200 CPU. Courtesy Siemens.

SIMATIC S7-300 Modular Controller

The S7-300 is a step above the S7-1200 and is designed to meet the needs of the low to mid-performance range of industrial PLC applications. Figure 1-29 shows an example of a S7-300 PLC. There are several varieties of CPUs available.

Figure 1-29. An example of a S7-300 PLC system. Courtesy Siemens.

Failsafe S7-300 CPUs are also available. These can be used in applications that demand higher safety requirements.

S7-300 stations use DIN rail for the rack. The modules connect to each other on the back of the module though a u-shaped connector. This is called a serial backplane bus. These serial backplane connectors connect each module to the next module. The slots for a S7-300 rack are numbered in consecutive order. Slot 1 is for the power supply (even if not used). Slot 2 is for the CPU. Slot 3 is for an interface module (IM), even if not installed. Slots 4-11 are reserved for I/O modules. There may not be any gaps between the modules in slots 4-11. There is no connection between slot number and module width.

S7-300 Expansion Racks

Expansion racks can be used to expand the number of I/O that are possible. Depending on the size of the CPU a two-row expansion or a 4-row expansion rack can be used.

To use expansion racks, an IM interface module (communications) must be installed between the between the CPU and the first I/O module. The receiver IM module in the expansion rack establishes the connection with the rack that the CPU is located in.

SIMATIC S7-1500 Controllers

The S7-1500 controller is the newest addition to the Siemens PLC models. The S7-1500 controller is intended to be the next generation of medium to high-end machine and plant automation controller. The S78-1500 is characterized by high performance and efficiency. Its features include many functions that are integrated as standard, including motion control, security, and safety. The S7-155 has configurable diagnostic functions for plant status and integration into the TIA Portal for simple engineering and reduced project costs.

The S7-1500 hardware is extremely compact. Highly integrated and versatile components and modules save space in the cabinet and reduce the spare parts inventory.

The S7-1500 has a pluggable display available for commissioning and diagnostics. The PLC in Figure 1-30 has the display plugged into the CPU. The display can be plugged in or unplugged during operation. The use of the display can reduce downtime. The display has plain-text diagnostics information available on it on central and decentralized modules.

A password assignment for display operation can be assigned by using TIA Portal. The S7-1500 also has high memory capacity with up to 2 GB card for project data, archives, recipes and documents.

The S7-1500 features a uniform display concept for STEP 7, HMI, Web server, and the display on the CPU. The S7-1500 enables access to all machine-relevant operating data using Office tools and also via Web server. It also has easy access to machine configuration data via Web browser or SD card reader (bidirectional data interchange to and from controller).

Figure 1-30. A S7-1500 PLC.

SIMATIC S7-400 Controllers

Next in the SIMATIC family is the S7-400 model PLC (see Figure 1-31). The S7-400 is used for larger tasks such as the coordination of entire systems. The S7-400 is especially suitable for data-intensive tasks such as are found in the process and manufacturing industry.

S7-400 stations can be used as multiprocessing controllers by adding additional CPUs. Up to 4 CPUs can be used.

The overall application execution can be shared by means of multicomputing. Complex tasks such as position control, process control, computation, or communication can be appropriately divided among different CPUs. Each CPU would be assigned its own, local I/O for this purpose.

When more than one CPU is installed, the PLC will automatically assume multiprocessing operation. All of the CPUs will be in the same operating mode. They will all start together, and all go to stop if any of the CPUs fails. Each CPU will execute its own program independently.

Figure 1-31. An example of a S7-400 PLC system. Courtesy Siemens.

Remote I/O

Remote I/O is a term used to describe the ability to locate I/O modules close to where the I/O is located and away from the PLC CPU. Imagine a very large automated system with many I/O devices. Many would be located a long distance from the PLC CPU. Instead of running wires for every device a long distance from the device to the PLC CPU, the devices can be connected to remote I/O modules that are positioned close to the I/O on the machines. Just one communication cable needs to be run from the remote I/O to the PLC CPU. The use of remote I/O can result in substantial cost savings.

Siemens has a wide range of distributed I/O systems available - for solutions in the control cabinet or without a control cabinet directly at the machine, as well as for applications in hazardous areas.

Figure 1-32 shows a Siemens ET 200S remote I/O system. SIMATIC ET 200S is a multifunctional modular I/O system. The ET 200 system's modular design makes it possible to easily expand the system.

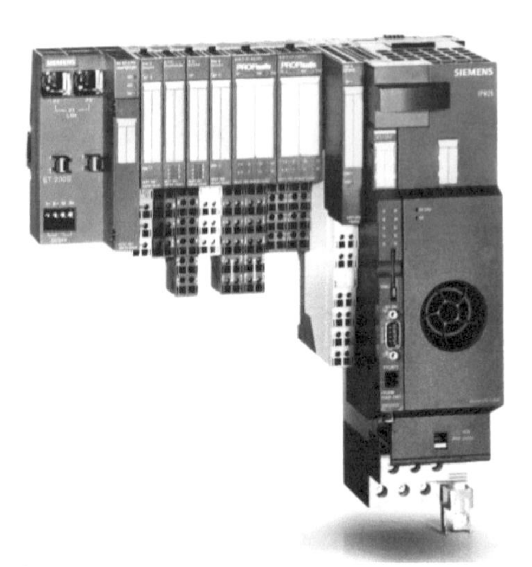

Figure 1-32. ET200S with motor starter.

The user can choose modules with many different combination options: digital and analog inputs/outputs, intelligent modules with CPU functionality, safety engineering, motor starters, pneumatic systems, frequency converters, and diverse technology modules.

Interface modules with integrated CPU and PROFINET/ PROFIBUS connection are available. They are available in standard and safety-oriented designs. The ET 200S model has a comprehensive module range available that includes, power modules, digital or analog input and output modules, technology modules, an IO-Link Master as well as motor starters or a pneumatic interface.

For cramped space conditions there is the ET 200S COMPACT a small block I/O which can also be expanded.

SIMATIC ET 200M Remote I/O

SIMATIC ET 200M is a modular I/O station for the control cabinet with high-density channel applications (see Figure 1-33). Connection to PROFIBUS and PROFINET is achieved with interface modules. The ET 200M can be used for standard as well as failsafe application. Up to 12 multi-channel signal modules (e.g., 64 digital inputs) and function modules as well as S7-300 communications processors can be used.

The ET 200M is particularly suitable for user-specific and complex automation tasks. Applicable analog input or output modules with HART protocol capability are also available for use in process applications.

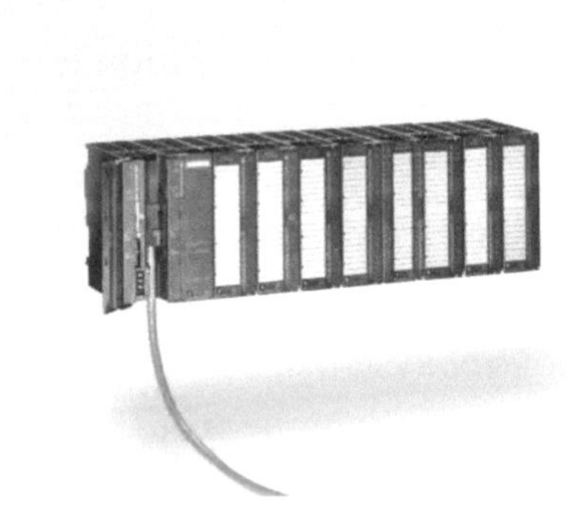

Figure 1-33. ET 200M remote I/O.

PROFINET

PROFINET is an open Industrial Ethernet Standard for automation. Existing fieldbus systems can be easily integrated into PROFINET networks. Information technology (IT) services such as web services, remote services, and TCP/IP communication can be used easily with PROFINET.

PROFINET has innovative diagnostics, new functions such as Shared Service, I-Device, MRP (Media Redundant Protocol), and high performance. These capabilities enable new, user-friendly applications, such as wireless automation.

There are over 4.3 million nodes of PROFINET installed. PROFINET is the world's leading Ethernet standard for automation.

PROFIBUS

PROFIBUS is a fieldbus communication protocol. Fieldbuses are control networks for industrial devices that are used in process control and industrial automation. Figure 1-34 shows the hierarchy of manufacturing communications. There are three levels of communication: the management level, cell level and device level. PROFIBUS is a family of protocols originally designed by Siemens to provide communications from real world sensors and actuators to controllers at the field level. Fieldbuses are bi-directional, digital serial networks for plant floor devices. PROFIBUS is the No. 1 fieldbus protocol.

There are more than 40 million PROFIBUS nodes installed worldwide.

PROFIBUS is not only implemented in the manufacturing area, but it can also be used throughout the process industry, even in hazardous areas. Standard interfaces support quick and easy connection of industrial I/O devices to other systems and integrated communication from the cell level down to the field level.

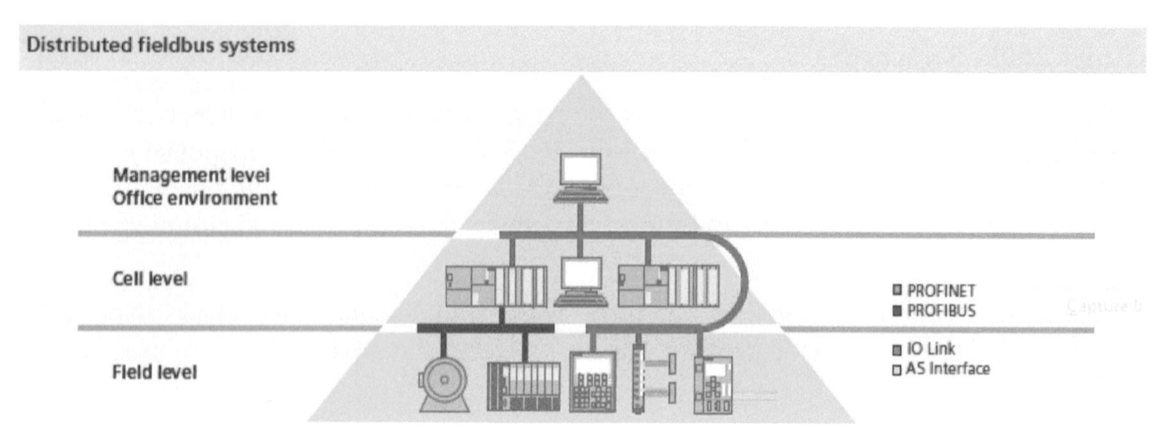

Figure 1-34. Communication levels.

Summary

PLCs are used in any imaginable application. They are used in manufacturing systems of course, but also in water treatment facilities, sewer treatment plants, bridge control, power generation and transmission, food production, prison control systems, building environmental control and security, and many others. They are used by inventors and entrepreneurs to start new companies producing newer and faster machines to do almost any task.

Chapter Questions

1. PLCs were originally intended to replace ________.
2. Name the main components that are found in a basic PLC system.
3. What does discrete I/O mean?
4. What is the difference between discrete and analog?
5. What is the name of the device that converts a physical change such as temperature to a standard analog signal?
   a. Sensor
   b. Analog sensor
   c. Transducer
   d. Analog converter
6. Before an analog value can be processed by the CPU it must be:
   a. Rectified
   b. Converted from a digital value
   c. Converted from analog to digital
   d. Scaled to 0-10 volts
7. Analog input channels can be configured as:
   a. Voltage input
   b. Current input
   c. AC
   d. DC
   e. A and B are correct
   f. C and D are correct
8. What is the most common PLC programming language?
9. Describe ladder logic.
10. What is FBD?
11. What is S7 an acronym for?
12. List the Siemens PLC models in order from the one that would be used for the smallest to the largest application.
13. The _________ connector is the most common when working with Industrial Ethernet.
    a. RS-232
    b. RS-505
    c. RJ45
    d. RJ12
14. A ___________ can be plugged into a S7-1200 CPU to add additional I/O.
    a. Signal module
    b. Signal board
    c. Auxiliary I/O module
    d. Mini I/O card
15. Ethernet devices have been assigned a fixed unique address by their manufacturer. The address is called a _________ address.
    a. Ethernet Access Control (EAC) address
    b. Global Access Control (GAC) address
    c. World Access Control (WAC) address
    d. Media Access Control (MAC) address

16. Networks can be divided into sub-networks by using a ________.
    a. RS-232 connector
    b. A subnet mask
    c. A subnetwork
    d. Ethernet bridge
17. If an Industrial Ethernet connection is to be established between the programming device and the PLC __________________.
    a. There must be termination resistors on each end of the cable
    b. Both devices IP addresses have to be in the same subnet
    c. Both devices must have the same Ethernet address
    d. Both devices must have their MAC addresses set the same
18. What typical output types are available in analog modules?
19. What is IEC 61131-3?

Chapter 2
Project Organization and Addressing

Objectives

Upon completion of this chapter, the reader will be able to:

Describe project organization in Siemens PLCs.

Explain the concept of modular programming.

Explain terms such as organizational blocks, function blocks, functions, data blocks, interrupts, and so on.

Explain how I/O is addressed on a Siemens PLC.

Introduction

To help in understanding how Siemens Step 7 organization works, it would be good to have an automated system to consider and see how a control program might be developed. It is difficult to find a system that all readers would be familiar with. We will use the space shuttle for our explanation. It was not chosen because we have any real knowledge of it, but because we can imagine the types of things that would be important in the control of the shuttle and there are many examples that will be helpful in explaining the many control features available in a Siemens control program. We can imagine that there are many mundane control tasks, many very complex and time sensitive tasks and, serious things that can go wrong that can be catastrophic. We can also imagine that under normal orbiting conditions we would monitor the temperatures and oxygen in the shuttle, the position of the shuttle, and the other systems on the shuttle. If anything unusual happens, the program would have to react instantly to evaluate and correct the problem.

We could develop the whole space shuttle application as one program. This could be done in what Siemens calls an organizational block (OB). It would be a long and complex program but we could accomplish the task in one organizational block. There would have to be some decision-making instructions so that if something goes wrong, the program would jump to different portions of the code within the one OB. Troubleshooting the code would be a nightmare being long, complex, all in one program. Putting all the code in one OB could be called called linear programing (see Figure 2-1).

Figure 2-1 also shows a simple example of what modular programming looks like. In modular programming we can divide the application into logical blocks of code that control portions of the application.

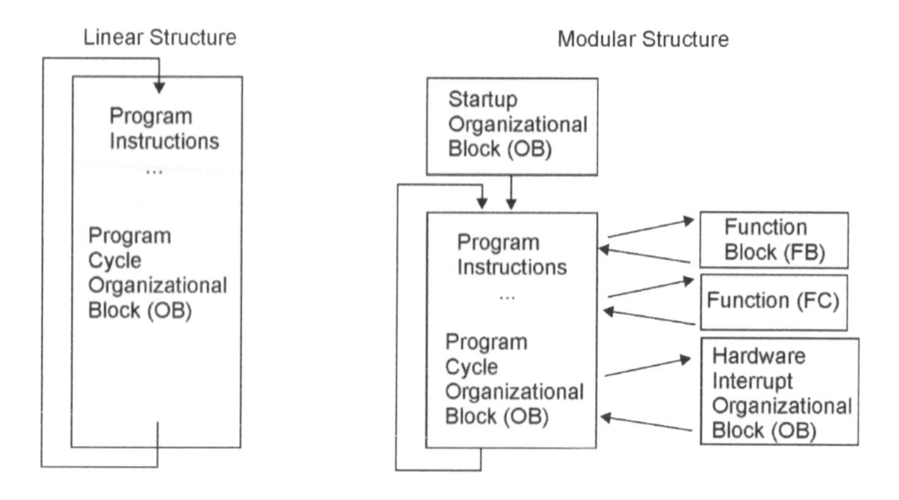

Figure 2-1. Linear versus Modular programming.

In Siemens PLCs program instructions are programmed in code blocks. When you create a user program for the automation tasks, you insert the instructions for the program into code blocks. There are three types of code blocks: organizational blocks (OBs), function blocks (FBs), and functions (FCs). We will look at organizational blocks first.

Organizational Blocks (OB)

Organizational blocks represent the interface between the operating system and the user program. The CPU's operating system calls the organizational block when specific events occur such as hardware or cycle events. The main program is in organizational block 1. An Organizational Block (OB) is a code block that is used to structure or organize a program. It is possible to just have one OB that contains all of the user's program logic.

The main organizational block is called a program cycle OB. You could put your whole program in the program cycle OB. OB1 is called cyclically by the operating system. In addition to the user's program cycle OB, there are other types of organizational blocks that can be used (see Figure 2-2). These other types of OBs are used for specific purposes.

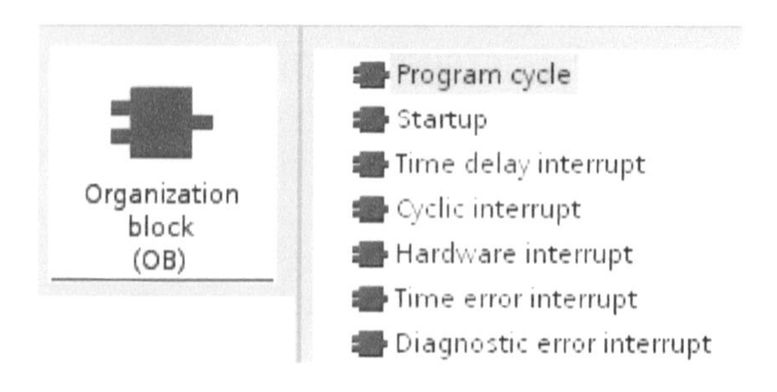

Figure 2-2. Types of organizational blocks (OBs).

One type of OB is a startup OB. Imagine the space shuttle again. There would be many things that need to be done at startup. There would be many values we would need to initialize, there would be sensors and inputs that would need to be checked to make sure everything was OK, before beginning the launch phase. A startup OB would be very helpful as this is the first thing the CPU would execute when put into run mode.

Other types of OBs are available to handle interrupts and errors, and also the ability to execute specific program code at a specific time interval (see Figure 2-2). For our shuttle example, we might want to check all of the life support systems on a regular time interval with a cyclic interrupt OB. We might execute a hardware interrupt OB to evaluate and resolve a problem with hardware.

Interrupt processing is event-driven program execution. When the event occurs, the operating system interrupts the execution of the main program and calls the routine that was assigned to this particular event. After the routine has been executed, the operating system will resume execution of the main program at the point where it was interrupted. A routine that is associated with an interrupt event is developed in an organizational block in which further blocks can be called. An event with a higher priority interrupts the execution of routine in an organizational block with a lower priority. The programmer can influence the interruption of a program by events of a higher priority by using program instructions such as Delay Interrupt Events (DIS_AIRT) and Enable Delayed Interrupt Events (EN_AIRT).

For now, we will only look at using various types of OBs in our modular program. Later in the chapter we will see there are other types of code blocks we can use. Study Figure 2-3. A modular program has been developed utilizing various types of OBs. In our space shuttle example, we could use a program cycle OB to handle the main tasks that need to be performed to launch, fly and land the shuttle. The startup OB (see Figure 2-3) would be used to handle all the start type tasks before the program cycle OB begins to execute. Note that the start tasks only need to be executed once before the actual operation begins. From then on, the program cycle OB will execute. In this application various types of interrupt OBs were also used. It appears in Figure 2-3 that a hardware interrupt routine is currently being called. This could be triggered by a hardware problem. Note that this has simplified the overall program. Startup tasks that only need to be done once during startup have been put into the startup OB. This shortens the code in the program cycle OB and may reduce its scan time. Note also that by using interrupt code blocks the program cycle OB code is shortened and the interrupt code blocks are only executed when needed. This also shortens the scan time and ensures that the code blocks will execute in a timely manner depending on their importance to the application.

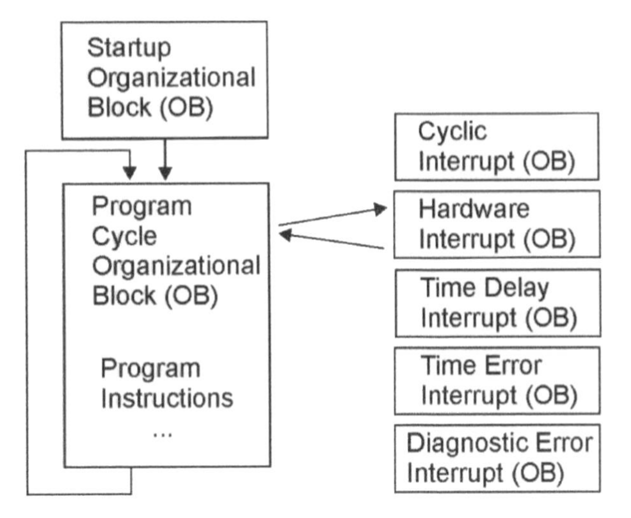

Figure 2-3. A modular program utilizing various types of OBs.

Error Handling

The CPU has the capability to detect and signal errors during program execution. These CPU detectable errors include:

- Errors when addressing peripheral inputs and outputs.
- Errors in arithmetic operations including overflow and floating point.
- Errors when calling blocks if the block does not exist or is still being processed or a program execution error.

If a serious error occurs such as a cycle monitoring time expiring twice in a program cycle, the CPU will immediately enter the stop state.
There are program instructions to help evaluate errors. The GET_ERR_ID can be used to read the program error number when a program error occurs. The GET_ERROR instruction can be used to read the program error information and provide it in a predefined data structure when an error occurs.

The Use of Organizational Blocks

Organizational blocks enable the programmer to divide a control program into individual sections. A block is a part of a control program. Organization blocks provide the structure for your program. They serve as the interface between the operating system and the user program. OBs are event-driven. An event, such as a diagnostic interrupt or a time interval, will cause the CPU to execute an OB. Some OBs have predefined start events and behavior.

Study Figure 2-4. Program execution by the CPU has been broken into operating system execution and user program execution.

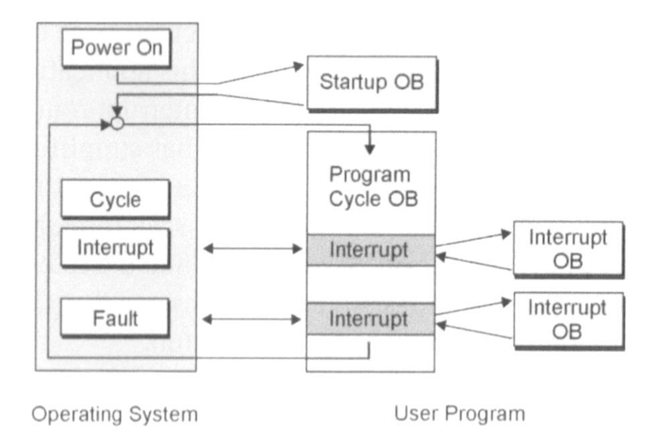

Figure 2-4. Example of program execution.

At power on and before going into Run mode the startup OB is executed. Next control goes back to the operating system which runs the program cycle OB. If any interrupts occur during the scan of the program cycle logic the interrupt OB is executed. When the interrupt OB is done executing it returns to the place in the program cycle OB where it was called. When the program cycle scan is complete, execution returns to the top of the program cycle for the next scan of logic.

The table in Figure 2-5 shows the types of OBs for a S7-1200 PLC. Note that additional OBs are available in S7-300 and S7-400 CPUs. The table also shows how many can be used, how they are numbered, what causes them to start execution, and their priority levels. For example, OB100 executes a complete restart when power is switched on or there is a transition from Stop to Run. A program cycle OB executes after the startup OB has executed or at the end of the last cyclic OB.

Event (OB)	Quantity Allowed	OB Number	Start Event	OB Priority	Priority Group
Program Cycle	>= 1	1 (default) 200 or greater	Starting or end of last cyclic OB.	1	1
Startup	>= 0	100 (default) 200 or greater	Transition from Stop to Run.	1	
Time Delay Interrupt	Up to 4	200 or greater	Delay time has expired.	3	2
Cyclic Interrupt	Up to 4	200 or greater	Constant bus cycle time expired.	4	
Hardware Interrupt	Up to 50 (more can be used with ATTACH and DETACH)	200 or greater	Rising edge (16 max.). Falling edge (16 max.).	5	
			• HSC: Count value = reference value (6 max.). • Count direction changed (6 max.). • HSC: External reset (6 max.).	6	
Diagnostic Error Interrupt	0 or 1	82 only	Module has detected an error.	9	
Time Error Interrupt	0 or 1	80 only	Maximum cycle time exceeded.	26	3
			• Called OB is still being executed. • Queue overflow. • Interrupt loss due to high interrupt load.		

Figure 2-5. S7-1200 OB Table.

The program cycle OB contains your main program. You can include more than one program cycle OB in your user program. During RUN mode, the program cycle OBs execute at the lowest priority level and can be interrupted by all other types of program processing. Note that startup OBs do not interrupt the program cycle OBs because the CPU executes the startup OBs before going to RUN mode.

After finishing the processing of the program cycle OBs, the CPU immediately executes the program cycle OB again. This cyclic processing is the normal type of processing used for PLCs. For many applications, the entire user program is located in a single OB, such as the default program cycle OB 1. You can create other OBs to perform specific functions, such as startup tasks, for handling interrupts and errors, or for executing specific program code at specific time intervals.

Interrupt Processing

Interrupt processing program execution is based on events. When an interrupt event occurs, the operating system interrupts the execution of the main program and calls the routine that was allocated to the particular interrupt event. After the interrupt routine has been processed the operating system resumes execution of the main program at the point where it was interrupted. Events can be interrupts or errors.

If multiple interrupt events occur simultaneously, an event scheduler controls the order of execution. A routine that is associated with an interrupt event is written in an organizational block from which other blocks can be called. An event of a higher priority will interrupt the execution of a routine in an organizational block with a lower priority. The programmer can delay and enable interrupts from routines with higher priority with the use of program functions that can delay and enable interrupts. The DIS_AIRT program instruction can be used to delay interrupt events and the EN_AIRT instruction can be used to enable delayed interrupt events.

An OB in a higher priority group will interrupt the execution of an OB in a lower priority group. The CPU stores any events that occur during the processing of an OB. After completing the execution of that OB, the CPU will then execute OBs in the queue according to their relative priority class within that priority group, processing the event with higher priority class first. However, the CPU completes the execution of each OB within that priority group before starting the execution of the next OB within the same priority group. After the CPU has executed all of the events for the interrupting priority group, the CPU then returns execution to the OB in the lower priority group that had been interrupted and resumes the execution of that OB at the point where it was interrupted.

A modular program organization calls specific code blocks to perform specific tasks. Each code block provides the program code for each subordinate task. Code blocks are executed by calling a code block from another block. This can make the main OB program more concise and easier to understand.

Function Blocks (FBs) and Functions (FCs)

In addition to OBs, Siemens has other types of code blocks available. These other types of code blocks could be thought of as subroutines.

Function blocks (FBs) and functions (FCs) can be thought of as subroutines. FCs and FBs can be used to organize, optimize, and simplify the programming of applications.

Uses for FBs and FCs:

- Divide and organize the program into logical tasks. This makes each task more manageable, easier to program, and easier to troubleshoot.
- FBs and FCs can be developed one time and can be used many times in the program. They can also be created and used as users' library for many projects.

FBs and FCs can have input and output parameters. This enables a FB or FC to be used many times, with different set of variables.

An FC uses the address of the given parameter (tag) to read/write directly from/to them. An FB copies its parameters (tags) to/from an associated data block (DBs), and works internally with the DB variables. There are two types of data blocks: global and instance. All program blocks in the user program can access the data in a global DB. An instance DB is used to store data for a specific function block (FB).

An FB can also have what is called an instance data block (DB) assigned to it. This enables a generic FB to use instance DBs for specific parameters for each individual device. For example, we might have several devices, such as valves, in an application. Each valve is similar but needs different parameters. We could write a generic FB to control valves and have a different instance DB for each valve.

Function Blocks (FBs)

A Function Block (FB) is a function that is able to remember its last operation. A function block (FB) is like a subroutine with memory. A Function Block (FB) is a subroutine that is executed when called from another code block (OB, FB, or FC). The calling block can pass parameters to the Function Block (FB) and also identifies a specific data block (DB) for the FB to use. A FB's calls can be programmed with block parameters. The FB stores the input (IN), output (OUT), and in/out (IN_OUT) parameters in variable memory that is located in a data block (DB). The DB is associated with that FB. Blocks can be monitored from the programming device if the device is online and the offline block and the online block are identical.

Data Blocks

Let's take a closer look at memory and data blocks (DBs). Data blocks contain your program's data. In addition to the data that are assigned to a function block, shared data can also be defined and used by any blocks. Data blocks can be protected against overwriting so that their data may only be read. Data blocks can be used as global data blocks, instance data blocks, or as a type data block. The data block (DB) is used to store the data for the specific call or instance of a function block (FB).

A data block (DB) is used to save the values that are written during execution of the program. In contrast to a code block, a data block only contains tag declarations. It contains no networks or instructions. The tag declarations define the structure of the data block.

Types of Data Blocks

There are two types of data blocks: global and instance. A global data block is not assigned to a particular code block (FB). The values of a global data block can be accessed from any code block. A global data block contains only static tags. The tags that a global data block contains are entered in its declaration table.

Instance data blocks are assigned directly to a function block (FB). The structure of an instance data block is determined by the declaration of the function block. An instance data block (DB) contains the block parameters and static tags that are declared there. The user can define instance-specific values in instance data blocks, such as initial values for the declared tags.

A data block (DB) can be used for fast access to data stored within the program itself. DBs can also be defined as being read-only. The data stored in a DB is not deleted when the data block is closed or the execution of the associated code block comes to an end. Data blocks can be configured to be accessible or non-accessible by the HMI.

Use of Instance DBs with FBs

FBs are used to control the operation of tasks or devices that do not finish their operation within one scan cycle. Each FB has one or more instance DBs that are used to store the operating parameters so that they can be quickly accessed from one scan to the next.
When an FB is called, an instance DB can also be opened to store the values of the block parameters and the static local data for that call or "instance" of the FB. The instance DB stores these values after the FB finishes.

The ability to have instance data blocks (DBs) is very powerful. This makes it possible to have one generic function block (FB) control several devices, each of which has its own instance data block (DB). For example, one FB can be used to control several pumps or valves, with different instance DBs containing the specific operational parameters for each pump or valve. Parameter assignable blocks are used for recurring program functions. This has several advantages. The program logic only has to be created once which reduces programing time. The block is only stored in user memory once which reduces the amount of memory used. The block can be called as often as needed. Function blocks (FBs) and Functions (FCs) can be programmed as parameter assignable.

By designing the FB for generic control tasks, you can reuse the FB for multiple devices by selecting different instance DBs for different calls of the FB.

Figure 2-6 shows an OB that could be used to control three different valves. The FB is called and told which instance DB to use for each call. This structure allows one generic FB to control several similar devices, such as valves, by assigning a different instance data block for each call for the different devices.
Each instance DB stores the data for an individual device. In this example, FB 20 controls three separate valves, with DB 201 storing the operational data for the first valve, DB 202 storing the operational data for the second valve, and DB 203 storing the operational data for the third valve.

Figure 2-6. This figure shows the use of one FB to control 3 different valves by using a different DB for each valve.

Initial values can be assigned to parameters in the FB. These values are then transferred to the instance DB that is associated with the FB. If parameters are not assigned, the values currently stored in the instance DB will be used.

The use of generic code blocks you can simplify the design and implementation of a user program. The code blocks can also be used multiple times in a program to help organize and structure the program and reduce redundant logic.

Reusable code blocks can be created for standard tasks, such as for motor control. The generic code blocks that you develop can also be stored in a library that can be used in different applications. This enables you to reuse your code in different applications.

Each instance DB maintains the values for the individual device FB between different or consecutive calls of that FB.

Figure 2-7 shows some logic that was used to create a FB. Note: it is not important to understand the logic at this point. The instructions will be covered in detail in later chapters. This logic takes input data, converts it to a scaled value and then an RPM value. The name of the FB is Input_Conversion. The progrmmer had to use this logic several times in the application for various motors. The programmer decided to create an FB to do the task.

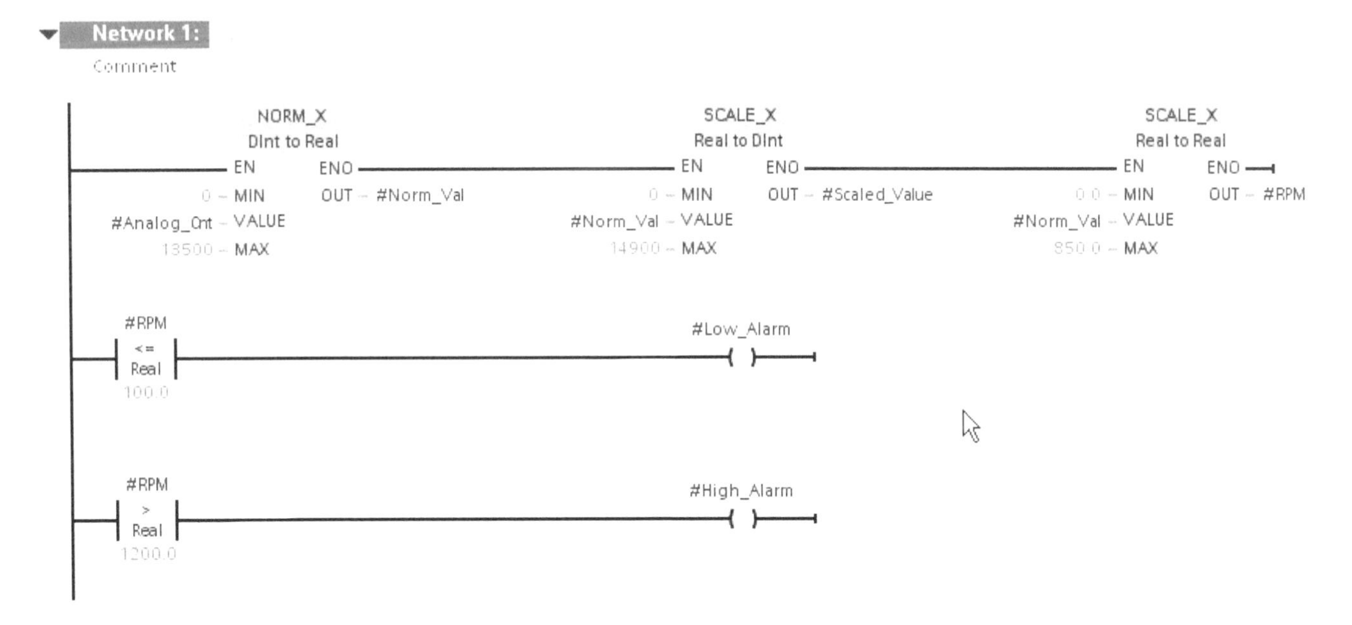

Figure 2-7. Logic for the Input_Conversion FB.

Figure 2-8 shows how tags are created for the instructions in a function block (FB). In this example a tag named Analog_Cnt is being created. The types of tag possible are shown in the figure. Local In was chosen for this tag. Choosing local for the type means that this tag will only be known to this FB. The table in Figure 2-9 shows the scope, uses, and naming for PLC and local type tags.

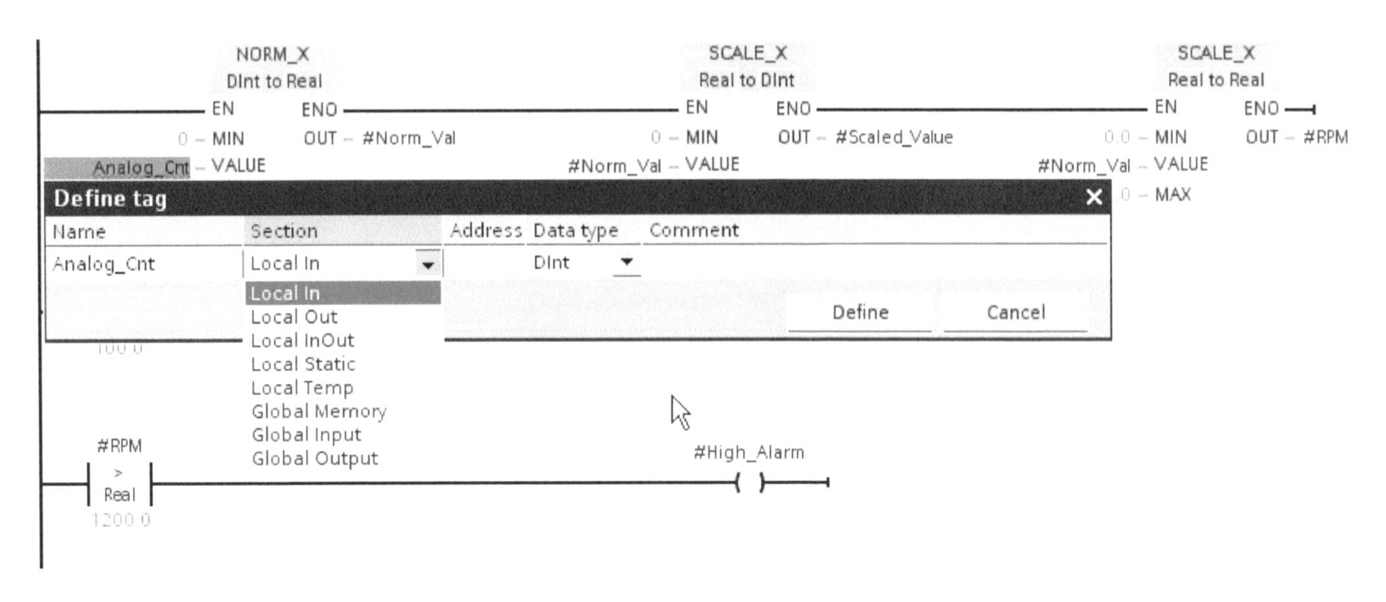

Figure 2-8. Definition of a tag for an FB.

	PLC Tags	Local Tags
Scope	Valid throughout entire CPU. Can be used by all blocks in the CPU. Have the same meaning in all blocks.	Known only in the block in which they are defined. The same tag name can be used in other blocks for different purposes.
Uses	PLC tags can be defined for: I/O signals (I, IB, IW, ID, Q, QB, QW, QD) Bit Memory (M, MB, MW, MD)	Local tags can be defined for: Block parameters (input, output, and inout parameters), Temporary local data of a block, Static data of a block.
Permitted Characters	Letters, numbers, and special characters. Quotation marks cannot be used.	Letters, numbers, and special characters. If special characters are used tags must be enclosed in quotation marks.
Location of the tag definition	PLC tag table	Block interface.
Representation	PLC tags are shown in quotes. "Analog_Count" for example.	Local tags from the block interface are shown with a # prefix. #In_Value.

Figure 2-9. Comparison of PLC tags and local tags.

Once a FB has been developed it can be used in logic. Figure 2-10 shows an example of what the Input_Conversion FB looks like in logic. Note that the FB is kind of a "black box". We cannot see the code that is in the FB. This can be beneficial in several ways. One is that it simplifies the logic for maintenance personnel. To use this FB the programmer only has to assign tags to the inputs and outputs. Note that this FB is using the data block (DB) named Input_Conversion_DB.

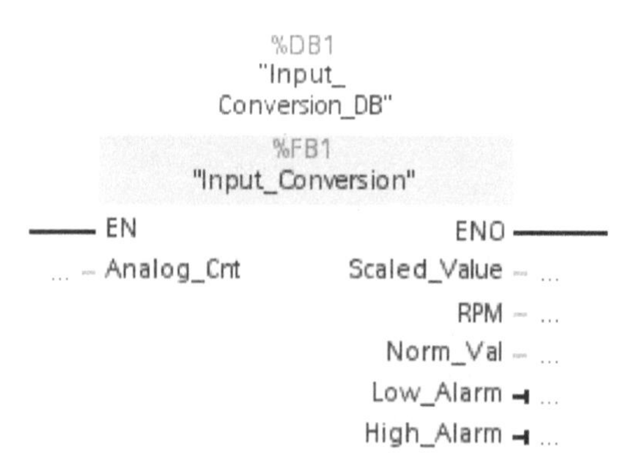

Figure 2-10. The Input_Conversion FB.

Figure 2-11 shows the program blocks for the project. Note that the Input_Conversion FB is in the list as well as a DB for the Input_Conversion FB.

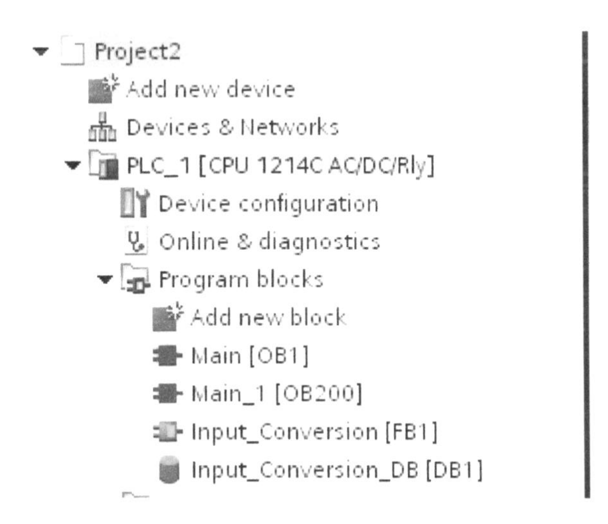

Figure 2-11. Program blocks showing the Input_Conversion FB and DB.

The associated DB for the Input_Conversion FB is shown in Figure 2-12. Note the tags and tag types. If you need to change types this is where it would be done. Note also that initial values could have been entered but all were left at 0 or False.

Input_Conversion_DB

	Name	Data type	Initial value	Monitor value	Retain	Comment
1	▾ Input					
2	Analog_Counts	Real	0.0			
3	Analog_Cnt	DInt	0			
4	▾ Output					
5	Scaled_Value	DInt	0			
6	RPM	Real	0.0			
7	Norm_Val	Real	0.0			
8	Low_Alarm	Bool	false			
9	High_Alarm	Bool	false			
10	▾ InOut					
11	▾ Static					

Figure 2-12. The Input_Conversion DB. This shows the tags that are used in the FB.

Now that the FB has been created it can be used in logic. Two of the Input_Conversion FBs were used in the logic shown in Figure 2-13. Note that different DBs were used for each of the instructions so that the data for each can be kept. Also note that while the same FB was used twice, the input and output tags were different. One FB was used for motor 1, and one FB was used for motor 2. Note something that can be very important. As you can see from Figure 2-13, we can hide the actual logic from others and only show them the tags for the FBs or FCs. This may make it easier for maintenance people to understand. It may also be used to protect proprietary logic that we would not want others to be able to copy or change.

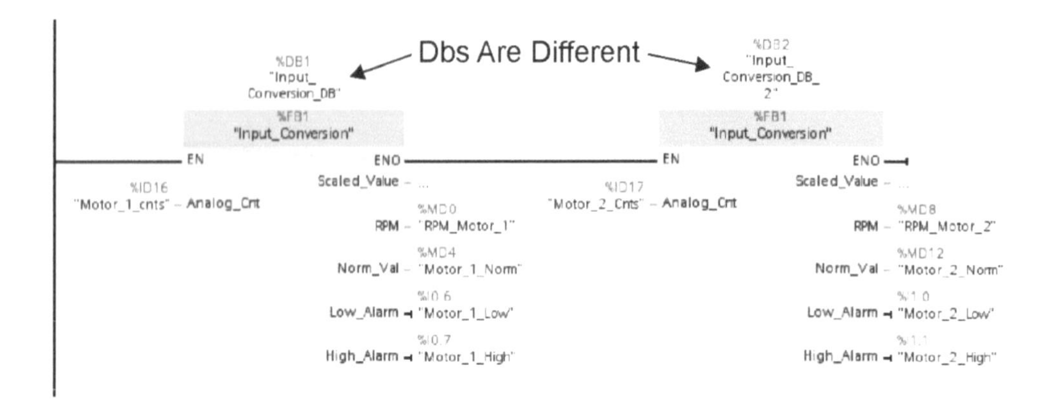

Figure 2-13. Two Input_Conversion FBs have been called in this logic. Each used different inputs and outputs and each used a different DB.

User created FBs and FCs can also be saved to the Project Library. Figure 2-14 shows that a folder was added to the Project Library (My_FBs) and the Input_Conversion FB was put into that folder. It could also have been added to the global library and have been available to all projects. Global libraries can be created or opened under the Options tab at the top of TIA.

Figure 2-14. The Input_Conversion FB has been added to the project library. Note that we could have added it to the global library.

Functions (FC)
An FC is a subroutine that is called from another code block. An FC can be called from an OB, FB, or a FC. A function (FC) is like a subroutine. An FC is a code block that typically performs a specific operation on a set of input values. The FC stores the results of this operation in memory locations. FCs are typically used to perform the following types of tasks:

- standard and reusable operations, such as calculations.
- functional tasks, such as for individual controls using bit logic operations.

An FC can also be called several times at different points in a program. This reuse simplifies the programming of tasks that are used multiple times. Unlike an FB, an FC does not have an associated instance DB. The program block that calls the FC passes parameters to the FC. The output values from an FC must be written to a memory address or to a global DB.
An FC uses its temporary memory (L) to do any required calculations. The temporary data is not saved. To store data for use after the execution of the FC has finished, the output value would have to be assigned to a global memory location, such as M memory or to a global DB.
The table in Figure 2-15 shows the uses and differences between FBs and FCs.

Use	FB	FC
Can be used as a subroutine.	Yes	Yes
Can use temporary variables.	Yes	Yes
Need an Instance DB for each call.	Yes	No
Can use parameters for Inputs, Outputs, and In/Out.	Yes	Yes
Parameters are passed as an address for internal use.	No	Yes
Can internally call a FB or FC.	Yes	Yes
Can call an FB as a multi-instance.	Yes	No
Can be called without filling all parameters.	Yes	No
Can use static variables.	Yes	No
Can use static variables retentively.	Yes	No
Parameters are copied to or from an- Instance DB for internal use.	Yes	No

Figure 2-15. Comparison of FBs and FCs.

An FC was developed to convert the analog signal from a load sensor to an actual weight in pounds that could be displayed on the display for the operator (see Figure 2-16). The actual instructions are not important at this point, they will be covered in later chapters. The inputs and outputs to these instructions were created as local inputs and outputs. Actual values were used in the NORM_X instruction for the high and low input values. Actual values were used in the SCALE_X instruction for the high and low input values. The #WGT would be the result of the logic and would be the weight in pounds. Note that when this FC executes these values will not be stored. They are only temporary and only last while the instruction executes.

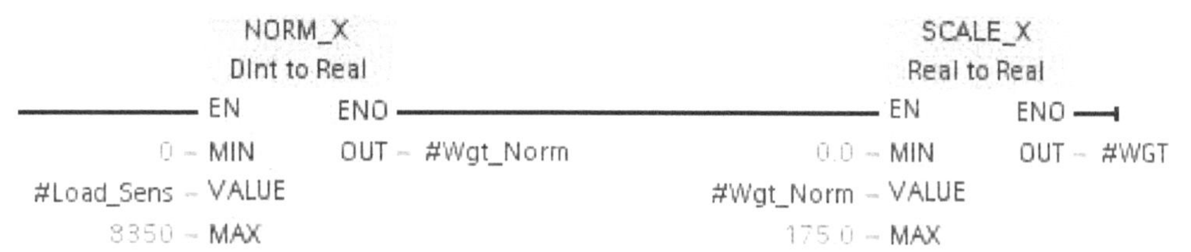

Figure 2-16. Logic for the Calc_Wgt Function (FC).

Figure 2-17 shows the variables that are used in the Calc_Wgt FC. Note that these are not PLC tags. They are only used while the FC executes.

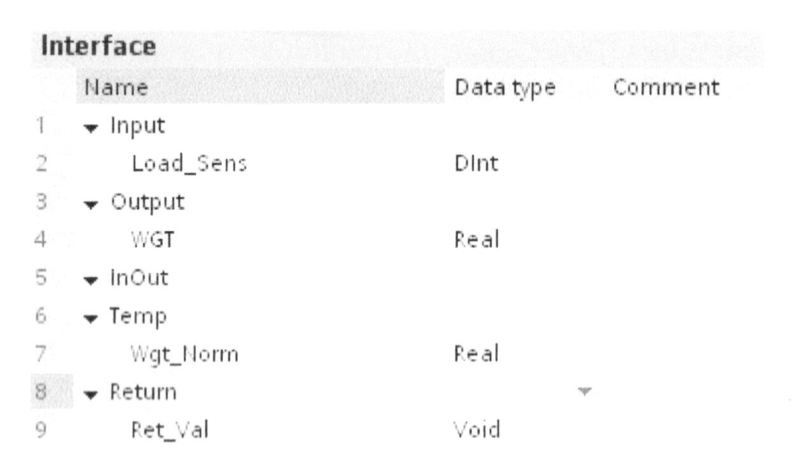
Interface

	Name	Data type	Comment
1	▾ Input		
2	Load_Sens	DInt	
3	▾ Output		
4	WGT	Real	
5	▾ InOut		
6	▾ Temp		
7	Wgt_Norm	Real	
8	▾ Return		
9	Ret_Val	Void	

Figure 2-17. Variables used in the FC.

Figure 2-18 shows what the Calc_Wgt FC looks like when placed in logic. Note that the user specifies the input and output tags.

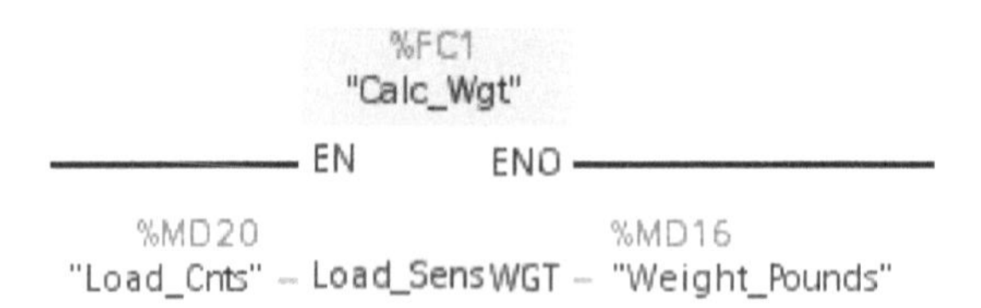

Figure 2-18. Calc_Wgt function.

Figure 2-19 shows the program blocks for the project. Note that the Calc_Wgt FC is now in the list. Note also that there is no DB for it. FCs do not have data blocks.

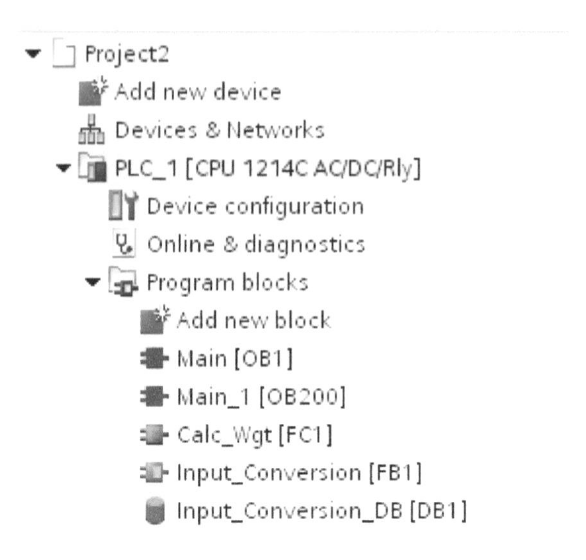

Figure 2-19. Calc_Wgt function added to the program blocks. Note that there is no associated data block (DB).

Figure 2-20 shows that the Calc_Wgt FC has been added to the project library. It could have been added to the global library. If it is added to the global library it would be available for use in all projects.

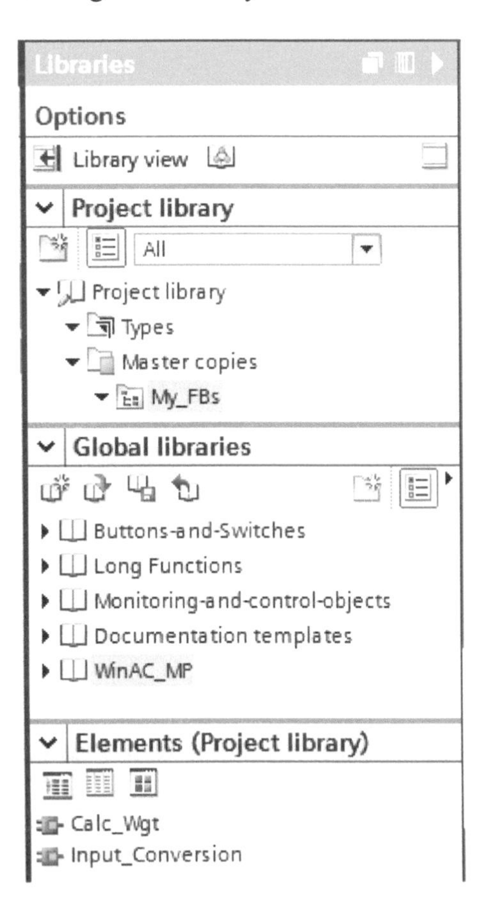

Figure 2-20. Function (FC) Calc_Wgt added to the project library. Note that we could have also added it to the global library.

Compiling Blocks

Blocks must be compiled to produce the code that the CPU executes. When you download a program, it starts the compile process before it can be downloaded to the CPU. You can also compile while you are developing your programs to find programming errors.

Know-How Protection

You may use Know-how protection for a block to prevent a program or its data from being read out or modified. You can still read block properties, parameters of the block interface, program structure, and global tags from a block that is know-how protected. You may also copy and paste the block (know-how protection is also copied), delete the block, call the block (FB or FC) from another block, compare online and offline version of the block, and modify the name and number in the block properties.

To configure Know-how protection, right click on the block you would like to protect and choose Know-how protection. If this is the first block you have protected you will be prompted to create a password. If it is not the first block you have protected you will have to enter the password. After a block has been protected a small padlock icon will appear on the block in the controller organizer. Protection can be removed by selecting the block in the controller organizer and then Edit/Manage undo protection for block (Block_Name), and entering the password.

If the password is lost, no access to the block is possible. You can only cancel protection of a block in its offline version. If you download a compiled version to the CPU, the recovery information is lost. A protected block that is uploaded from the CPU will overwrite the offline version and cannot be opened, even with the password.

Advantages of Modular Programming

When an application is broken into logical tasks and code blocks are used to accomplish each of the tasks, the program is easier to understand, troubleshoot, and manage.

The programmer can create their own FBs and FCs. The user can also configure them so that the logic cannot be seen. This can be used to protect proprietary logic for original equipment manufacturers or to make the instruction more user-friendly and understandable by "hiding" logic that a maintenance person does not need to see. FBs and FCs can be represented as “black boxes”. It is often not necessary for the user to know how the FB's or FC's functions were accomplished.

Nesting Blocks

The main and startup program have a nesting depth limit of 16. Interrupt blocks have a limit of 4. If more blocks are called the CPU will generate a program execution error.

Number Systems

Next, we will take a look at Siemens addressing. We must start by reviewing number systems. We are most familiar with the decimal number system (base 10). In the decimal system there are ten possible digits: 0 – 9. If we have to represent a number larger than 9, we then have to have two digits. The right-most digit represents the number of 1s in the number and the next digit to the left represents the number of 10s in the number. So, if we need to represent the number 14, there would be 1 ten and 4 ones (see Figure 2-21). There are other number systems. Computers work in the binary system.

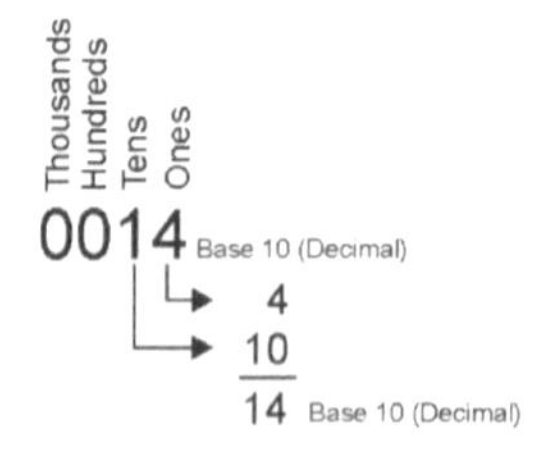

Figure 2-21. The number 14 (base 10) consists of 1 ten and 4 ones.

Binary Number System

The binary number system has two digits: 0 and 1. It is also called base 2.

In the binary system a number is represented by a number of bits (see Figure 2-22). There are 16 bits in the binary number shown in Figure 2-22. Eight bits are called a byte. There are two bytes in this example. Four bits are called a nibble. In common usage the length of a word is commonly thought of as 16 bits. Eight bits make a byte and 2 bytes make a word. 32 bits would be considered a double word.

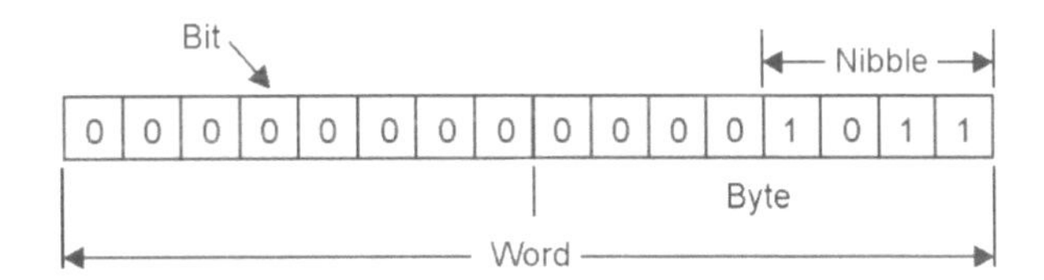

Figure 2-22. Bit, byte, and word.

Each position, left-to-right, in a binary number has a value (see Figure 2-23). The right most bit (bit 0) is the one's bit. The value of this bit can be zero or one. The next bit is the 2's bit. It can have the value of zero or two. The next bit is the 4's bit. Each bit is associated with a power of 2 based on its position in the number. Study Figure 2-23. The further to the left, the higher the power of 2. Note also that the bit in the right-most position is called the least-significant bit and the one on the left is called the most-significant bit. This is because the value of each bit gets larger as we move from right to left. In fact, the value of each bit position doubles as we move from right to left. Study the bit values on the top of Figure 2-23.

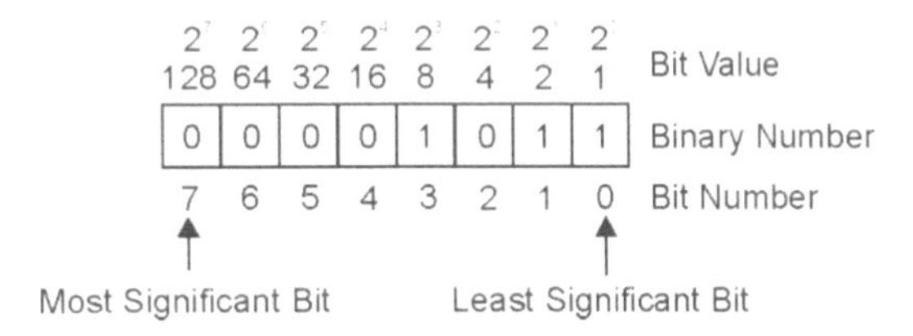

Figure 2-23. Binary system.

To convert a binary number to its decimal value you add the equivalent decimal value for each position in the binary number where a 1 is shown. Binary bit positions with a 0 do not add to the number value.

Figure 2-24 has an example of the conversion of a 4-bit binary number to its decimal equivalent.

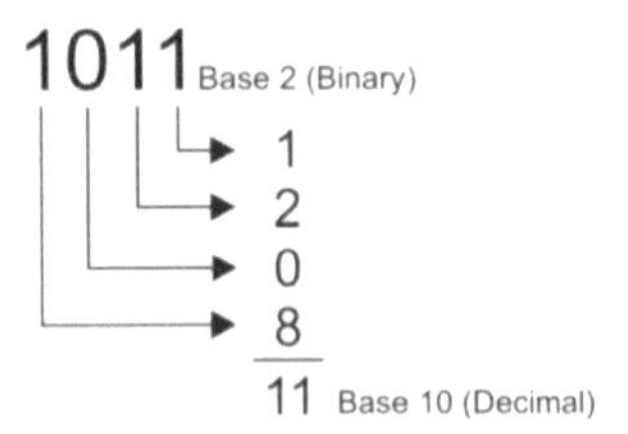

Figure 2-24. Conversion of binary to decimal.

One will often hear the term K when talking about computer or PLC memory. K is the abbreviation for Kilo. K usually means 1000. When K is used to describe the amount of PLC or computer memory, however, 1k means 1024. Remember that computers work in the binary system. In the binary number system (2^{10}=1024). So, if one is talking about memory 1k means 1024. 16K would be (16 * 1024 = 16,384). The term K is often used to refer to bits, bytes, or words.

BCD Number System

While PLCs use binary values, people want to see decimal numbers. Some input and output devices provide a decimal display that convert 4 binary digits to one decimal digit. In BCD numbering each decimal digit corresponds to four PLC binary inputs or outputs.

One example of a BCD device is a type of four-digit thumbwheel switch. Each thumbwheel digit is connected to 4 PLC inputs. A four-digit BCD thumbwheel would require 16 inputs. Each thumbwheel digit would have 10 available values: 0-9. Binary would require 4 bits to represent 0-9. So, the operator chooses a digit and the PLC would look at the values of 4 inputs. If the operator chose 7 the PLC would see 0111 at the 4 inputs connected to this digit of the thumbwheel switch.

Hexadecimal

Hexadecimal is another very commonly used number system in industrial controls. Hexadecimal (hex) is a base 16 system. There need to be 16 digits to represent each possible hex digit.

The numbers 0-9 are used to represent the first 10 hex digits. Then the letters A-F are used to represent the values of 10-15. Study Figure 2-25. The columns on the right of the figure show a comparison of the binary, hex and decimal system. Note the decimal values of the hex digits (0-15).

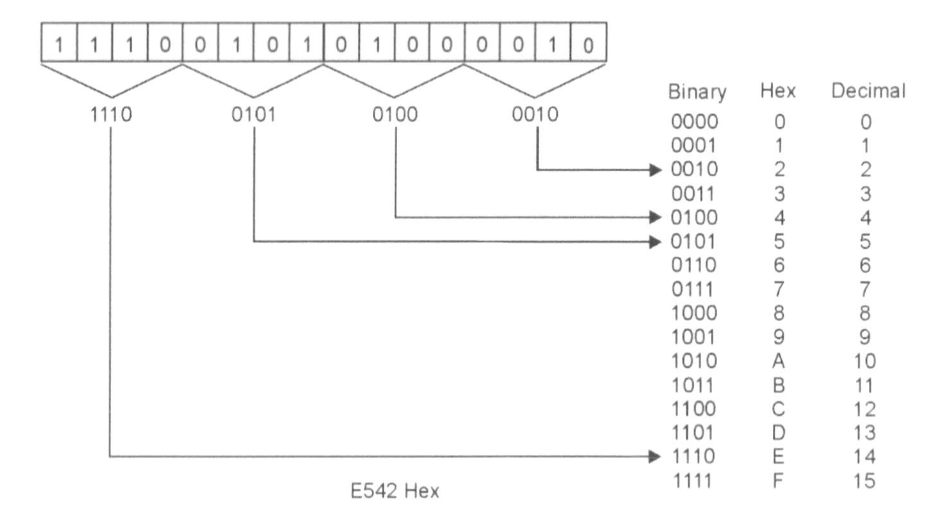

Figure 2-25. Conversion of Binary to Hexadecimal.

The hexadecimal system is commonly used when working with computers or PLCs because it allows the status of a large number of binary bits to be represented by a much shorter hex number. For example, a 16-digit binary number can be represented by 4 hex digits. Each hexadecimal character represents four binary bits. Figure 2-25 shows an example the conversion of a 16-bit binary number to hex and decimal. First each 4 bits of the binary number were converted to their equivalent hex digit. In this example the first 4 bits value was 2. The value of the next 4 bits was 4, The third three binary digit value is 5 and the last 4 (most significant) binary bits had the decimal value of 14 which is E in hex.

Next the value of the 4 hex digits will be converted to their decimal equivalent. Just as the value of each binary digit increased as a power of 2 as we moved from right to left, each digit of a hex number increases by a power of 16 as we move from right to left. Study Figure 2-26. The first hex (right-most) digit represents the number of ones (1^0 power) in the number. This can have the value of 0-F (base 16) or 0-15 (base 10). The second hex digit represents the number of 16's (16^{1st} power). The third hex digit represents the number of 256s (16^{2nd} power) in the number. The fourth hex digit would represent the number of 4096s (16^3 power) in the number.

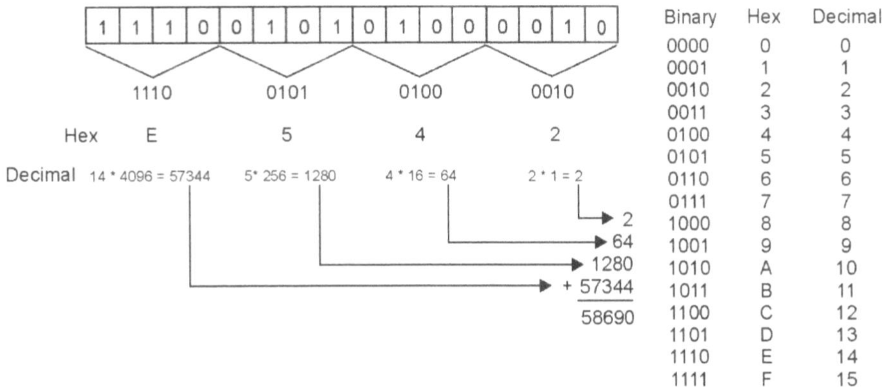

Figure 2-26. Hex digits and weights.

You use the binary number system (base 2) for counting the numbers. A word can represent an integer number from -32768 to +32767. The bit with the value 215 is used to denote a negative number (when position 215 has the value "1" the number is negative).

The CPU organizes memory into groups. Figure 2-27 shows the data types. Note that the CPU also supports a grouping of 8 bytes that form a "long real" data type (LReal) for storing very large or very precise values. The range for the LReal is: +/-2.23 x 10^{-308} to +/- 1.79 x 10^{308}

Type of Data	Size	Range
Bool (Boolean)	1 bit (0, 1, TRUE, FALSE)	0 to 1
Byte	8 bits (1 byte)	0-255
Word	16 bits (2 bytes)	0-65,535
DWord	32 bits	0- 4,294,967,295
Char	8 bits (1 byte)	0-255
SInt	8 bits (1 byte)	-128 to 127
USint	8 bits (1 byte)	0 to 255
Int	16 bits (2 bytes)	-32,768 to 32767
UInt	16 bits (2 bytes)	0 to 65535
DInt	32 bits (4 bytes)	-2,147,483,648 to 2,147,483,647
UDInt	32 bits (4 bytes)	0 to 4,294,967,295
Real	32 bits (4 bytes)	+/-1.18 x 10^{-38} to +/-3.40 x 10^{38}
LReal	64 bits (8 bytes)	+/-2.23 x 10^{-308} to +/-1.79 x 10^{308}
Time	32 bits (4 bytes)	T#-24d_20h_31m_23s_648ms to T#24d_20h_31m_23s_647ms Stored as: -2,147,483,648 ms to +2,147,483,647 ms
String	Variable	0 to 254 byte-size characters
DTL (Date and Time)	12 bytes	Minimum: DTL#1970-01-01-00:00:00.0 Maximum: DTL#2554-12-31-23:59:59.999 999 999

Figure 2-27. Data types.

Note that a user can also create their own data types. Under "PLC data types" in the project tree there is a "Add new data type" choice. The user can create their own data type including simple data types and arrays or structures. For example, if a user had several of the same type of machines, they might want to make a new data type that held the common types of data for a generic machine and then use it for each of their machines. These are sometimes called User-defined Data Types (UDTs).

Figure 2-28 shows how memory bits, bytes, words and double words are addressed. A word is 16 bits long. A double word is 32 bits long. A byte is 8 bits long. Notice that the whole 32 bits could be addressed as MD24 (memory, double-word 24). There are two words in the double word: %MW24 and %MW26. There are four bytes in the double word: %MB24, %MB25, %MB26, and %MB27). Individual bits can be addressed as shown. The two examples of bit addressing are %M25.6 and %M27.1. The next double word address would begin at %MD28.

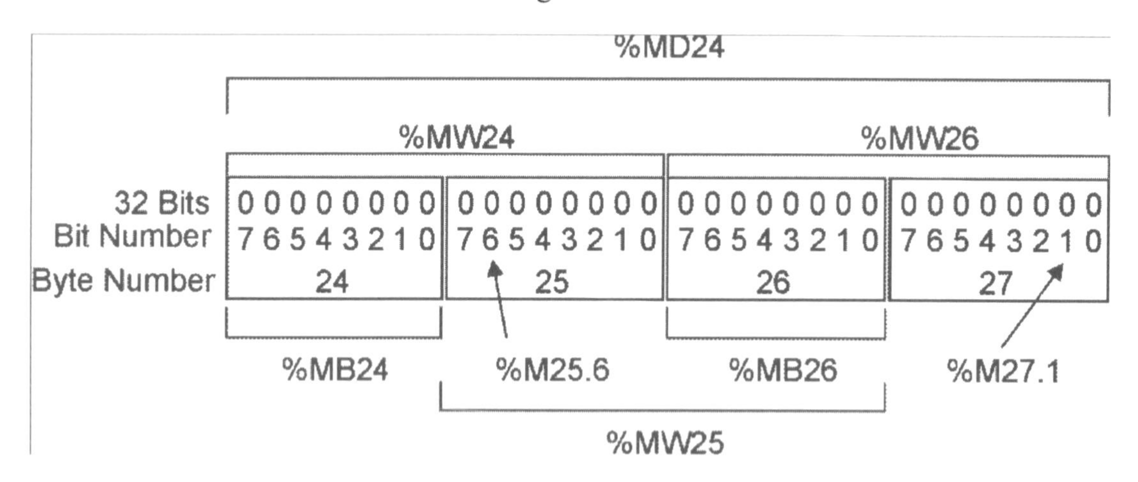

Figure 2-28. Memory addressing.

Figure 2-29 shows how different types of memory are addressed. Note that x = the bit address; y = the byte address; and z = the data block number.

Type	Type ID	Bit (1 Bit)	Byte (8 Bits)	Word (16 Bits)	Doubleword (32 Bits)
Input	I	%Iy.x	%IBy	%IWy	%IDy
Peripheral	:P added to the	%Iy.x:P	%IBy:P	%IWy:P	%IDy:P
Output	Q	%Qy.x	%QBy	%QWy	%QDy
Peripheral	:P added to the	%Qy.x:P	%QBy:P	%QWy:P	%QDy:P
Bit Memory	M	%My.x	%MBy	%MWy	%MDy
Data	DB	%DBz.DBXy.x	%DBz.DBBy	%DBz.DBWy	%DBz.DBDy
z = Data Block Number, x = Bit Address, y = Byte Address					

Figure 2-29. Addressing.

Absolute I/O Addressing

When absolute addressing is used a signal or value is addressed directly using the memory address. The address of an input or output of a digital module consists of a byte address and a bit address. The byte address depends on the module start address. The bit address is the number printed on the module. Figure 2-30 shows how the addresses of the individual channels of a digital module are obtained. Every module will have a start byte address. The first 8 discrete inputs or outputs would be addressed based on the starting byte address. The next 8 inputs or outputs would use an address that would use the start byte + 1.

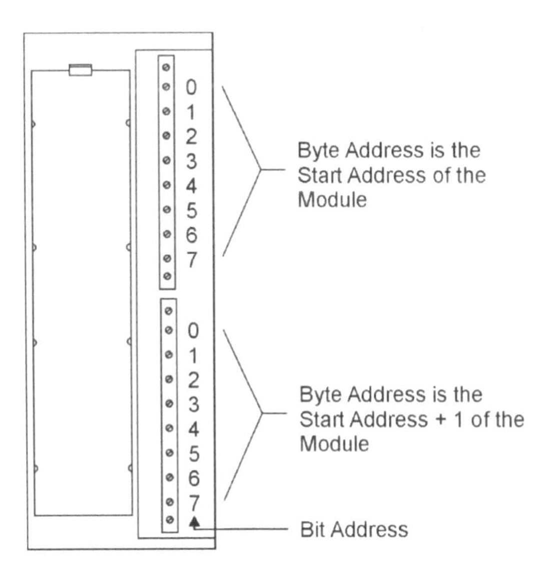

Figure 2-30. Byte addressing on a digital module.

Figure 2-31 shows two examples of addressing. The first part of the I/O address is whether it is an input (I) or an output (Q). The second part of the address is the byte address on the module where the input or output is located. The next part of the address is the actual bit position within the byte. The first example in the figure would be a discrete input located on byte 1, bit 5. Bit 5 would be the sixth input in the byte (0, 1, 2, 3, 4, 5). The second example would be an output located in an output module with a byte address of 4 and a bit address of 5. This would be the 6th output in the module with that byte address.

Figure 2-31. Digital input and output addressing.

Addressing for a S7-300

Figure 2-32 shows an example of module byte addressing. This module in slot 4 has a starting byte address of 0. The module is a 16-point module. It has 2 byte addresses: 0 and 1. Note that the actual I/O addresses for the first 8 are 0.0, 0.1, 0.2, 0.3, 0.4, 0.5, 0.6, and 0.7. The second 8 I/O would be addresses 1.0 through 1.7. If this were an input module the address would be I0.0 – I0.7 and I1.0 – I1.7.

The starting addresses for a S7-300 PLC is shown in Figure 2-33. Note the module in slot 4 in Rack 0. The starting address byte for this module would be 0. There are 4 bytes that can be used for this slot for a digital module. They would be bytes 0-3. Each byte has 8 bits. The module in this figure is a 16-point module so it would use 2 bytes. If a module with 8 inputs was installed in this slot it would only use byte 0. The addresses of inputs in this module would be I0.0 – I0.7. A 16-point input module in this slot would use 2 bytes for addresses, I0.0 – I0.7 and I1.0 – I1.7. If a 32-input module were used it would use 4 bytes (0-3).

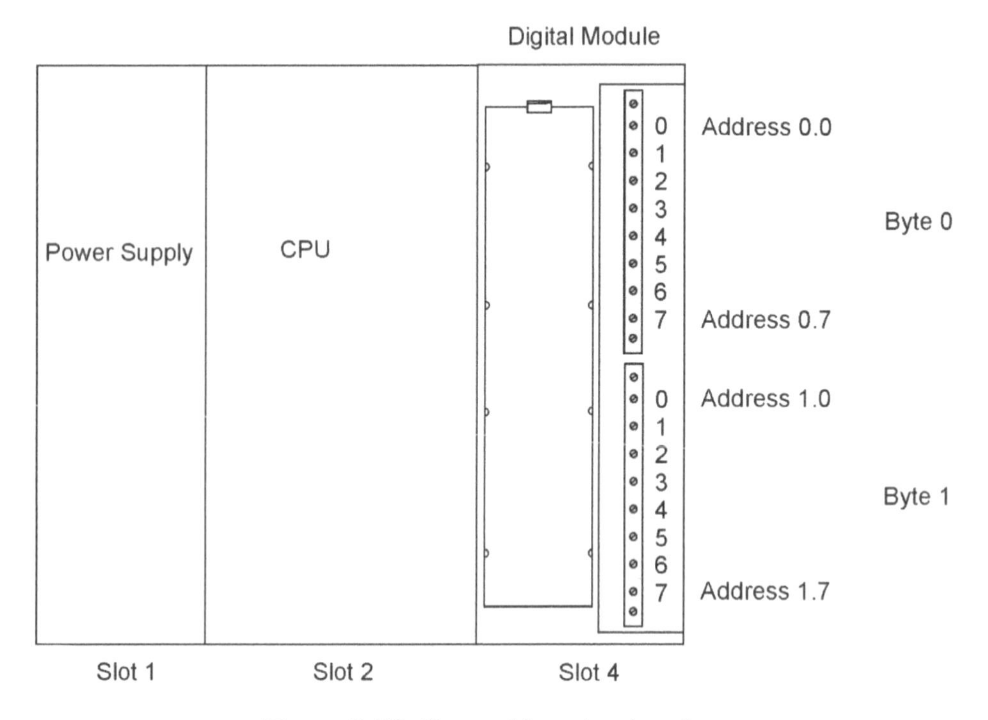

Figure 2-32. Byte addressing by slot.

If we added additional racks the addressing would be similar. Examine Figure 2-33. Look at the addressing for Rack 1 slot 4. If we installed a digital module in slot 4 the module could use bytes 32-35 (4 bytes). If a module with 16 outputs is installed, the addresses would be Q32.0-Q32.7 and Q33.0 – Q33.7. If an analog card was installed instead, the addresses would start at 384.

Note that in some models of the S7-300 it is possible for the user to define their own addresses for the modules. User-defined address allocation means that the user is free to allocate modules an address of their choice. The addresses are allocated in STEP 7 software.

The user simply defines the start address of the module, and then the addresses of this module are based on the user defined start address.

Rack	**I/O Type**	**Slot Number**										
		1	2	3	4	5	6	7	8	9	10	11
0	Digital	Power	CPU	Interface	0	4	8	12	16	20	24	28
1	Digital	-		Interface	32	36	40	44	48	52	56	60

Figure 2-33. Byte addressing for digital and analog modules for rack 0 and rack 1.

Study the example in Figure 2-34. If we used an analog module in slot 4 the address of the analog inputs or outputs would start at address 256 and go to 271 (16 words). Note that the next possible analog byte address would be 272.

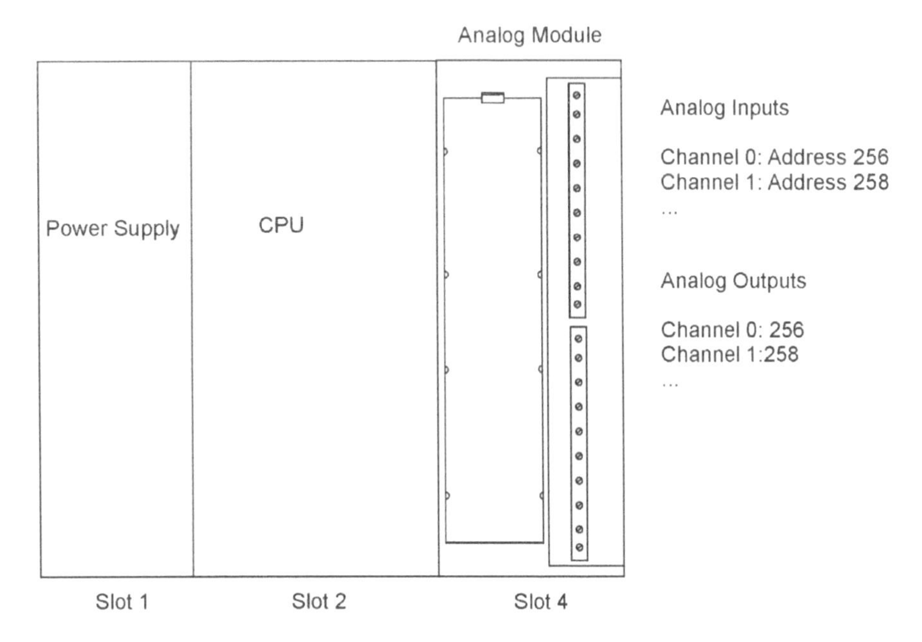

Figure 2-34. Analog byte addressing.

Integrated I/O addressing for the 312 and the 314 CPU

The 300 CPUs have some I/O available that are integrated into the CPU module. Figure 2-35 shows the integrated I/O for a 312 model CPU. There are 10 digital inputs and 6 digital outputs available. 4 of the inputs are special purpose and can be used for high-speed counter or frequency functions.

Inputs/Outputs	Addresses	Description
10 Discrete Inputs	124.0 – 125.1 4 of these are special (124.6-125.1).	The 4 special channels can be used for counter and frequency functions or as interrupt inputs.
6 Discrete Outputs	124.0 – 124.5	

Figure 2-35. Integrated I/O for a 312 IFM CPU.

Figure 2-36 shows the integrated I/O for a 314 model CPU. There are 20 digital inputs and 16 digital outputs available. There are also 4 analog inputs and 1 analog output available. 4 of the inputs are special purpose and can be used for high-speed counter or frequency functions.

Inputs/Outputs	Addresses	Description
20 Discrete Inputs	124.0 – 126.3 4 of these are special (126.0-126.3).	The 4 special channels can be used for counter, frequency, positioning functions or as interrupt inputs.
16 Discrete Outputs	124.0 – 125.7	
4 Analog Inputs	128 - 135	
1 Analog Output	128 - 129	

Figure 2-36. Integrated I/O for a 314 IFM CPU

Addressing for a S-7 1200

Figure 2-37 shows a typical S7-1200 PLC. The PLC station has 4 digital modules installed and 1 analog module installed. The CPU also has a signal module (SM) installed in the CPU.

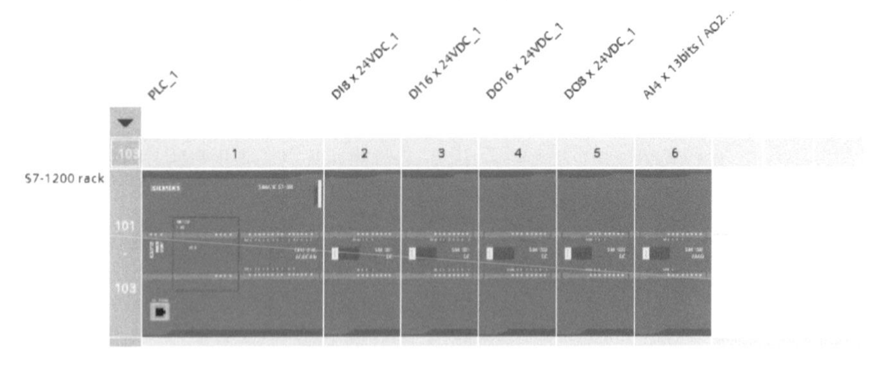

Figure 2-37. An example of a S7-1200 station. Courtesy Siemens.

Figure 2-38 shows the I/O addressing for the S7-1200 PLC shown in Figure 2-34. Note that the first external discrete module is listed as being in slot 2. Slot 2 has an 8 input DC module installed in it. The starting byte for this module is byte 8. The module has bytes 8, 9, 10, and 11 reserved for it. There are only 8 inputs so only byte 8 will be used. The addresses for this module would be I8.0 – I8.7.

Device overview

Module	Slot	I address	Q addre...	Type	Order no.	Firmware
▾ PLC_1	1			CPU 1214C AC/DC/Rly	6ES7 214-1BE30-0XB0	V1.0
DI14/DO10	1.1	0...1	0...1	DI14/DO10		
AI2	1.2	64...67		AI2		
AO1 x 12bit...	1.3		80...81	AO1 signal board	6ES7 232-4HA30-0XB0	V1.0
HSC_1	1.16			High speed counters (HSC)		
HSC_2	1.17			High speed counters (HSC)		
HSC_3	1.18			High speed counters (HSC)		
HSC_4	1.19			High speed counters (HSC)		
HSC_5	1.20			High speed counters (HSC)		
HSC_6	1.21			High speed counters (HSC)		
Pulse_1	1.32			Pulse generator (PTO/PWM)		
Pulse_2	1.33			Pulse generator (PTO/PWM)		
▸ PROFINET in...	X1			PROFINET interface		
DI8 x 24VDC_1	2	8		SM 1221 DI8 x 24VDC	6ES7 221-1BF30-0XB0	V1.0
DI16 x 24VDC...	3	12...13		SM 1221 DI16 x 24VDC	6ES7 221-1BH30-0XB0	V1.0
DO16 x 24VD...	4		16...17	SM 1222 DO16 x 24VDC	6ES7 222-1BH30-0XB0	V1.0
DO8 x 24VDC...	5		20	SM 1222 DO8 x 24VDC	6ES7 222-1BF30-0XB0	V1.0
AI4 x 13bits /...	6	160...167	160...163	SM 1234 AI4/AO2	6ES7 234-4HE30-0XB0	V1.0
	7					
	8					
	9					

Figure 2-38. Addressing for a S7-1200 station.

Slot 3 has a 16 input DC module. The starting byte address for his module is byte 12. Four-byte addresses are reserved for this module: 12, 13, 14, and 15. The module has 16 inputs so only bytes 12 and 13 will be used. The addressing for the module will be I12.0-I12-7 and I13.0- I13.7.

The module in slot 4 is a DC output module with 16 outputs. The starting byte address for this module is 16. The module has 4 bytes reserved for it, bytes 16-19. The module only has 16 outputs so only bytes 16 and 17 will be used for addressing this module. The addresses would be Q16.0-Q16.7 and Q17.0-Q17.7. Slot 5 has an 8 output DC module installed. Its byte address will be 20. The outputs will be addressed as Q20.0-Q20.7.

Lastly there is an analog combination module installed in slot 6. It has 4 analog inputs and 2 analog outputs on it. Its analog input addresses will be IW160, IW162, IW164, and IW166. The analog outputs addresses will be QW160 and QW162. The W in the analog addresses means word. Each analog input or output uses 2 bytes (one word).

S7-400 Addressing

In the S7-400 CPU starting address can be calculated by the module's slot location. The starting byte is determined by taking the slot location – 1 and then multiplying by 4. In the example shown in Figure 2-39 the module is located in slot 6. In this example (6-1) * 4 = 20. The starting byte address for this module would be byte 20. The actual input addresses for the module are shown in the figure. This is a 32-bit module. The addresses would be: I20.0- I20.7, I21.0-I21.7, I22.0-I22.7, and I23.0-I23.7.

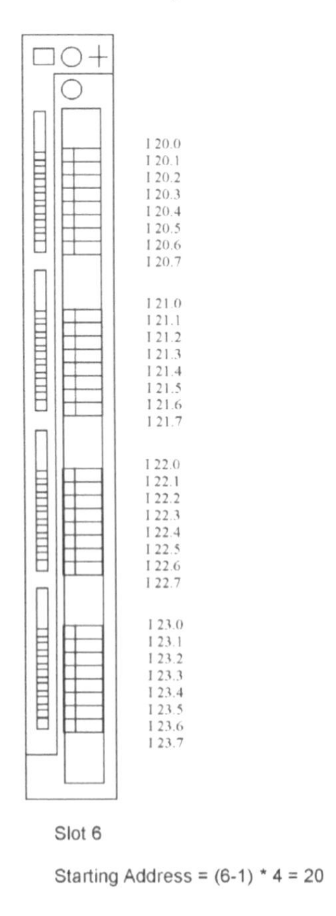

Figure 2-39. Input byte addressing for a S7-400.

Figure 2-40 shows an analog module. The starting byte is determined by taking the slot location – 1 and then multiplying by 64. In the example shown in Figure 2-40 the module is located in slot 4. In this example (4-1) * 64 = 192. The starting byte address for this module would be byte 192. The actual input addresses for the module are shown in the figure. The addresses for the 8 analog outputs would be: QW192, QW194, QW196, QW198, QW200, QW202, QW204, and QW206. Note that each analog output (or input) uses 2 bytes. Note also the W in the QW means word or 2 bytes.

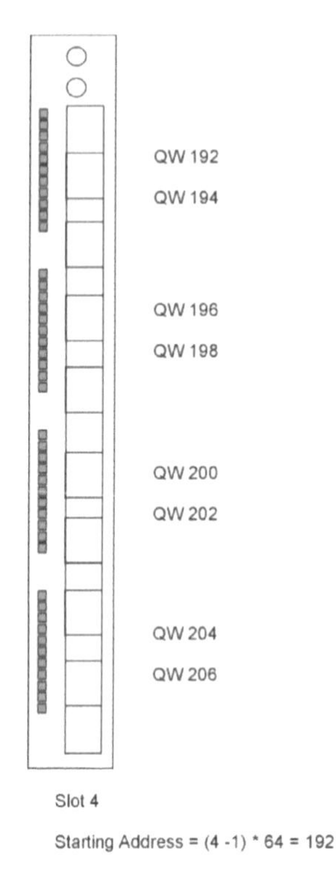

Figure 2-40. Analog byte addressing for a S7-400.

Symbolic Addresses (Tagnames)

STEP 7 facilitates symbolic programming. You create symbolic names or "tags" for the addresses of the data, whether as PLC tags relating to memory addresses and I/O points or as local variables used within a code block. These tagnames should be descriptive of use of the I/O point. The use of symbolic addressing (tagnames) makes the logic much easier to understand. To use these tags in your user program, simply enter the tag name for the instruction parameter. For a better understanding of how the CPU structures and addresses the memory areas, the following explains the "absolute" addressing that is referenced by the PLC tags.

Local and Global Tags

A local tag is known only in the block in which it was created. You can use the same name for a local tag in different blocks for different purposes. A global tag is known throughout the entire program.

Memory Areas of the S7-1200

The CPU provides the following memory areas to store the user program, data, and configuration data: load and work memory.

Load Memory

Load memory is non-volatile storage for the user program, data and configuration. When a project is downloaded to the CPU, it is first stored in the Load memory area. This area is located either in a memory card (if present) or in the CPU. This non-volatile memory area is maintained through a power loss. The memory card supports a larger storage space than that built-in to the CPU.

Work Memory

Work memory is volatile storage for some elements of the user project while executing the user program. The CPU copies some elements of the project from load memory into work memory. This volatile area is lost when power is removed, and is restored by the CPU when power is restored.

Retentive Memory
Retentive memory is non-volatile storage for a limited quantity of work memory values. The retentive memory area is used to store the values of selected user memory locations during a power loss. When a power down occurs, the CPU has enough hold-up time to retain the values of a limited number of specified locations. These retentive values are then restored upon power up.

Retentivity of Data Values
To prevent data loss if power is lost, data values can be stored in a retentive memory area.

Retentive Memory Areas
Data loss after power failure can be prevented by making data retentive. Data set up to be retentive is stored in a retentive memory area. A retentive memory area is an area that retains its content during a STOP to RUN transition.

The values of retentive data are deleted during a cold restart.

The following data can be made retentive:

- Bit memory: The precise width of the memory can be defined for bit memory in the PLC tag table or in the assignment list.
- Function block (FB) tags: In the interface of an FB, individual tags can be defined as being retentive if symbolic tag addressing has been enabled. Retentivity settings can only be defined in the assigned instance data block if symbolic addressing is not activated for the FB.
- Global data block tags: depending on the settings for symbolic addressing retentivity can be defined either for individual or for all of the tags of a global data block. If the "Symbolic access only" attribute of the DB is enabled, retentivity can be set for each individual tag. If the "Symbolic access only" attribute of the DB is disabled, the retentivity setting applies to all of the tags of the DB; either all tags are retentive or no tags are retentive.

Retentivity can be set in software. You select the Retain button and the screen shown in Figure 2-41 appears. Then you enter the number of memory bytes you would like to make retentive.

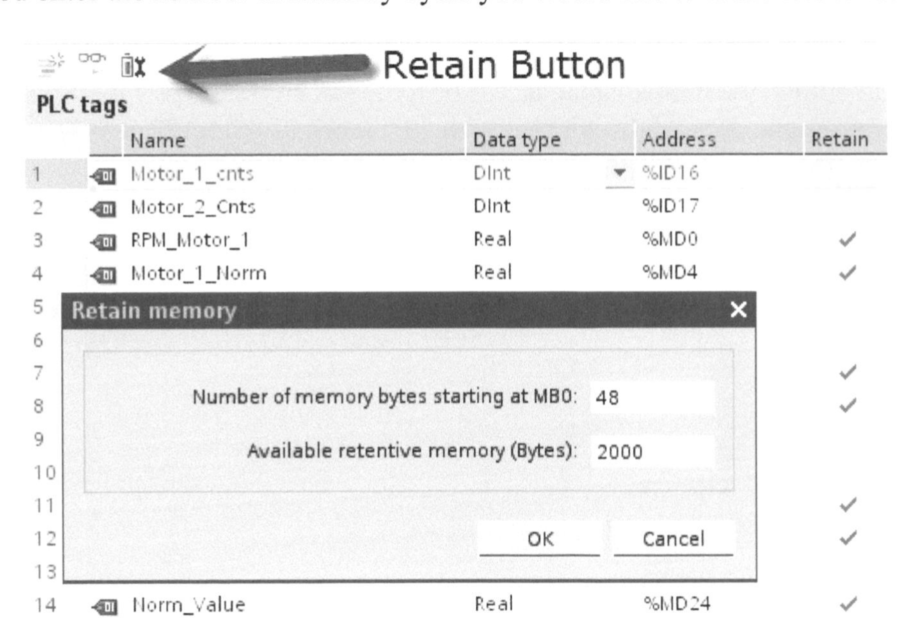

Figure 2-41. Setting retentivity.

An optional SIMATIC memory card provides an alternative memory for storing the user program. A memory card can also be used to transfer programs. If you use the memory card, the CPU runs the program from the memory card and not from the memory in the CPU. The CPU supports only a preformatted SIMATIC memory card.

The CPU provides a variety of specialized memory areas (see Figure 2-42), including inputs (I), outputs (Q), bit memory (M), data block (DB), and local or temporary memory (L). Your user program accesses (reads from and writes to) the data stored in these memory areas. Each different memory location has a unique address. Your user program uses these addresses to access the information in the memory location. The Process-Image Input table (PII) is the area of memory in which the states of all digital inputs are stored. The states of all digital outputs are stored in the Process-Image Output table (PIQ) memory area.

Memory Type	**Description**
I Process image input	The CPU copies the state of the physical inputs to I memory at the beginning of the scan cycle. To immediately access or to force the physical inputs, append a ":P" to the address or tag (such as "Start:P" or I0.3:P).
Q Process image output	The CPU copies the state of Q memory to the physical outputs at the beginning of the scan cycle. To immediately access or to force the physical outputs, append a ":P" to the address or tag (such as "Stop:P" or Q0.3:P).
M Bit memory	The user program reads and writes the data stored in M memory. Any code block can access the M memory. You can configure addresses within M memory to retain the values of the data after a power cycle.
L "Temp" memory	Whenever a code block is called, the CPU allocates the temporary, or local, memory (L) to be used during the execution of the block. When the execution of the code block finishes, the CPU reallocates the local memory for the execution of other code blocks.

Figure 2-42. Memory types.

Whether you use a tag (such as "Start" or "Stop") or an absolute address (such as "I0.3" or "Q1.7"), a reference to the input (I) or output (Q) memory areas accesses the process image and not the physical output. To immediately access or force the physical (peripheral) input or output in your user program, append the address with ":P" (such as "Stop:P" or "Q0.3:P"). Figure 2-43 shows the memory types and their identifiers.

Operand Area	Description	Size	S7 Identifier
Process Image Output	The CPU writes the values from the process image output table to the output modules at the start of the cycle.	Output Bit	Q
		Output Byte	QB
		Output Word	QW
		Output Double Word	QD
Process Image Input	The CPU reads the inputs from the input modules and saves the values to the process image input table at the start of the cycle.	Input Bit	I
		Input Byte	IB
		Input Word	IW
		Input Double Word	ID
Bit Memory	This area provides storage for intermediate results calculated in the program.	Memory Bit	M
		Memory Byte	MB
		Memory Word	MW
		Memory Double Word	MD
Data Block	Data blocks store information for the program. They can either be defined so that all code blocks can access them (global DBs) or assigned to a specific FB or SFB (instance DB). Requirement: The block attribute "Symbolic access only" is not enabled.	Data Bit	DBX
		Data Byte	DBB
		Data Word	DBW
		Data Double Word	DBD
Local Data	This area contains the temporary data of a block while the block is being processed. Requirement: The block attribute "Symbolic access only" is not enabled. Recommendation: Only access local data (temp) symbolically.	Local Data Bit	L
		Local Data Byte	LB
		Local Data Word	LW
		Local Data Double Word	LD
I/O Area	The I/O input area permits direct access to central and distributed input and output modules.	I/O Bit	<Tag>:P
		I/O Byte	
		I/O Word	
		I/O Double Word	

Figure 2-43. Memory areas and identifiers.

Examples of Tag Addressing

Figure 2-44 shows some examples of tag addressing. The first column is the symbolic tagname. The second column shows the type of data and the third column shows the actual address that was used for each tag. The first tag (Analog_In_1) is an input from an analog input card. Int was chosen for the data type for this input. The address for the input is %IW96. The I means input, the W means word, and the 96 means byte 96 in memory. The 96 is the starting byte address of an input on an analog input card. The tag named Temp_5 is not a real input; it is just a tag value stored in memory. The type for this tag is Real and the actual address in memory is %MD28. The MD means Memory Double word (32 bits - 4 bytes). The starting byte for this tag is 28. The tag uses 4 bytes so the next tag (Temp_6) was started at memory byte address 32, as Temp_5 used bytes 28, 29, 30, and 31. Sensor_2 is a discrete input. The type is Bool. The address of the tag is %I0.2. The I means that it is an input. The 0 in the address would describe which byte it would be based on the card the input is connected to. The next number after the period (2) would specify that this tag represents the 3rd input in the byte. The first input would be 0, second 1, and third would be 2.

Output_4 is a discrete output. The type is Bool. The actual address in memory is %Q2.4. The Q means output. The 2 specifies that it is located in the module whose starting address is byte 2. It is output bit 4, the 4th output (0, 1, 2, 3, 4). The tag named Analog_Out_1 is an output on an analog output card. The card's starting memory byte address is 112. This analog output is a type Int so it will use two bytes in memory 112 and 113.

Analog_In_1	Int	%IW96
Temp_5	Real	%MD28
Temp_6	Real	%MD32
Part_Cnt_1	DInt	%MD36
Part_Cnt_2	DInt	%MD40
Value_1	DWord	%MD44
Value_2	DWord	%MD48
Sensor_2	Bool	%I0.2
Output_4	Bool	%Q2.4
Analog_Out_1	Int	%QW112

Figure 2-44. Examples of tag addressing

Function

When the user program addresses the input (I) and output (O) operand areas, it does not query or change the signal states directly on the digital signal modules. Instead, it accesses a memory area in the system memory of the CPU. This memory area is referred to as the process image.

The process image is part of the CPUs system memory. The process image consists of the input process (inputs (I)) and output process image (outputs (Q)). The input area of the process image is called the Process-Image Input (PII) table. The output area of the process memory is called the Process-Image Output (PIQ) table. The process image in the S7-1200 is 1024 bytes. This would be addresses 0-1023. All of the digital and analog I/O channels are in this memory range.

After a CPU restart and before the CPU executes the main program, the operating system will transfer the signal states of the output process image to the outputs modules and take the input states from the input modules and write them into the input process image. After that has occurred, the CPU will begin execution of the main program. After the main program is executed a new cycle starts with the update of the process image, followed by program execution, etc.

Advantages of the Process Image

Compared with direct access to input and output modules, the main advantage of accessing the process image is that the CPU has a consistent image of the process signals for the duration of one program cycle. If a signal state on an input module changes during program execution, the signal state in the process image is retained until the process image is updated again in the next cycle. The process of repeatedly scanning an input signal within a user program ensures that consistent input information is always available.

Access to the process image also requires far less time than direct access to the signal modules since the process image is located in the internal memory of the CPU.

Append the suffix ":P" to the I/O address if you want the program to access user data directly instead of using the process image. This could be necessary, for example, during execution of a time-sensitive program which also has to control the outputs within the same cycle.

Creating Arrays

An array is a group of data of the same data type. Imagine you need to keep a record of the process temperature every 30 minutes for a 5-hour process. This would be 10 temperatures. We could create 10 individual tags to keep the temperatures in. We could also create a tag array so that all 10 could have the same tag name and just be addressed individually by an index value. Let's examine how to create an array in TIA Portal.

Arrays are created in Data Blocks. The Add new Block under the Program Blocks area of TIA Portal is chosen See Figure 2-45.

Figure 2-45. Adding a new data block for a tag array.

The new data block was named Block_1 in Figure 2-46.

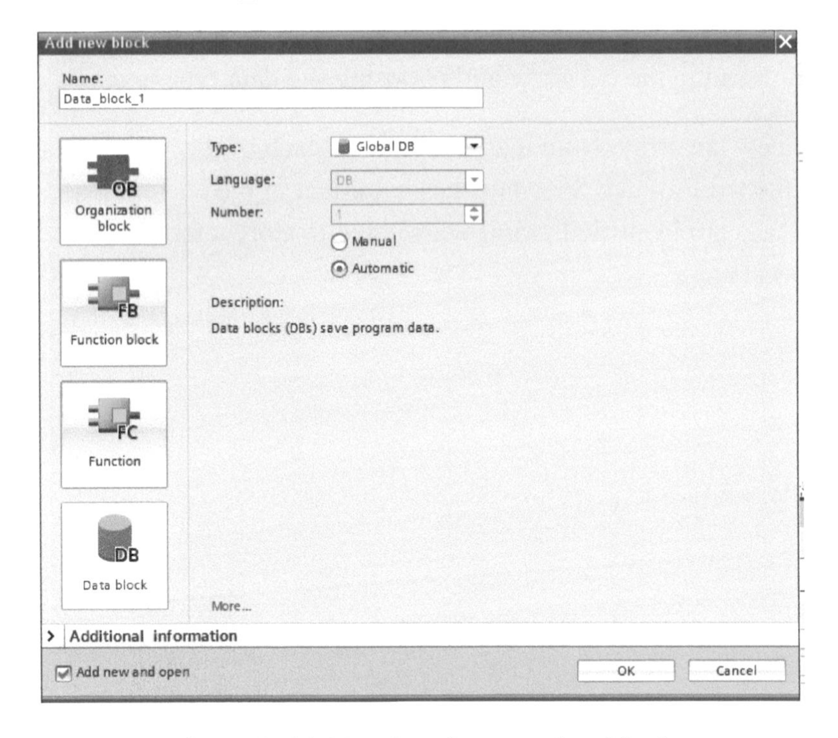

Figure 2-46. Naming the new data block.

After you click on the OK button the screen in Figure 2-47 will appear. Note the Add new box. This is where we will name the array tag.

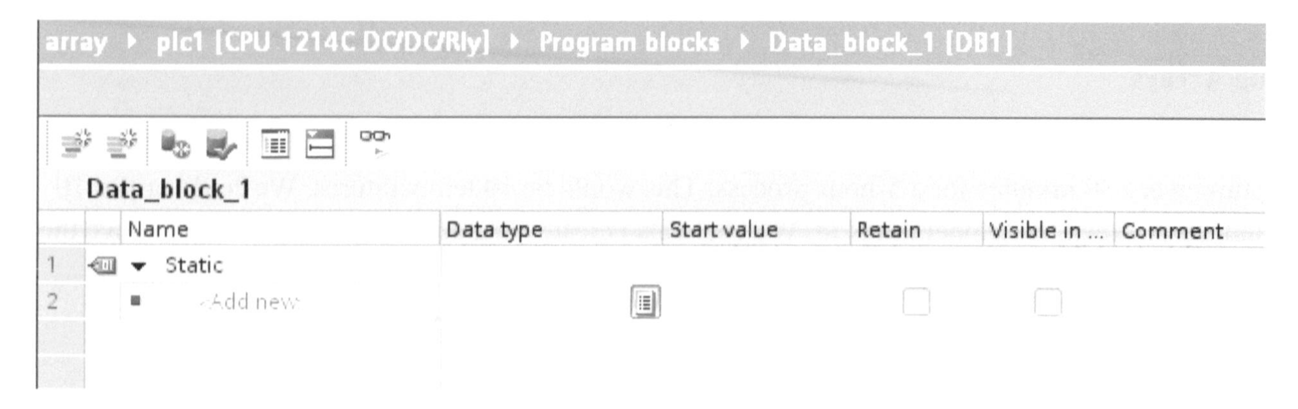

Figure 2-47. Adding an array tag.

The array tag was named Product_Spec in Figure 2-48 and array was chosen for the type. The type of array to be created was Int and the array size will be 10.

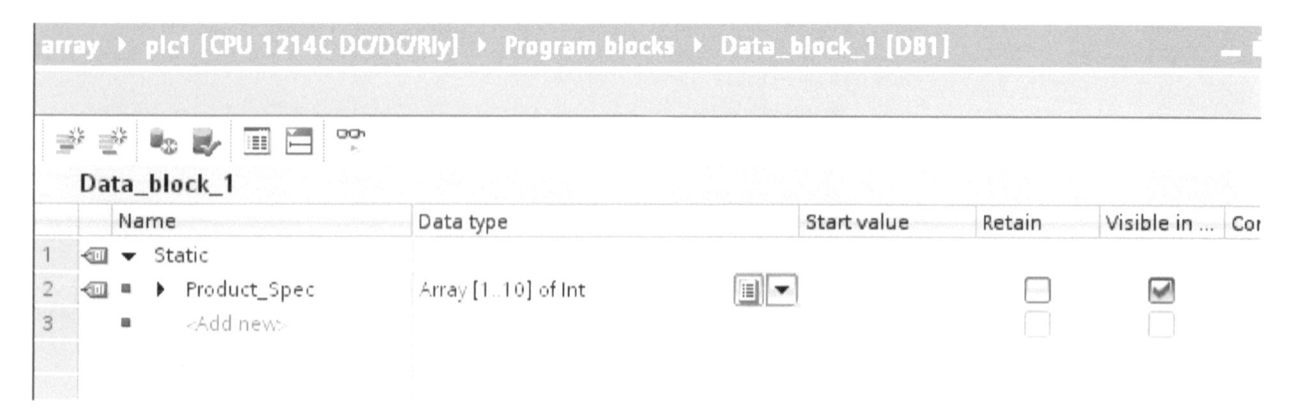

Figure 2-48. Naming the tag array and choosing the data type and size of the array.

Figure 2-49 shows the new tag array. Note that the name of each of the 10 array tags is Product_Spec. Note also that each of the ten Product_Spec tags has a number at the end enclosed by square brackets. This is how individual tags are identified. So, if we wanted to store a temperature in the 10th tag we would use Product_Spec[10].

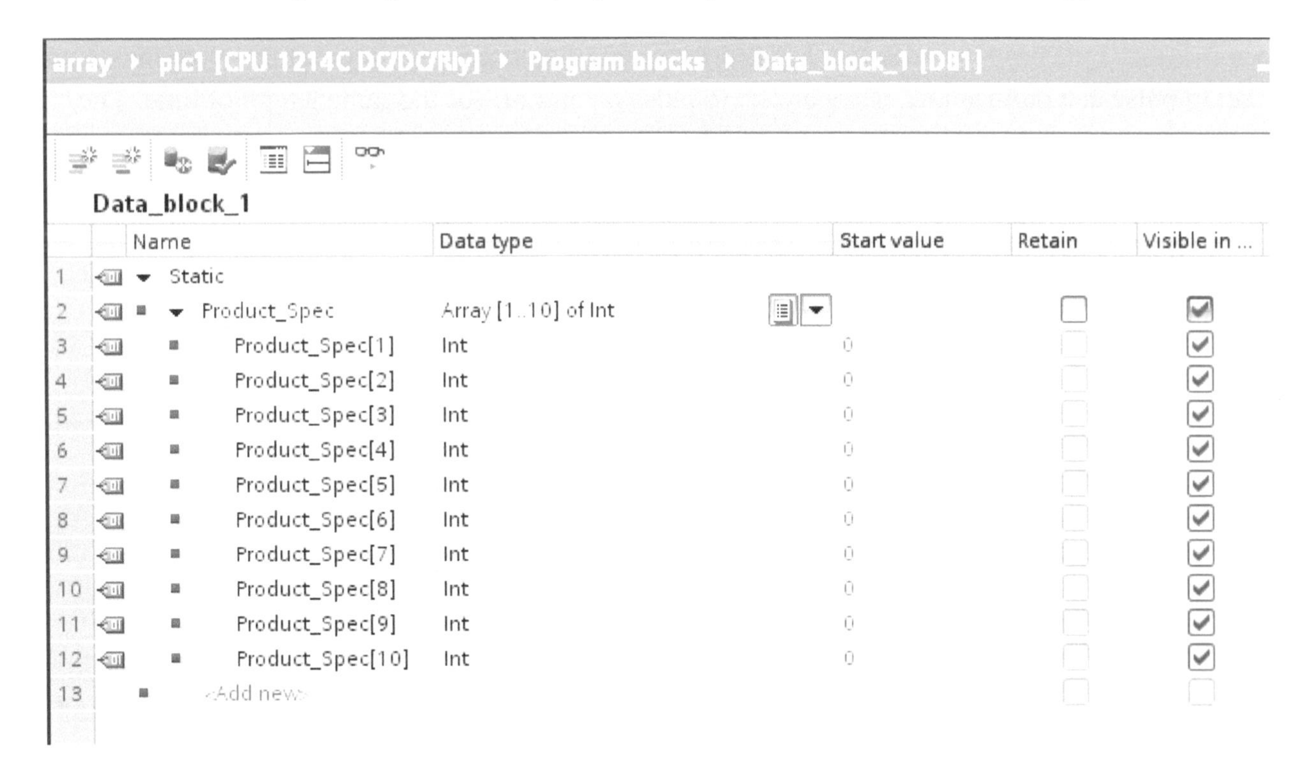

Figure 2-49. Completed array of 10 tags.

Structure Tags

An array-type tag can only have one data type for all of the tags. A structure-type tag can have multiple data types for the tags. For example, we may have similar machines in our automated system that all have similar data available. We could create a structure to hold those types of data. Figure 2-50 shows an example of a structure-type DB named machine that was created to have 5 tag members. Note that 4 different data types were used for the tag members.

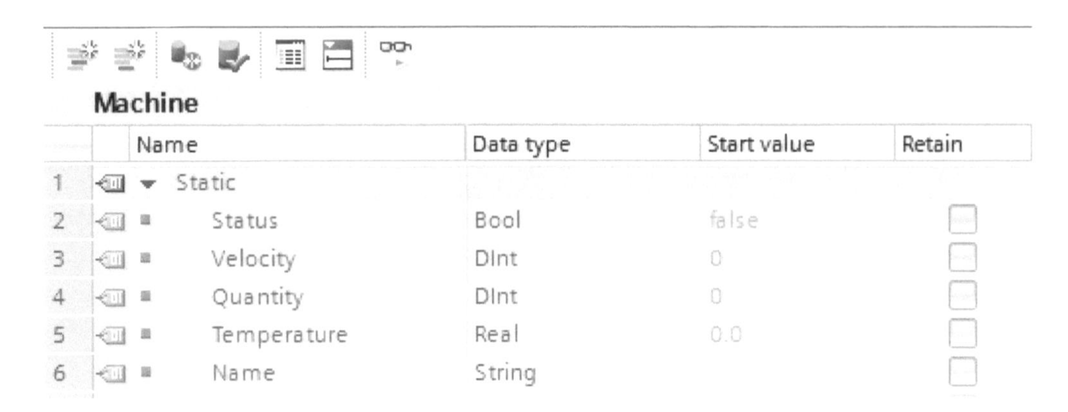

Figure 2-50. A structure-type DB.

System Memory Bits

You can define a clock and a system memory byte with fixed purpose, individual bits whose states are controlled by the operating system. These bits can be very useful to the programmer. Figure 2-51 shows the two types of memory bytes: clock and system. You can enable the clock and system bytes and define their addresses in the CPU properties. You can also look up the purpose of each of the bits in the CPU properties under the system and clock memory configuration.

The clock memory byte has 8 bits that can be used for timing functions in your program. Note that these bits are set to pulse at a defined rate. They are set to pulse on and off for the same length of time. The system memory byte has 4 bits that can be used for various functions.

Bit 0 is set to a 1 when the main program is processed for the first time following the CPU being switched on. It is set to 0 in all other processing cycles.

Bit 1 is set to a 1 when the diagnostics sate has changed compared to the previous program cycle, except during STARTUP and not in the first RUN cycle. Otherwise, it has a state of 0.

Bit 2 is always true.

Bit 3 is always 0.

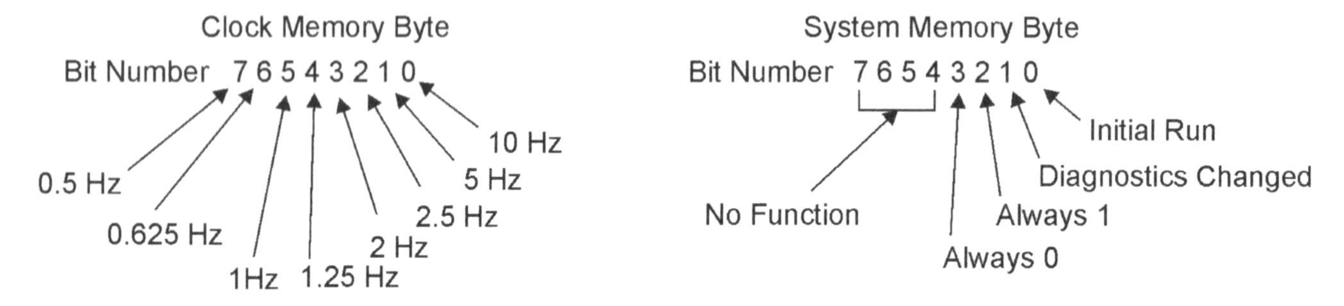

Figure 2-51. Clock and system bytes.

Chapter Questions

1. Describe the differences between linear and modular programming.
2. What is an OB and what types are available?
3. Which of the following are the interface between the operating system and the user program?
    a. Function blocks (FBs)
    b. Functions (FCs)
    c. Data blocks (DBs)
    d. Organizational blocks (OBs)
4. What is an FB and what are they used for?
5. What is an FC and what are they used for?
6. Which of the following is cyclically called by the operating system?
    a. OB155
    b. DB1
    c. FC25
    d. OB1
7. When power is switched from Stop to Run the CPU executes a complete restart by executing _______.
    a. OB1
    b. FC1
    c. DB100
    d. OB100
8. An Organizational Block's (OB) execution can be interrupted by an event (OB) with a _____________.
    a. Priority level 100.
    b. Higher priority
    c. Lower priority
    d. An OB's execution cannot be interrupted by another OB.
9. What is the main difference between a FB and an FC?
10. What is a DB?
11. What types of DBs are available and what would they be used for?
12. __________ blocks contain information that can be accessed by all logic blocks in the user program.
    a. Universal
    b. Access
    c. Instance
    d. Global
13. Which of the following are always assigned to a specific Function Block (FB)?
    a. Global Data
    b. Organizational Block
    c. Instance Data
    d. Function Call
14. Function Blocks (FBs) are user program blocks and can be called by:
    a. OBs
    b. FBs
    c. FCs
    d. All of the above

15. If data are to be retained even after a block is executed, _____________ variables should be used.
   a. Temporary
   b. Static
   c. Permanent
   d. Stored
16. If data are to be stored only while the block is being executed, ____________ variables should be used within the block.
   a. Static
   b. Temporary
   c. Unstored
   d. Short duration
17. Parameter-assignable blocks are created for frequently recurring functions. Which of the following is true regarding this type of block?
   a. The block only has to be created once, which significantly reduces programming time.
   b. The block is only stored once in user memory, which significantly reduces the amount of memory used.
   c. The block can be called as often as needed.
   d. All of the above.
18. Which of the following can be programmed as parameter-assignable?
   a. Organizational blocks (OBs)
   b. Functions (FCs)
   c. Function blocks (FBs)
   d. Both B and C.
19. Unlike Functions (FCs), Function blocks (FBs) have
20. Convert the following numbers to binary, decimal, or hex as required.

Binary	Decimal	Hex
0011		
	13	
10001		
011011		
	37	
	256	
10001001		
	73	
00100111		
		B7
		AF
		C6

21. Explain the concept of byte addressing for a discrete module.

22. The memory area in which the states of all digital inputs are stored is called the ________.
   a. Digital Input Image table (DII)
   b. Process-Image Input table (PII)
   c. Input Image table (II)
   d. I/O Process Image table (IOPI)

23. The memory area in which the states of all digital outputs are stored is called the ________.
    a. I/O Process Image table (IOPI)
    b. Output Image table (OI)
    c. Process-Image Output table (PIQ)
    d. Digital Output Image table (DOI)

24. Explain the concept of byte addressing for an analog module.

25. UDT refers to:
    a. User-Defined Data Type
    b. Undefined Data Type
    c. Underwater Demolition Technology
    d. None of the above

26. Which data type is used for data that is 8 bits long?
    a. BOOL
    b. SINT
    c. INT
    d. DINT
27. Which data type is used for data that is 32 bits long?
    a. BOOL
    b. SINT
    c. INT
    d. DINT
28. Which data type is used to store numbers that include a decimal?
    a. REAL
    b. WORD
    c. DWORD
    d. DINT

29. The default unit of measure for time is:
    a. Seconds
    b. Milliseconds
    c. Hundredths of seconds
    d. Minutes

30. Which of the following contain the number of milliseconds since the beginning of the day?
    a. Milliseconds_Since_Midnight (MSM)
    b. Milliseconds_Since_Yesterday (MSY)
    c. Time_Of_Day (TOD)
    d. Ontime_MS (OM)

31. Which of the following STRING variables?
    a. They are one-bit arrays
    b. They hold characters
    c. They hold REALs
    d. There are no string variables in a PLC

32. This data type can be used to store groups of values that are of the same data type.
   a. DINT
   b. STRUCT
   c. STRING
   d. ARRAY

33. This data type can be used to store groups of values that are of different data types.
   a. DINT
   b. STRUCT
   c. STRING
   d. ARRAY

34. When creating or editing a Data Block (DB) the user can determine whether the data can be:
   a. Visible on the HMI
   b. Accessed by the HMI
   c. Both A and B
   d. None of the above

35. Describe each of the following memory data types:

W

D

I

Q

IW

IB

QW

M

MB

MD

DBX

LW

36. Thoroughly describe each part of the following tag addresses.

I0.6

Q2.3

I12.6

QW112

IW96

MD48

I8.5:P

Q2.6:P

IW0

QW2

Temp[5]

Chapter 3

Ladder Logic Programming

Objectives

Upon completion of this chapter, the reader will be able to:

Define terminology such as rung, contact, coil, scan, normally-open, and normally-closed.

Explain the difference in operation between normally-open and normally-closed logic contacts.

Write basic ladder logic.

Describe the relationship between the states of real-world switches and normally-open versus normally-closed logic contacts.

You can download a trial version of Siemens Step 7 at the following link - www.siemens.com/sce/trial

Ladder Logic

There are multiple languages that can be used to program industrial controllers. Ladder logic is still the most commonly used language. Ladder logic was designed to be easy for electricians to use and understand. Symbols were chosen that look similar to schematic symbols of electrical devices so that a program would look like an electrical circuit. An electrician who has no idea how to program a computer can understand the basics of a ladder diagram.

The instructions in ladder logic programming can be divided into two broad categories: input and output instructions. The most common input instruction is a contact and the most common output is the coil. Figure 3-1 shows input and output instructions in logic. Input instructions are on the left and output instructions are to the right of the logic.

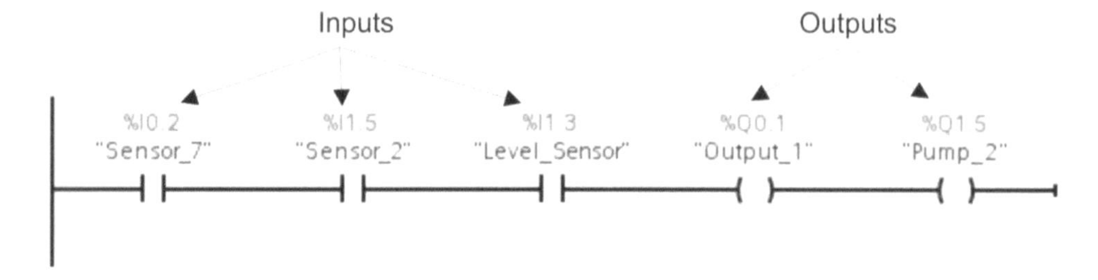

Figure 3-1. A rung of ladder logic.

Contacts

Most inputs to a PLC are discrete. Discrete devices only have two states: on or off. Contacts are used in ladder logic to represent input devices with two states. Symbols for the two types of contacts are shown in Figure 3-2. The one on the left is called a normally-open contact. The one on the right is called a normally-closed contact. Contacts are like discrete inputs.

Figure 3-2. A normally-open and a normally-closed contact.

Real-World Switches

To understand how contact will work in ladder logic we need to take a look at the two main types of mechanical switches: normally-open and normally-closed. Figure 3-3 shows the two types of switches: normally-open and normally-closed. A door bell switch is a normally-open switch. A normally-open switch will not allow current flow until it is pressed. A normally-closed switch will allow current to flow until it is pressed.

Figure 3-3. Diagram of a real-world normally-open switch and a normally-closed switch.

Normally-Open Logic Contacts

Think about a door bell switch again. If you use a normally-open, real-world switch, the bell will be off until someone pushes the switch and it allows current to flow. If you used a normally-closed switch, the bell would be on until someone pushes the switch to stop the current flow. A normally-open logic contact is true if the real-world input associated with it is true. Think about a home door bell again. There is pushbutton switch (momentary) next to the front door. The actual real-world switch is a normally-open switch. When someone pushes the switch the bell in the house rings.

The normally-open contact in ladder logic is similar to the normally-open door bell switch. Consider a sensor that we would like to control a pump. A normally-open contact in a ladder diagram could be used to monitor the real-world level sensor that controls the pump. When the level sensor becomes true, the normally-open contact in the rung of logic would allow current flow and the output (pump) would be on while the sensor is true (see Figures 3-4 and 3-5).

%I0.0
"Sensor_2"
%Q0.4
"Pump_2"

Figure 3-4. A normally-open contact. Real-world input Sensor_2 is false so the normally-open contact is not energized. The rung is false and the output is off.

%I0.0
"Sensor_2"

%Q0.4
"Pump_2"

Figure 3-5. A normally-open contact. Real-world input Sensor_2 is true so normally-open contact is energized so the rung is true and the output is on.

Figure 3-6 shows a conceptual diagram of a PLC with an input and output. The logic is shown in the middle in the block that represents the CPU. Note the input and output image table. The sensor attached to input 0 is true and there is a 1 in the input table with the bit associated with input 0. In the rung of the ladder logic the normally-open contact is true because the bit in memory for input 0 is a 1 (true). This makes the rung conditions true, so the output is energized. Note the 1 in the output table associated with output 11.

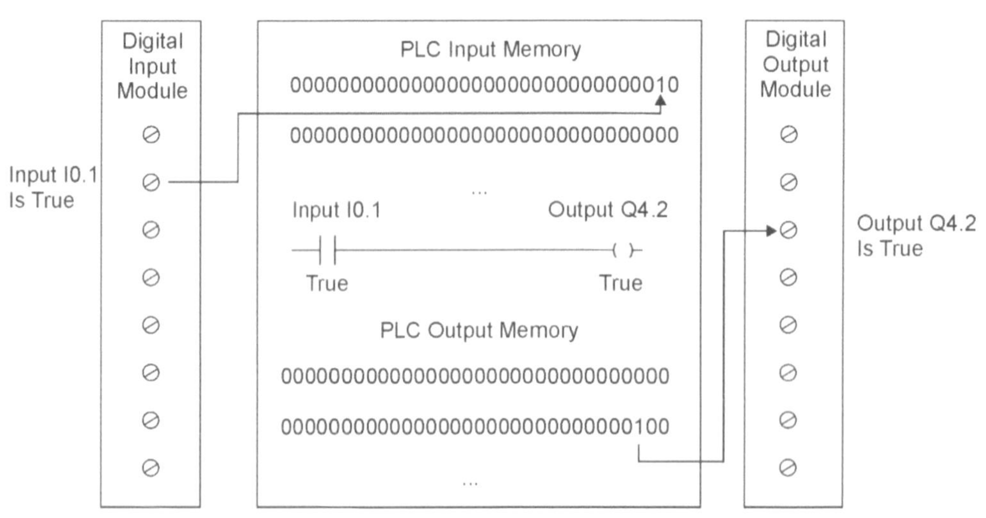

Figure 3-6. Simple conceptual diagram of a PLC.

Normally-Closed Contacts

The other type of discrete contact is the normally-closed contact. A normally-closed contact is also called an examine-if-open contact. This type of contact can be confusing at first. A normally-closed contact will pass power until the real-world condition associated with it is energized. A normally-closed contact in a ladder diagram is only energized if the real-world input associated with it is false. Figure 3-7 shows an example of a normally-closed contact in a rung of logic. If real-world Sensor_2 is false, the normally-closed contact in the rung is true and the output (Pump_2) will be on. If Sensor_2 in the real-world is on (true) this contact will open and the output (Pump_2) will be off.

Figure 3-7. A normally-closed contact is used in the rung. If the input associated with that contact is closed, it forces the normally-closed contact open. No current can flow through the rung to the output, so the output is off.

If the bit associated with a normally-closed instruction is a zero (off), the instruction is true and passes power. If the bit associated with the instruction is a one (true), the instruction is false and does not allow current flow. Note that the normally-closed contact has the opposite effect of the normally-open contact.

Figure 3-8 shows two rungs. The first rung has a normally-open contact. The second rung has a normally-closed contact. The tag is the same for each: Sensor_2. Sensor_2 is the tag name for a real-world switch that is off. Note that the normally-open contact in the first rung is open and will not pass power. In the second rung the normally-closed contact is closed and does energize the output because the real-world switch is off.

%I0.0 "Sensor_2" — %Q0.5 "Pump_1"

%I0.0 "Sensor_2" — %Q0.6 "Pump_2"

Figure 3-8. Logic example

Figure 3-9 shows the same logic that was shown in Figure 3-8, but in this example the real-world switch (Sensor_2) is true. Note that the normally-open contact in the first rung is true. Note that the normally-closed contact in the second rung is now false because the switch is true.

%I0.0 "Sensor_2" — %Q0.5 "Pump_1"

%I0.0 "Sensor_2" — %Q0.6 "Pump_2"

Figure 3-9. Logic example.

Note that contacts can cause confusion for the beginning programmer. You must have a clear understanding of the relationship between real-world conditions and their effect on logic contacts.

Adding Comments

Comments can be added to your logic by right clicking on the element you would like to attach a comment to. Two comments were added to the logic in Figure 3-10.

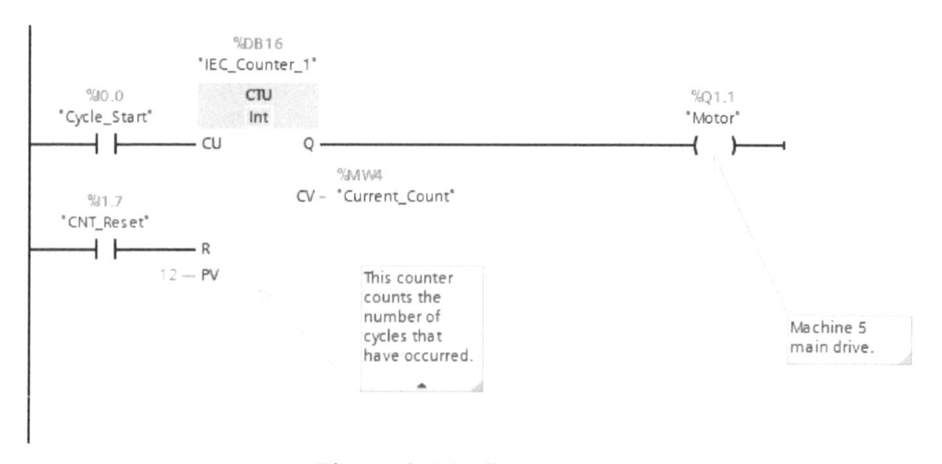

Figure 3-10. Comments.

Coils

A coil is symbol for an output. A coil instruction sets a bit in memory. If the input logic in its rung is true, the output bit will be set to a 1. If the logic of its rung is false, the output bit is reset to a zero. Outputs are things such as: lights, signals to other devices, motors contactors, pumps, counters, timers, valves, and so on. Coils are only used on the right side of a rung. Contacts are conditions on rungs. If all of the conditions are true, the rung is true and the coil (output) will be true. The symbol for a coil is shown in Figure 3-11. A specific output coil should only be used once in logic. Note that there can be parallel conditions to control the coil on the rung. A specific coil should ever be used in more than one rung as an output coil. The output that the coil represents can however be used as rung input contact(s) as many times as is useful.

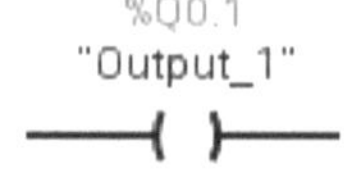

Figure 3-11. The symbol for a coil.

Figure 3-12 shows Siemens toolbars for ladder logic programming.

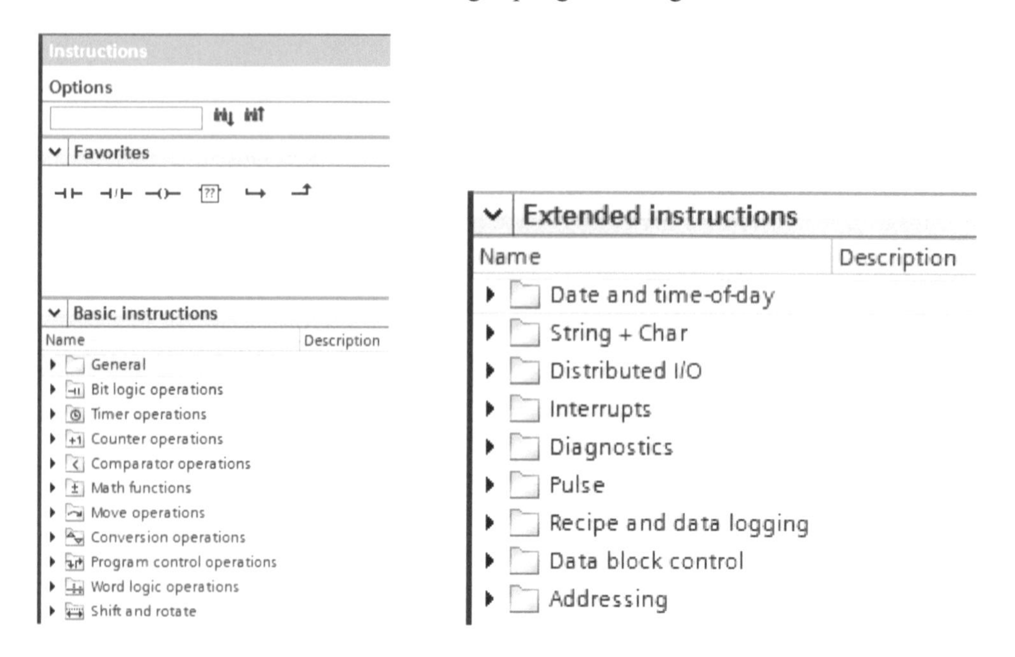

Figure 3-12. Ladder logic instruction folders.

Figure 3-13 shows some of the types of coils that are available for programming. Each of these will be covered later.

Simple Coil	—( )—	Negated Coil	—(/)—
Set Coil	—(S)—	Reset Coil	—(R)—
Pulse on Positive Edge	—(P)—	Pulse on Negative Edge	—(N)—
Multiple Bit Set	—(SET_BF)—	Multiple Bit Reset	—(RESET_BF)—
Jump If 1	—(JMP)—	Jump If 0	—(JMPN)—
Conditional Block End	—(RET)—		

Figure 3-13. Types of coils.

Ladder Diagrams

The basic structure of a ladder diagram looks similar to a ladder (see Figure 3-14). There are two rails (uprights) and there are rungs. The left and right rails represent power. If the logic on a rung is true, power can flow through the rung from the left upright to the right upright.

Figure 3-14 is a simple application involving a heat sensor and a fan. There is one input and one output for this application. A heat sensor is connected to a PLC input and the fan is connected to a PLC output. The left and right rails represent voltage that will be used to power the fan if the sensor state is true.

Figure 3-14. Ladder logic example.

When a PLC is in run mode it monitors the inputs and controls the states of the outputs. This is called scanning. The amount of time it takes for the PLC to evaluate the logic and update the I/O table each time is called the scan time. The more complex the ladder logic, the more time it takes to scan.

Figure 3-15 shows a conceptual view of a PLC system. The real-world inputs are attached to an input module (left side of the figure). Outputs are attached to an output module (right side of the figure). The center of the figure shows that the CPU evaluates the logic. The CPU evaluates user logic by looking at the inputs and then turns on outputs based on the logic.

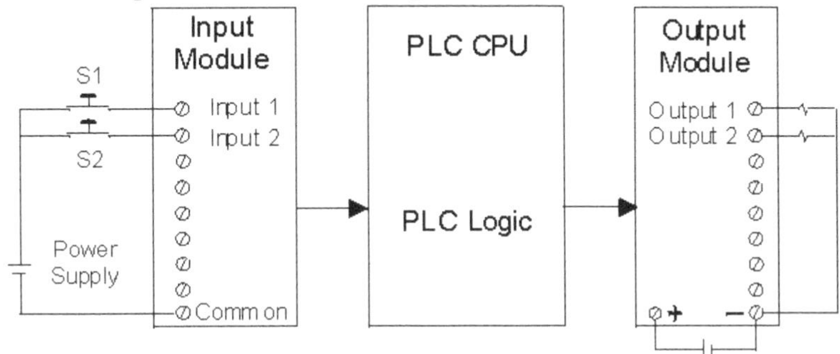

Figure 3-15. Conceptual view of a PLC system.

Multiple Contacts

Series (AND) Logic

Contacts can be combined to form logic on a rung. A two-hand switch on a punch press is a good example. For safety the punch press should only operate if both of the operator's hands are on the switches.

Figure 3-16 shows an example of a two-hand safety switch and series logic. The hand switches are on the left and right. The switch in the middle is a stop switch. A two-hand switch can help assure that the operator's hands cannot be in a dangerous part of the machine when it punches a part. The switches represent an AND condition. The ladder program for the PLC is also shown in Figure 3-16. Note that the real-world switches were programmed as normally-open contacts in the logic. They are in series in the rung. Both must be true for the machine to punch. This is for illustrative purposes only.

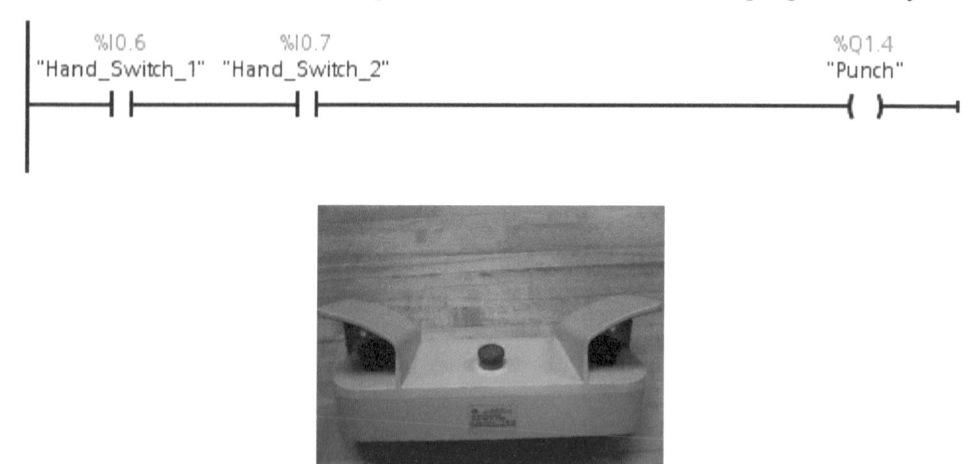

Figure 3-16. A two-hand safety switch. The two hand switches are on the left and the right of the photo.

If the operator removes one hand, the punch press will not operate. In fact, with newer safety relay technology, both switches have to be turned on at almost exactly the same time to make the machine run. Safety relays also prevent an operator from taping one switch closed. Contacts in series are logical AND conditions. In this example, the left-hand switch AND the right-hand switch would have to be true to run the punch press.

Figure 3-17 shows a series circuit. Sensor_1 AND Sensor_2 AND NOT Sensor_3 AND Sensor_4 must be true to turn the output named Fan_Motor on. Note that NOT represents the normally-closed contact named Sensor_3. This means that in the real-world Sensor_1, AND Sensor_2 AND Sensor_4 must be true, AND Sensor_3 must be false in the real-world for the normally-closed contact (Sensor_3) to be true in logic.

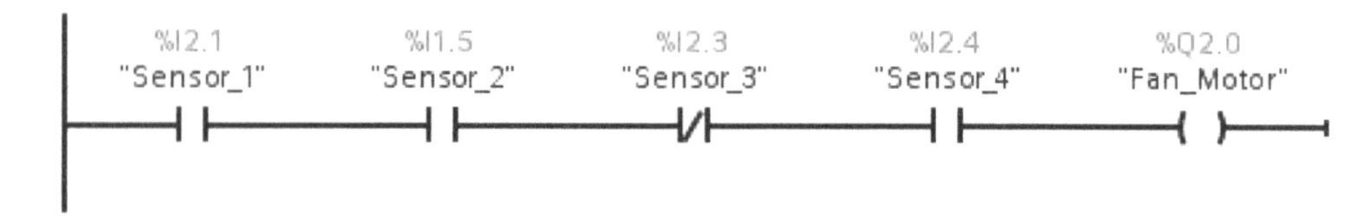

Figure 3-17. This figure shows a series circuit.

Figure 3-18 shows a robotic cell with light curtain protection. The light curtain has two safety outputs. If someone or something blocks any of the light between the light curtain transmitter and receiver the light curtain outputs become false. If everything is normal and no light is blocked, the light curtain outputs are both true. There are two outputs from the light curtain to make the system safer. Assume the two outputs from the light curtain are connected to inputs on an input module. The logic in the PLC would be written so that the two light curtain inputs would be used in series (AND) to make sure both were true for safety. This example was an oversimplification of safety technology and logic in order to make contacts more understandable.

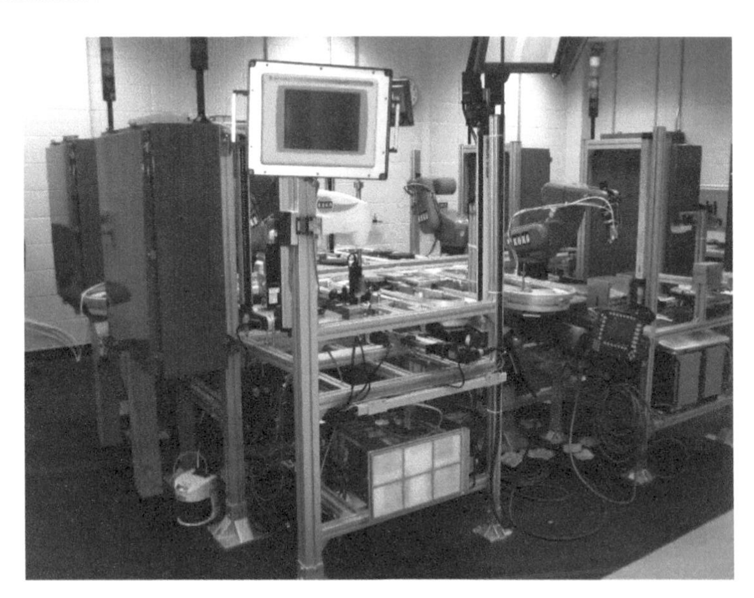

Figure 3-18. Robotic cell with light curtains and other safety devices.

OR Logic

Series logic was used to create AND conditions. Parallel logic is used to create OR conditions. These are often called branches. Branching can be thought of as an OR situation. One branch OR another can control the output.

Study Figure 3-19. This logic uses two different inputs to control a door bell. If either switch is on, we would like the bell to sound. A branch is used to create this logic. If the front door switch is closed, electricity can flow to the bell. OR if the rear door switch is closed, electricity can flow through the bottom branch to the bell.

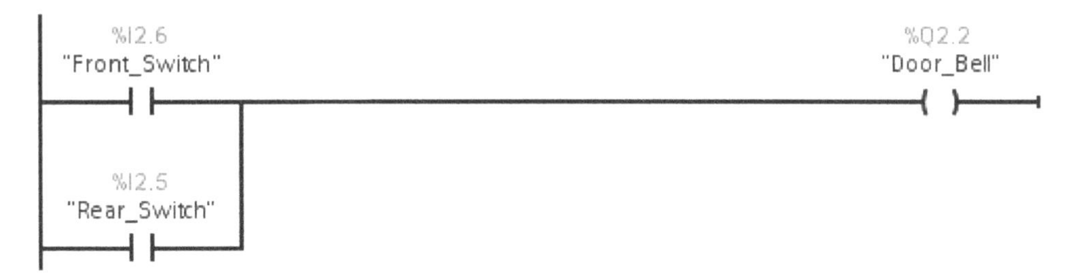

Figure 3-19. Ladder logic to control a home door bell. This figure shows a parallel condition. If the front door switch is closed the doorbell will sound, OR if the rear door switch is closed the doorbell will sound. These parallel conditions are also called OR conditions.

ORs allow multiple conditions to control an output. This is very important in industrial control of systems. Think of a motor that is used to move the table of a machine. There are usually two switches to control table movement: a jog switch and a feed switch. Either switch must be capable of turning the same motor on. This is an OR condition. The jog switch OR the feed switch can turn on the table feed motor.

Series and parallel conditions can be combined in logic. Figure 3-20 shows a simple example. If Inp_1 AND Inp_2 are true, Out_1 will be turned on. OR, if Inp_3 AND Inp_2 are true, Out_1 will be turned on.

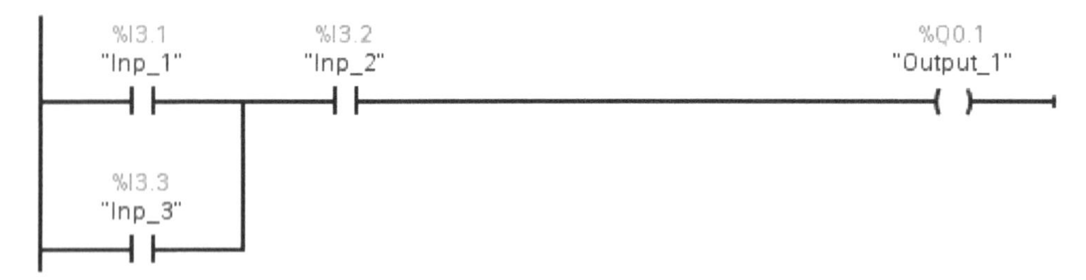

Figure 3-20. Rung using parallel (OR) and series (AND) logic.

Outputs may also be branched. Figure 3-21 shows an example of two outputs being branched. In this example, if the rung is true, both outputs will be turned on.

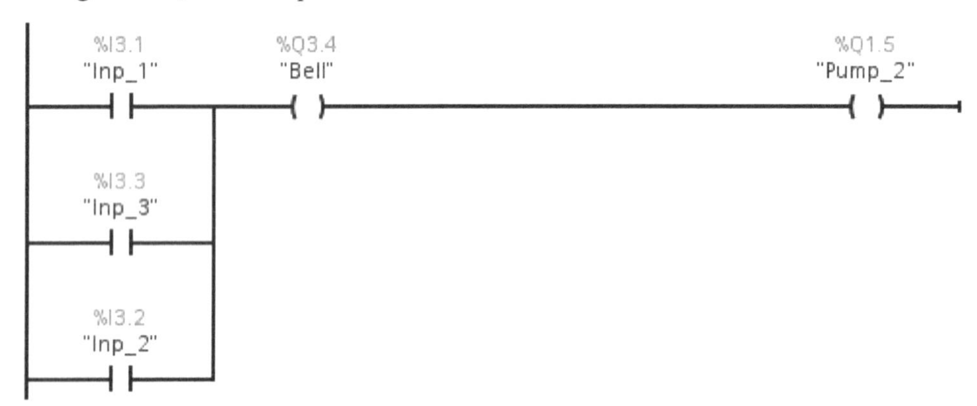

Figure 3-21. Example of parallel outputs.

Start/Stop Circuits

Machines typically have a start and a stop switch. The start switch is typically a normally-open, momentary type. The stop switch is typically a normally-closed momentary switch. The ladder logic for a simple start/stop circuit can help understand several important concepts about ladder logic. It can also help us understand the effect of real-world switch conditions on normally-open and normally-closed logic contacts. The normally-closed logic contact is often confusing to many who are beginning to program ladder logic.

Examine Figure 3-22. Notice that a real-world start switch is a normally-open momentary push-button. When it is pressed, it closes the switch. When the button is released, the switch opens. A real-world stop switch is a normally-closed switch. When pressed, it opens the contacts and stops current flow.

Figure 3-22. Start/stop circuit.

If the start switch is pressed momentarily, the normally-open start contact will become true. The Stop contact is also true, because the real-world stop switch is a normally-closed switch. Output coil Run will be true. The Run contact bit is then used in the branch around the Start contact. Notice that an output (Run in this example) can be used as an input condition (contact). The Run bit would be true so this bit would "latch" around the Start bit.

The output (Run) remains latched when the momentary start switch opens. The output (Run) will shut off only if the normally-closed stop switch (Stop) is pressed. If normally-closed contact Stop opens, then the rung would be false and the Run bit would be turned off. To restart the system, the Start button must be pushed. Note that the real-world stop switch is a normally-closed switch, but that in the ladder, it is programmed as a normally-open contact. This is done for safety. It is called fail-safe. If the stop switch fails, we want the machine to stop. If a wire to the stop switch is cut, we want the machine to stop. By using a normally-closed real-world switch with a normally-open contact in the logic we fail-safe the logic.

There are many ways to program start/stop circuits and ladder logic. Figures 3-23 and 3-24 show examples of the wiring of start/stop circuits. Note the types of switches that are used and how they are used to latch the circuit. Safety is always the main consideration in start/stop circuits.

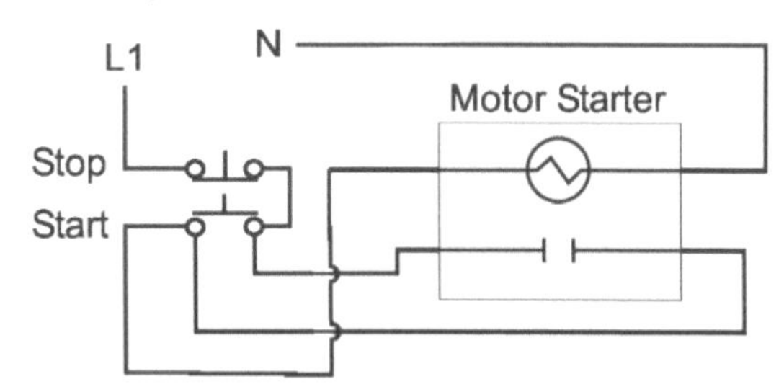

Figure 3-23. Single-phase motor start/stop circuit example.

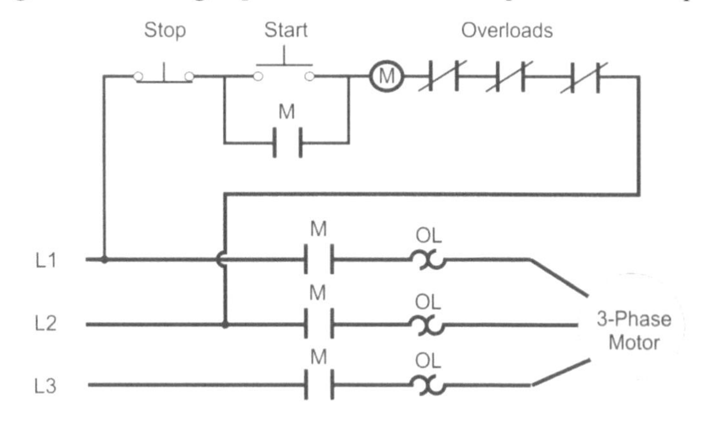

Figure 3-24. Start/stop circuit example for a 3-phase motor.

Use of Outputs in Logic

Output instructions can be placed in series on a rung (see Figure 3-25). This is equivalent to outputs in parallel. The more common and accepted method however is shown in Figure 3-26.

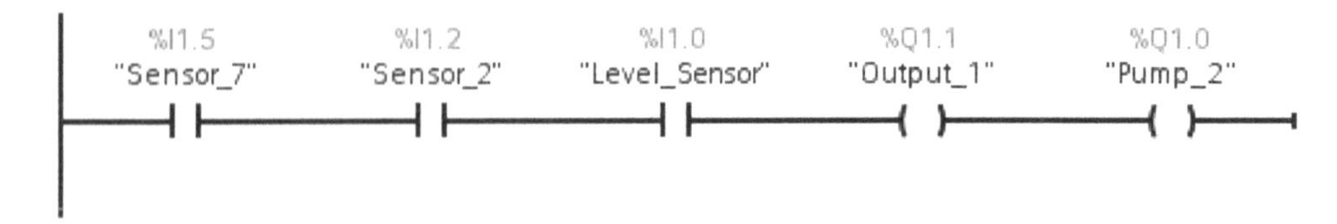

Figure 3-25. Coils used in series at the end of a rung.

Output instructions can be placed in branches (parallel). Figure 3-26 shows an example of output instructions used in parallel in logic. This is much more common than programming outputs in series.

%I1.5 "Sensor_7" %I1.2 "Sensor_2" %I1.0 "Level_Sensor" %Q1.0 "Pump_2" %Q1.1 "Output_1"

Figure 3-26. Example of output instructions used in parallel.

Output branches can also be nested. Figure 3-27 shows an example of a nested branch. Note that if Sensor_3 is true, the Bell will be turned on. The branched rung below that uses Sensor_2 to control the remaining branches. If Sensor_3 is true and Sensor_2 is true, the Bell output and the Pump_2 output will be turned on. The next branch adds another condition (Sensor_7).

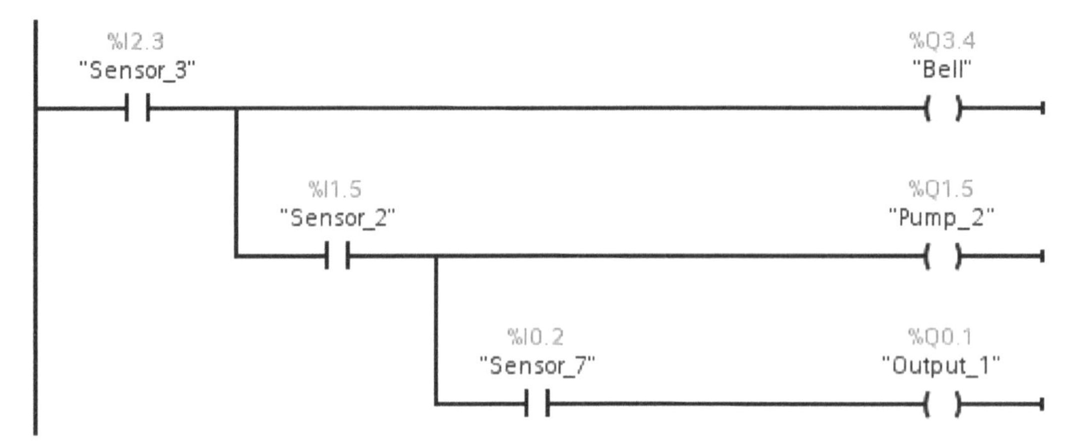

Figure 3-27. Nested output branch.

Output instructions can be placed between input contacts as long as the last instruction on the rung is an output instruction (see Figure 3-28).

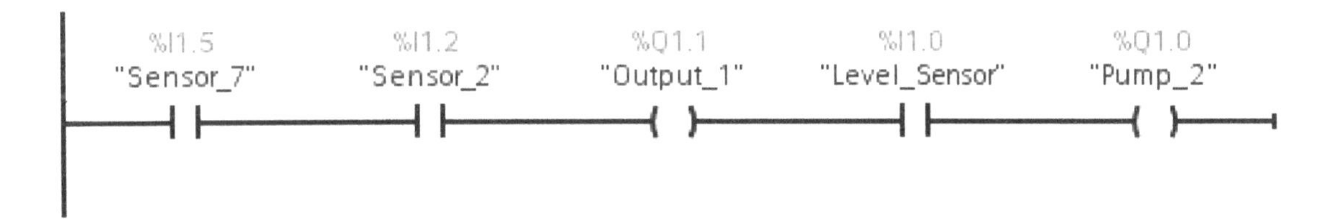

Figure 3-28. Example of OTE coils being used between contacts in logic.

Logic Examples

A rung of ladder logic and the states of the real-world inputs are shown in Figure 3-29. Study the logic in Figure 3-29. The output is on. Why (note that the answer is in the caption for the figure)?

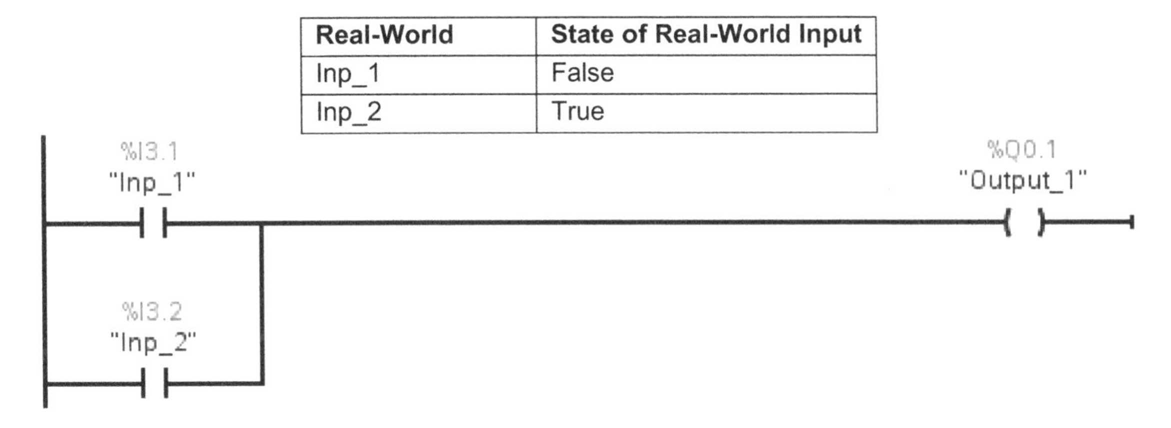

Real-World	State of Real-World Input
Inp_1	False
Inp_2	True

Figure 3-29. Output is true in this example. The output is on because Inp_2 is true making the rung true. This is parallel (OR) logic.

A rung of ladder logic and the states of the real-world inputs are shown in Figure 3-30. Study the logic in Figure 3-30. The output is energized. Why?

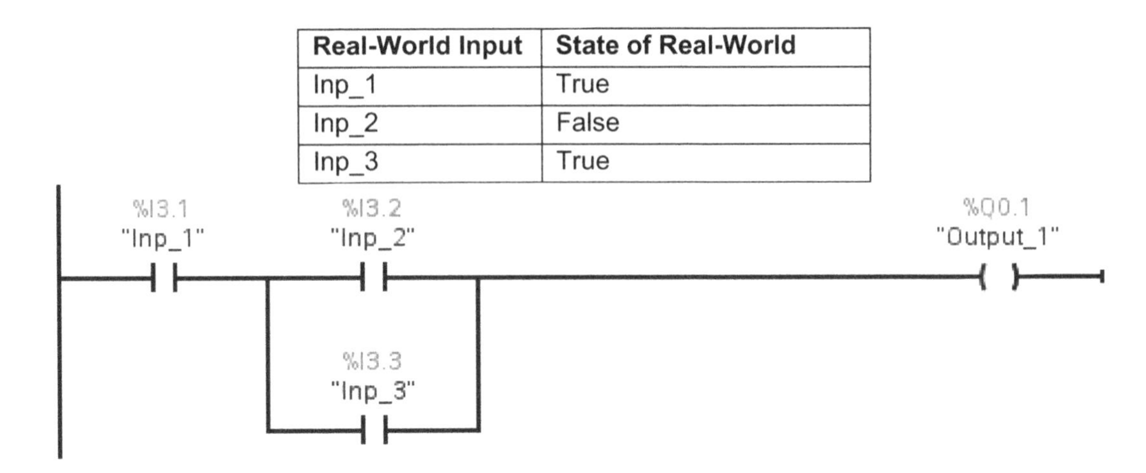

Real-World Input	State of Real-World
Inp_1	True
Inp_2	False
Inp_3	True

Figure 3-30. Out_1 is true in this example because Inp_1 and Inp_3 are true.

A rung of ladder logic and the states of the real-world inputs are shown in Figure 3-31. Study the logic in Figure 3-31. The output is energized. Why? Hint: remember how normally–closed contacts work.

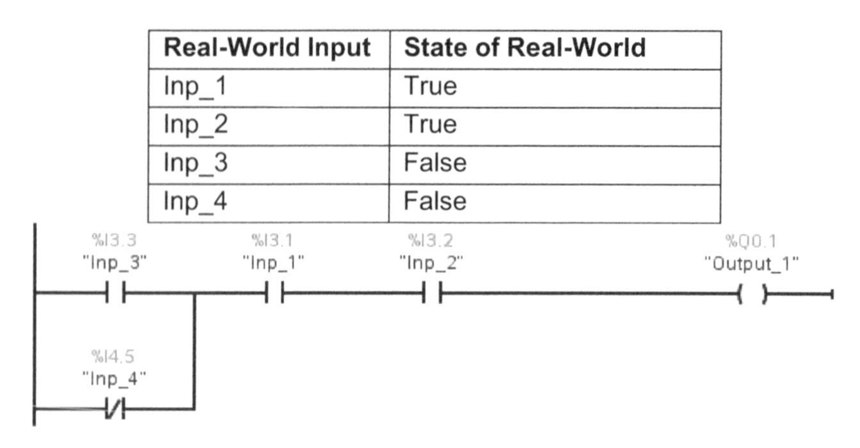

Real-World Input	State of Real-World
Inp_1	True
Inp_2	True
Inp_3	False
Inp_4	False

Figure 3-31. Output is energized in this example because Inp_4 contact is true because its real-world input is false and Inp_1 and Inp_2 are true.

A rung of ladder logic and the states of the real-world inputs are shown in Figure 3-32. Study the logic in Figure 3-32. The output is off. Why?

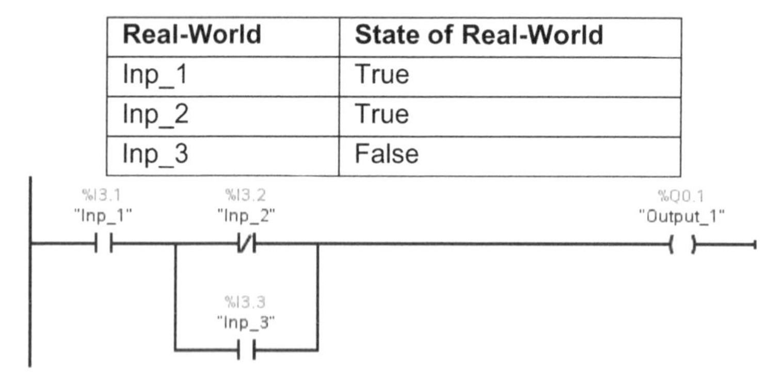

Real-World	State of Real-World
Inp_1	True
Inp_2	True
Inp_3	False

Figure 3-32. Output is off in this example because the normally-closed contact (Inp_2) in logic is false because the real-world input is true, and Inp_3 is also false.

A rung of ladder logic and the states of the real-world inputs are shown in Figure 3-33. Study the logic in Figure 3-33. The output is off. Why?

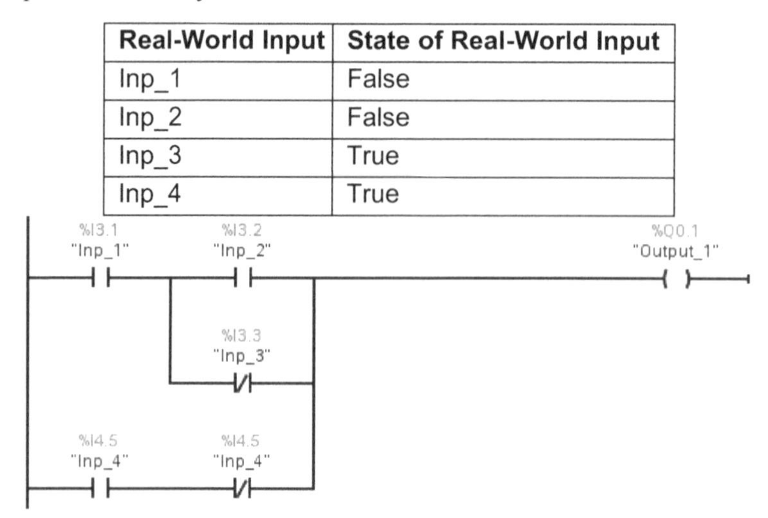

Real-World Input	State of Real-World Input
Inp_1	False
Inp_2	False
Inp_3	True
Inp_4	True

Figure 3-33. Output is off in this example because contacts Inp_2, Inp_3, and Inp_4 are false in logic. Note that there is no way that the branch with normally-open, Inp_4 and normally-closed Inp_4 can ever be true, because one or the other will always be false.

NOT Instruction

NOT instructions are used to invert the status of bits. A one is made a zero and a zero is made a 1. Figure 3-34 shows the result of a NOT instruction on bits.

Input Value to the	Result
0	1
1	0

Figure 3-34. Result of a NOT operation. Note that a 0 becomes a 1 and a 1 becomes a 0.

Figure 3-35 shows the use of a NOT instruction. If Inp_5 is true, the NOT instruction executes.

The NOT instruction can be used to invert the state of the result of logic operation. When the state is true at the input of the NOT instruction, the output of the NOT instruction will be false. When the signal state is "0" at the input of the instruction, the output of the instruction provides the signal state "1".

Figure 3-35. Use of a NOT instruction.

Latching Instructions

Latches are used to lock in a condition. For example, if an input contact is on for only a short time, the output coil would be on for the same short time. If it were desired to keep the output on even if the input goes low, a latch could be used. This can be done by using the output coil to latch itself on (see Figure 3-36).

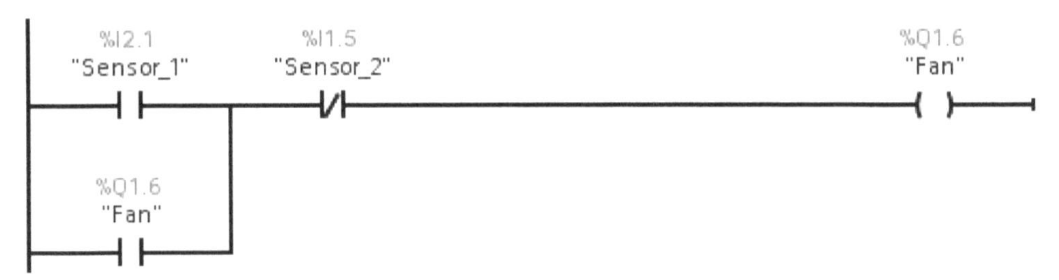

Figure 3-36. Note that even if coil Sensor_1 is only on for a very short time the output (Fan) will latch around the Sensor_1 contact and keep itself on. It would stay on in this example until the normally closed contact would open. There are instructions available that perform the task of latching.

Set Output Instruction

You can use the Set output instruction to set the signal state of an output to true. Figure 3-37 shows an example of a Set output instruction.

If Inp_1 is true in this example the Set output instruction will execute and Output_5 will be turned on and latched on even if Inp_1 becomes false. Output_5 will remain true until a Reset output instruction is executed. In this example Output_5 will be true until Inp_2 is true and the Reset output instruction executes. Note that the name of the output we want to reset must be used for the Reset output instruction.

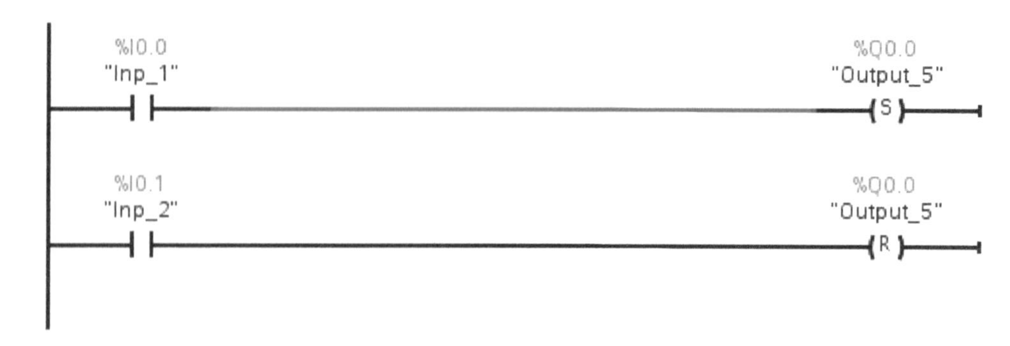

Figure 3-37. A Set output instruction and a Reset output instruction in logic.

Special Purpose Contact and Coils

Negated Coil Instruction

The Negated Coil instruction (see Figure 3-38) inverts the result of logic operation and assigns it to the specified operand. When the rung logic before the input of the coil is true, the output is set to false. When the RLO at the input of the coil is false, the operand is set to true.

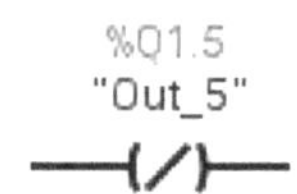

Figure 3-38. Negated coil

Edge Evaluation Instructions

An edge instruction records the change in a signal state. A positive (rising) edge is present if a signal changes state from 0 to 1. A negative (falling) edge is present if a signal changes state from 1 to 0. The program records the change of signal state. During the processing the CPU compares the actual result of the logic operation with the saved result of the previous logic evaluation. A signal edge is present if the two signal states (current and saved state) are different.

Set Output on Negative Signal Edge Instruction

The set operand on negative signal edge instruction can be used to set an output when there is a true to false change in rung logic (see Figure 3-39). The instruction compares the current result of the rung logic to the result of operation from the previous result. The previous result is saved in the edge bit memory. When the instruction detects a change in rung logic from true to false, there is a negative, falling edge.

When a negative edge is detected, the output is set to signal state true for one scan. Otherwise, the output has signal state false.

The address of the edge memory bit (a tag named State in this example) must not be used more than once in the program, otherwise the memory bit would be overwritten. This would influence the edge evaluation and might result in an unwanted result. Note that the State tag is just a bit in memory (M0.0). It is not a real output. Memory bits are very useful in logic.

Figure 3-39. Negative transition coil.

The Scan Negative Signal Edge instruction is used to query a change in the signal state from true to false of an operand (Inp_5) in this example (see Figure 3-40). The instruction compares the current signal state of Inp_1 (operand 1) to the signal state of the previous query saved in inp_State (operand 2). If the instruction detects a change in from true to false, there is a negative, falling edge.

If a falling edge is detected, the output of the instruction has signal state true.

If there is no falling edge, the signal state to the right power rail is reset to false.

The address of the edge memory bit (operand 2) must not be used more than once in the program, otherwise the memory bit would be overwritten. This would influence the edge evaluation and might result in an unwanted result.

Figure 3-40. Negative transition contact.

Scan Positive Signal Edge Instruction

Figure 3-41 shows the use of a Scan Positive Signal Edge instruction. The Scan Positive Signal Edge instruction can be used to determine if there is a false to true change in the signal state of operand 1 (Output 2 in this example). The instruction compares the current signal state of operand 1 to the signal state of the previous query saved in operand 2 (State in this example). If the instruction detects a change in the result of logic operation from false to true, there is a positive, rising edge.

If a rising edge is detected, the output of the instruction has signal state true. In all other cases, the signal state at the output of the instruction is false.

The address of the edge memory bit (operand 2) must not be used more than once in the program, otherwise the memory bit would be overwritten. This would influence the edge evaluation and might result in an unwanted result.

Figure 3-41. An example of the Scan Positive Signal Edge instruction.

The Scan Positive Signal Edge instruction can be used to determine if there is a false to true change in the signal state of operand 1, Output_2 in this example (see Figure 3-42). The instruction compares the current signal state of operand 1 (Output_2) to the signal state of the previous logic result saved in Operand 2 (State in this example). Operand 2 is also called the edge memory bit. If the instruction detects a change in the result of logic operation from false to true, there is a positive, rising edge.

If a rising edge is detected, the output of the instruction will be set to true. Otherwise, the signal state of the output is false.

The address of the edge memory bit (operand 2) must not be used more than once in the program, otherwise the memory bit would be overwritten. This would influence the edge evaluation and might result in an unwanted result.

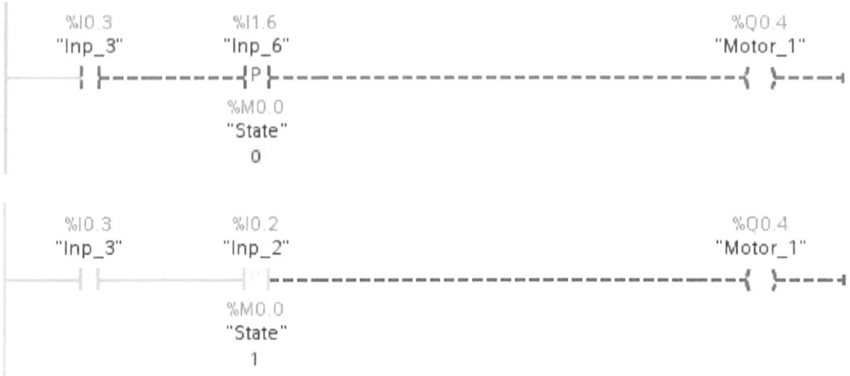

%I0.3 "Inp_3" | %I1.6 "Inp_6" P %M0.0 "State" 0 | %Q0.4 "Motor_1"

%I0.3 "Inp_3" | %I0.2 "Inp_2" P %M0.0 "State" 1 | %Q0.4 "Motor_1"

Figure 3-42. Use of a Scan Positive Signal Edge instruction.

Q Boxes in Ladder Logic

Siemens has some instructions they call Q boxes. Q boxes have a discrete output named Q that can be used in logic. There are Q boxes available to represent memory functions, edge evaluations, and timer and counter functions. Figure 3-43 shows Q boxes available in ladder programming. The first discrete input on a Q box must be connected and the connection of other inputs and outputs is optional. The inputs of a Q box cannot be directly connected to the left power rail.

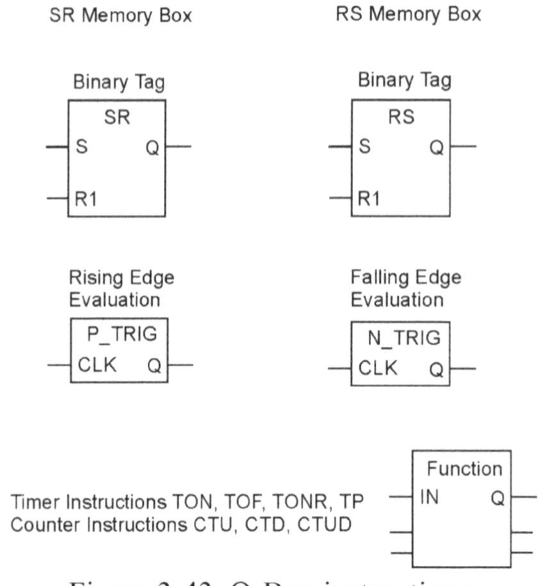

Figure 3-43. Q-Box instructions.

Edge Trigger Q boxes

Edge evaluation Q boxes are used to register a change in flow prior to the box. A change of flow from 0 to 1 would be a rising edge. A change of flow from 1 to 0 would be a falling edge. If the signal state changes from a 0 to a 1(rising edge) at the CLK input of a positive (P-TRIG) instruction there will be a 1 at the Q output for one program (scan) cycle. If the input to the CLK input of a negative trigger (N_TRIG) instruction changes from a 1 to a 0 (falling edge) the Q output will be set for one program (scan) cycle.

Negative Trigger Instruction (N-TRIG)

The set output on negative signal edge instruction can be used to detect a true to false change in the signal state of the result of logic (see Figure 3-44). The instruction compares the current signal state of the logic to the signal state of the previous scan saved in the edge memory bit (State in this example). If the instruction detects a change in the logic from true to false, there is a negative, falling edge.
If a falling edge is detected, the output of the instruction is true. In all other cases, the signal state at the output of the instruction is false.

%I0.3
"Inp_3"
N_TRIG
CLK Q
%M0.0
"State"
1
%Q0.4
"Motor_1"

Figure 3-44. Use of a Negative Trigger instruction.

Positive Trigger Instruction (P-TRIG)

The set output on positive signal edge instruction can be used to determine if a change in the signal state of the result of logic from false to true has occurred (see Figure 3-45). The instruction compares the current signal state of the result of logic to the signal state from the previous scan, which is saved in the edge bit memory. If the instruction detects a change in rung logic from false to true, there is a positive, rising edge.
If a rising edge is detected, the output of the instruction is set to true. Otherwise, the signal state at the output of the instruction is false.

%I0.3
"Inp_3"
P_TRIG
CLK Q
%M0.0
"State"
1
%Q0.4
"Motor_1"

Figure 3-45. Use of a Positive Trigger instruction.

Memory Q Boxes (SR and RS)

There are two types of Q memory boxes available a reset dominant (SR) and a set dominant (RS) box. Figure 3-46 shows an example of a SR and an RS memory box. The discrete tag above the memory box is set when the set input is true and the reset input is false. The discrete tag is reset when the reset input is true and the set input is false. A signal state of false at both inputs has no effect on memory functions. If both inputs are true, the SR and RS respond differently. If both inputs are true for a SR instruction the memory bit tag will be reset. If both inputs are true for a RS instruction, the memory bit will be set.

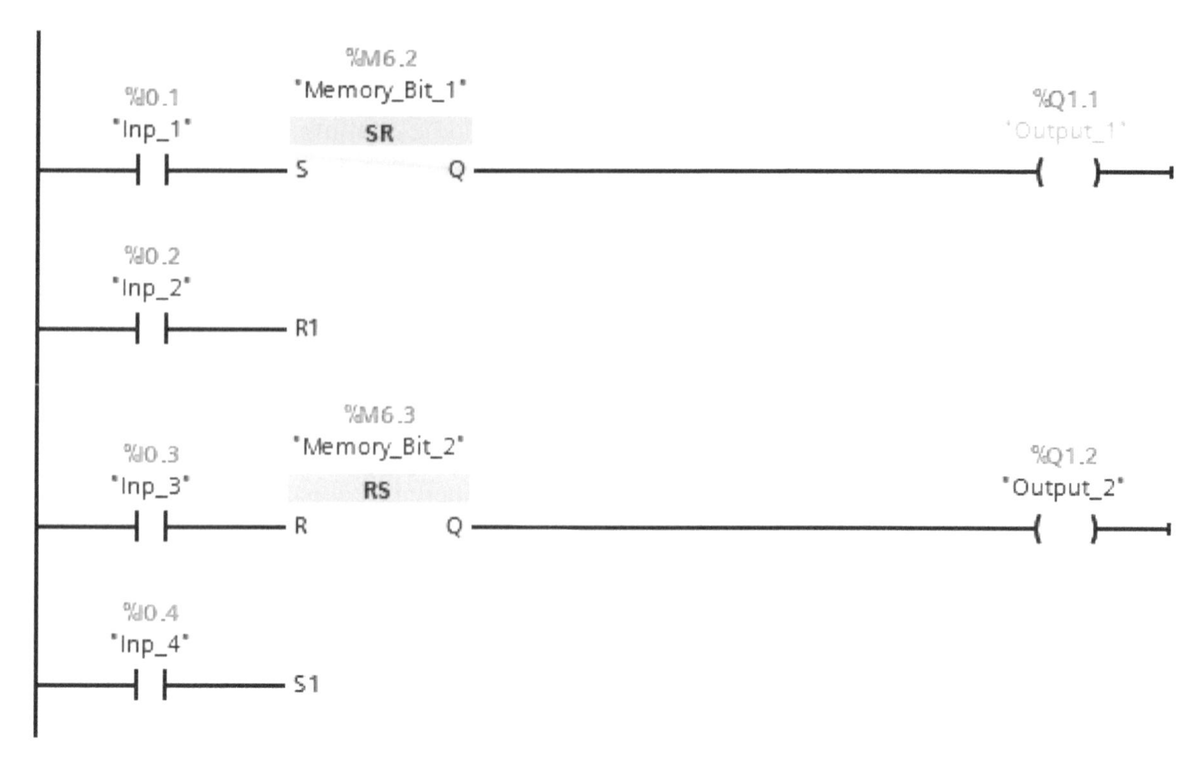

Figure 3-46. Example of the use of a SR and RS memory box instruction.

Set/Reset (SR) Instruction

The set reset set flip-flop (SR) instruction can be used to set or reset the bit of a tag (Output_Bit in this example) based on the signal state of inputs S and R1 (see Figure 3-47). If the signal state at input S is 1 and is 0 at input R1, the specified tag (Output_Bit) is set to 1. If the signal state at input S is 0 and is 1 at input R1, the specified operand is reset to 0. Input R1 takes priority over input S. If the signal state is 1 at the two inputs S and R1, the signal state of the specified tag is reset to 0.

If the signal state at the two inputs R1 and S is 0, the instruction is not executed and the signal state of the operand is unchanged.

The operands Output_Bit and Out_5 are set when the following conditions are fulfilled:

Sensor_5 (Input S) is true.

Sensor_6 (Input R1) is false

The operands Output_Bit and Out_5 (Q bit) are reset when one of the following conditions is fulfilled:

Sensor_5 (Input S) is false and the operand Sensor_6 (Input R1) is true.

Sensor_5 (Input S) and Sensor_6 (Input R1) are both true.

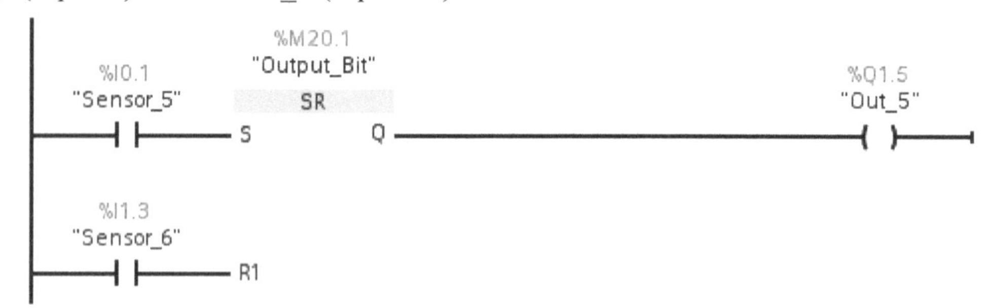

Figure 3-47. Example of a set/reset flip/flop (SR) instruction.

Reset/Set (RS) Flip-Flop Instruction

The reset/set flip-flop instruction can be used to reset or set based on the signal state of the inputs at R and S1 (see Figure 3-48). If the signal state at input R is 1 and is 0 at input S1, the specified tag (Output_Bit) is reset to 0. If the signal state at input R is 0 and is 1 at input S1, the specified tag (Output_Bit) is set to 1.

Input S1 takes priority over input R. If the signal state is 1 at inputs R and S1, the signal state of the specified tag is set to 1.

If the signal state at the two inputs R and S1 is 0, the instruction is not executed. The state of the output (Q) then remains unchanged.

Bit memory Output_Bit and output Output_5 are reset under the following conditions:

Inp_1 is true.

Inp_2 is false.

Tags Output_Bit and Output_5 are set under the following conditions:

Input Inp_1 is false and Inp_2 is true.

Both inputs Inp_1 and Inp_2 are true.

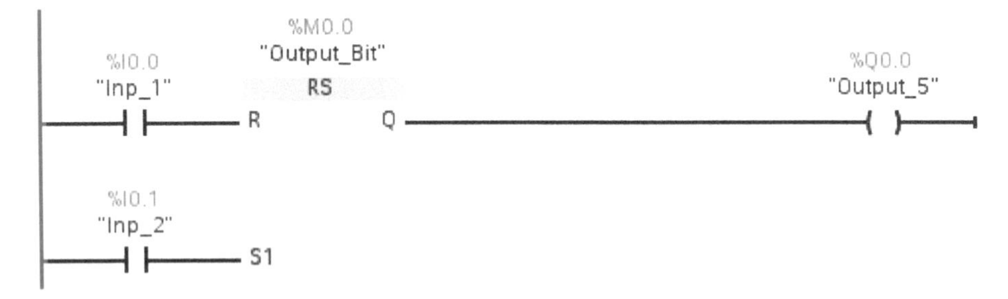

Figure 3-48. Example of a set/reset (RS) instruction.

Program Flow Instructions

There are many types of flow control instructions available for PLCs. Program flow instructions can be used to control the sequence in which your program is executed. Program flow instructions allow the programmer to control the order in which the CPU scans the ladder diagram. These instructions can be used to minimize scan time and/or develop more efficient programs. They can be used to troubleshoot ladder logic as well. Program flow instructions can be used to jump around sections of logic for testing. Program flow instructions must be used carefully. Serious consequences can occur if they are improperly used because their use can cause portions of the ladder logic to be skipped.

Jump Instructions

Siemens has jump (JMP) and label (LBL) instructions available (see Figure 3-49). These can be used to reduce program scan time by not executing a section of program until it is needed.

The Jump in block instruction can be used to interrupt the linear execution of the program and resume it in another network. The network that is to be jumped to is identified by a label. The jump label is located above the instruction.

The specified jump label must be in the same block in which the instruction is executed. The label can only be used once in the block.

If the logic up to the input of the instruction is true, the jump instruction is executed. The jump direction can be forward or backward in networks. If the logic up to the input of the jump instruction is not true, execution of the program will continue in the next network.

In the example shown in Figure 3-49, if Inp_5 is true, the JMP instruction will execute and program execution will go to label Loc_2 in Network 3. The logic in Network 2 will be skipped.

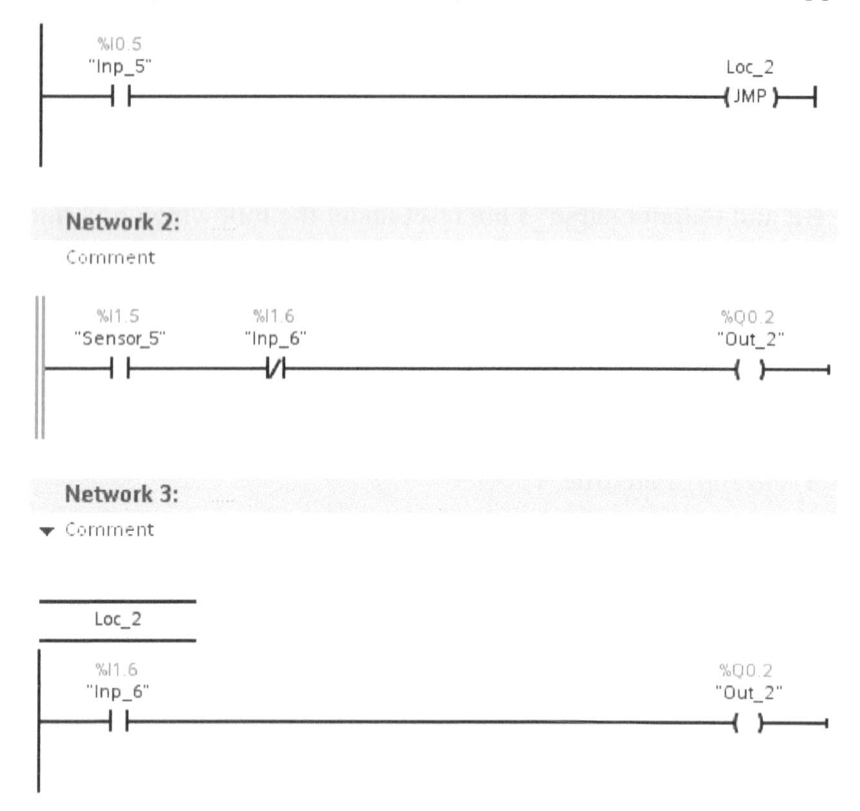

Figure 3-49. A jump if true instruction.

Jump if Zero (JMPN)

The Jump in block if false (JMPN) instruction can be used to interrupt the execution of the program and resume it in another network, when the logic up to the input of the JMPN instruction is false (see Figure 3-50). The target network must be identified by a jump label. The specified jump label is located above the JMPN instruction.

The specified jump label must be in the same block as the JMPN instruction. The label can only be used once in the block.

If the logic up to the instruction is false, the jump instruction (JMPN) is executed. The jump can be forward or backward in networks.

If the logic up to the input of the JMPN instruction is true, the instruction will not execute and execution will continue in the next network.

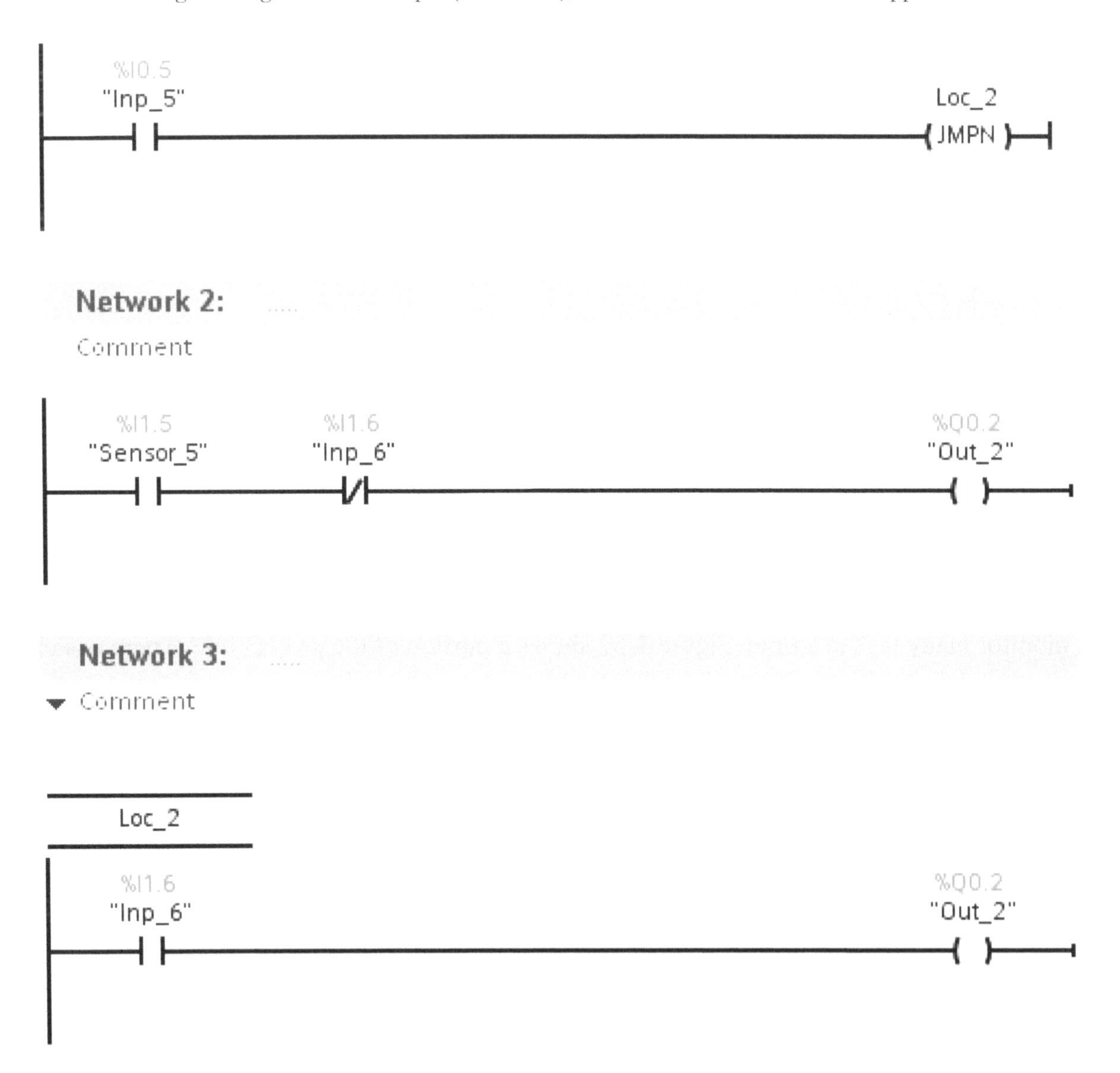

Figure 3-50. Use of a JUMPN (jump if zero) instruction.

Return Instruction

*The return instruction can be used to prematurely stop the execution of a bloc*k (see Figure 3-51). The instruction is then only executed if the signal state at the left connector is true. If this condition is fulfilled, program execution is terminated in the currently called block and continued after the call function in the calling block (for example in the calling OB). The status of the call function is determined by the parameter of the Return instruction. If it is true, the Output ENO of the call function is set to true. If it is false, Output ENO of the call function is reset to false.

The Output ENO of the call function is determined by the signal state of the specified operand.

If an organization block is terminated by the Return instruction, the CPU will continue in the system program. If the signal state at the input of the Return instruction is false, the instruction is not executed. In this example, program execution continues in the next network of the called block.

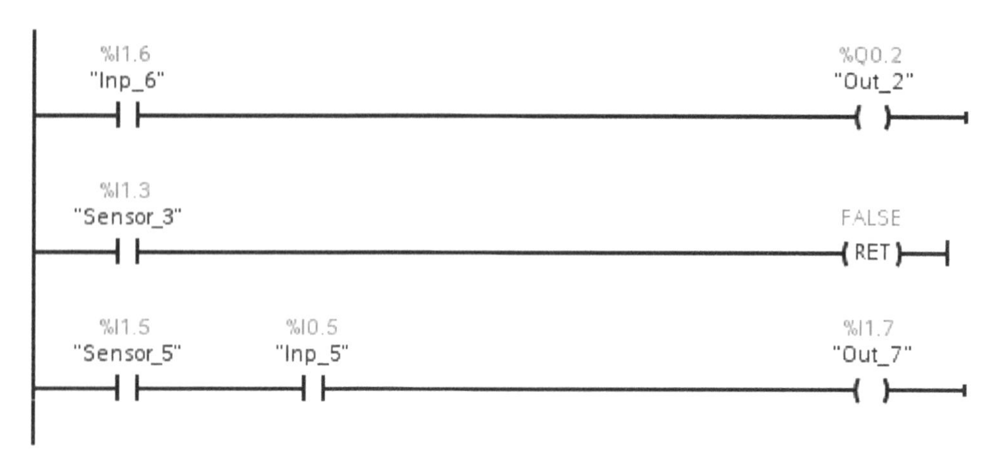

Figure 3-51. Use of a RET (return) instruction.

Watch and Force Tables

TIA software enables you to create watch tables. Watch tables can be used to create a table of tags that you would like to monitor in a table format. This can be very useful for troubleshooting as it enables the user to monitor many tags at a time. Figure 3-52 shows a portion of the project tree. There is a choice to Add new watch table under the Watch and force tables tab.

Force Tables

Force tables can be used to force inputs and outputs to a false or true state. This is usually done when troubleshooting. Note that forcing can be dangerous as you are overriding logic and actually turning real inputs and outputs on or off. Figure 3-52 also shows a Force table under the Watch and force table tab. If you double click on the Force table choice the screen in Figure 3-53 will appear.

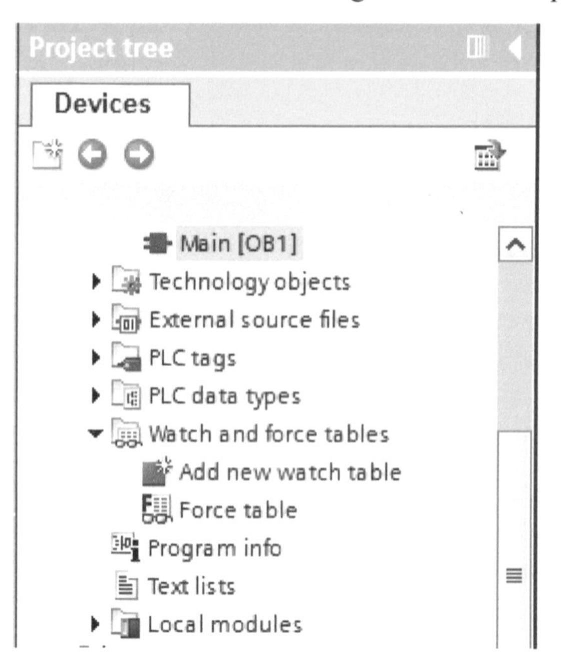

Figure 3-52. Project tree showing watch and force tables.

You can see the Add new choice in the Address column of Figure 3-53. This is where you would add tags that you would like to force on or off.

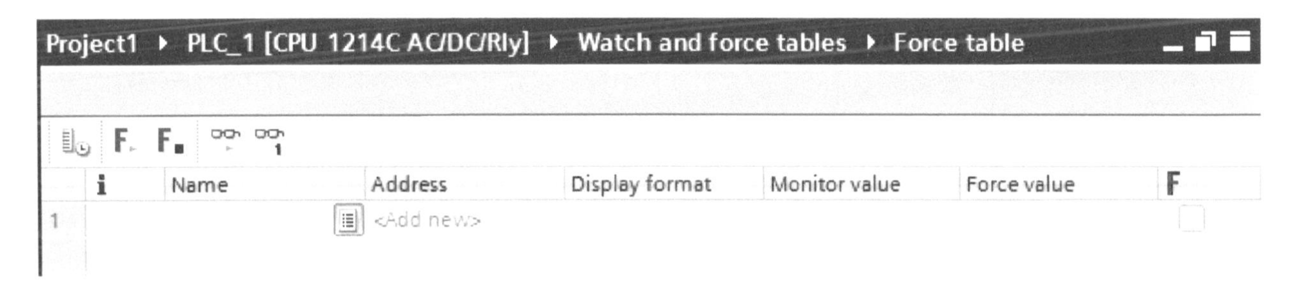

Figure 3-53. Force table configuration screen.

Figure 3-54 shows that 4 tags were added to the Force table. Note the symbols in the Monitor value column. Note that the symbols are both in the Monitor value column for outputs. It is not possible to use the monitor for peripheral outputs.

Project1 ▸ PLC_1 [CPU 1214C AC/DC/Rly] ▸ Watch and force tables ▸ Force table

	i	Name	Address	Display format	Monitor value	Force value	F
1		"Tag_1":P	%I0.0:P	Bool			
2		"Tag_2":P	%I0.1:P	Bool			
3		"Tag_3":P	%Q0.0:P	Bool			
4		"Tag_4":P	%Q0.1:P	Bool			

Figure 3-54. Force table with 4 tags added.

You can right-click on the tag you would like to force in the F column in Figure 3-54. You then have 4 choices: Force to 0; Force to 1; Force all; and Stop forcing. Tag_1 has been forced on in Figure 3-55. Note the F that appears above the tag name showing that it is in a forced condition.

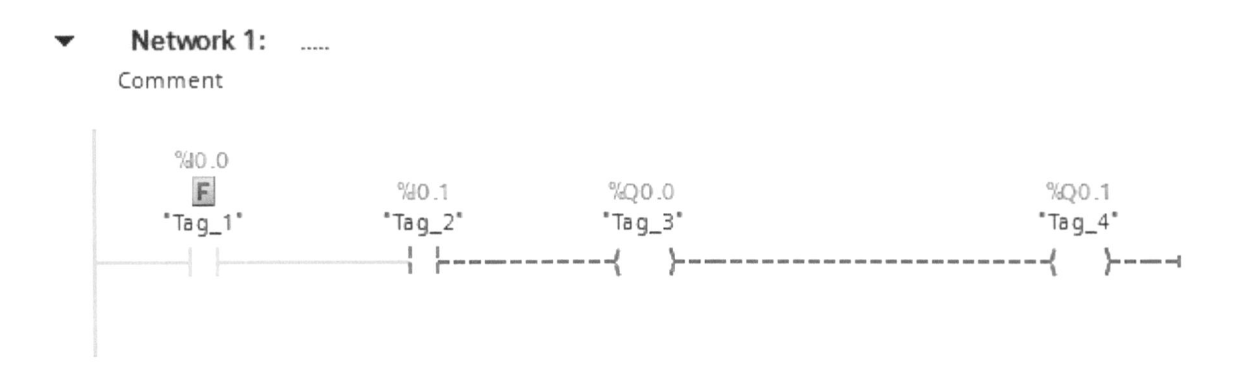

Figure 3-55. Tag_1 has been forced on.

Two inputs have been forced on in Figure 3-56. Note that these inputs are now true in the rung which turns on the two outputs in this logic.

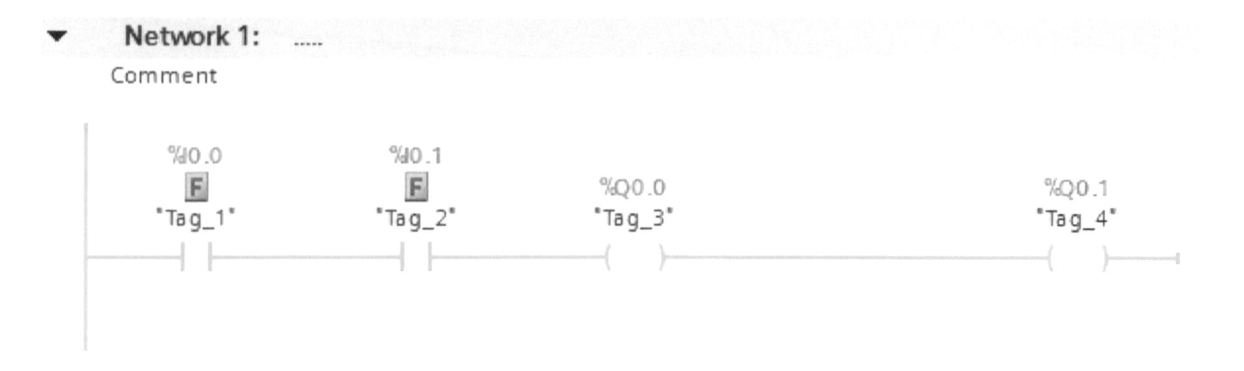

Figure 3-56. Tag_1 and Tag_2 have been forced on.

Figure 3-57 shows that output coils can also be forced on or off. Output tag Tag_3 was forced on (TRUE) in Figure 3-57.

	i	Name	Address	Display format	Monitor value	Force value	F
1		"Tag_1":P	%I0.0:P	Bool		FALSE	☐
2		"Tag_2":P	%I0.1:P	Bool		FALSE	☐
3	F	"Tag_3":P	%Q0.0:P	Bool		TRUE	☑
4		"Tag_4":P	%Q0.1:P	Bool			

Figure 3-57. Output forced to TRUE.

If forces are active, the Maintenance LED will blink (See Figure 3-58).

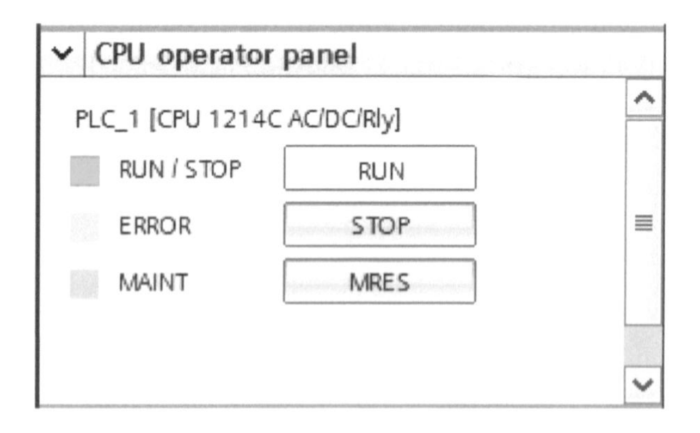

Figure 3-58. Maintenance LED.

TIA Portal Online Status Indicators

When the computer is online with the PLC the Project Tree has indicators that show the current condition of the PLC and program components (See Figure 3-59). A green circle or a green square with a checkmark means the item is OK. Everything shown in Figure 3-59 is OK.

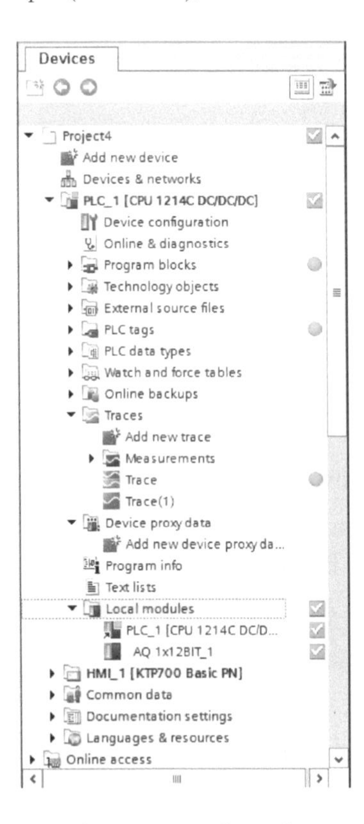

Figure 3-59. Project Tree when the PLC is online.

Figure 3-60 shows a situation where there is a problem. In this example the programmer added a contact to the logic but has not compiled or downloaded it yet . There is an orange circle with an exclamation point in it next to the Project name. This means that there is a problem in this project.

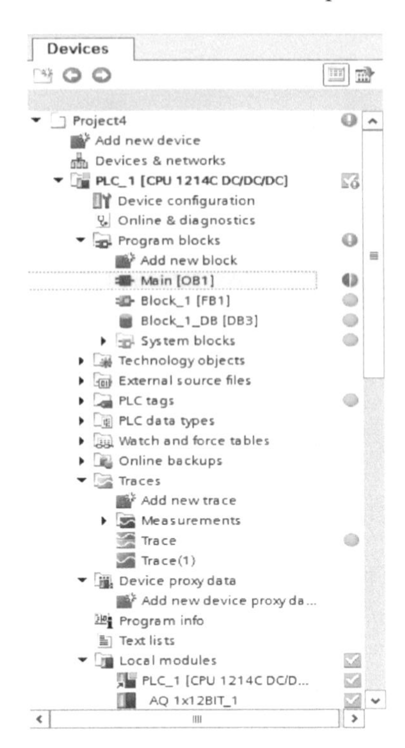

Figure 3-60. Project Tree when the PLC is online.

The orange circle next to Program Blocks in the Project Tree indicates that there is a problem in one of the program block. Note the circle that is half blue and half orange next to Main[OB1]. This one means that the online program and the offline program are different. This message is telling the programmer to compile and download the changed program. Note that if you hover the mouse over the indicator, while online, it will give you a description of what the indicator means.

Diagnostics

TIA Portal has a rich set of diagnostics available. Note that the Diagnostics section is located under the PLC in the Project Tree. Figure 3-61 shows the main Diagnostics screen and choices of the types of diagnostics. This screen shows the General Diagnostic information. It shows the specifics about the make and model of the PLC as well as Firmware version and serial number of the PLC as well as other information.

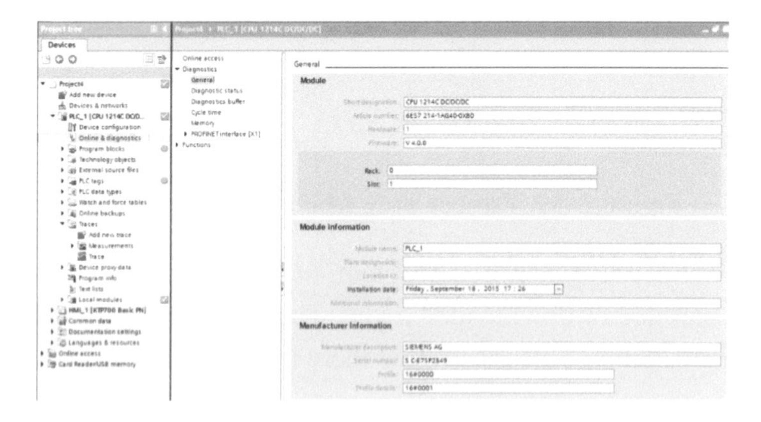

Figure 3-61. General Diagnostic screen.

Figure 3-62 shows the Diagnostic status of the PLC. In this example it shows that the PLC (module) exists and is OK.

Figure 3-62. PLC status screen.

Figure 3-63 shows the first portion of the Diagnostics buffer screen. This screen shows the events that have occurred and the time and date when they happened. Note that by clicking on them the specifics of that event are shown.

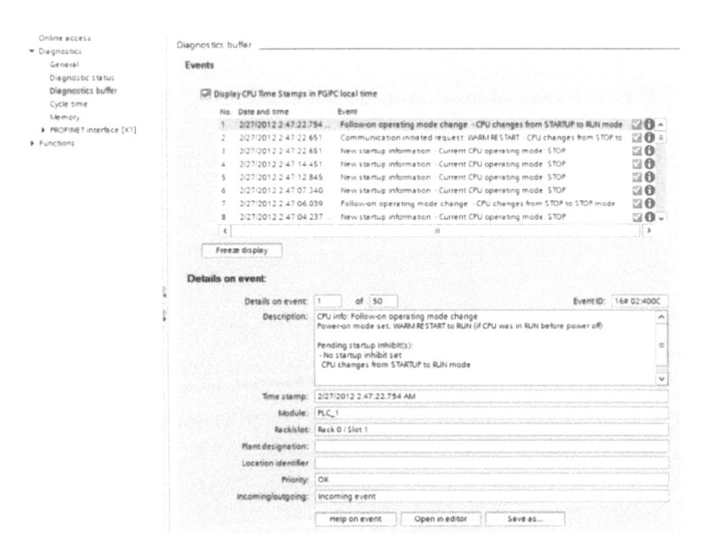

Figure 3-63. Diagnostics Buffer.

Figure 3-74 shows the Settings that control which information is shown in the Diagnostics Buffer Events list.

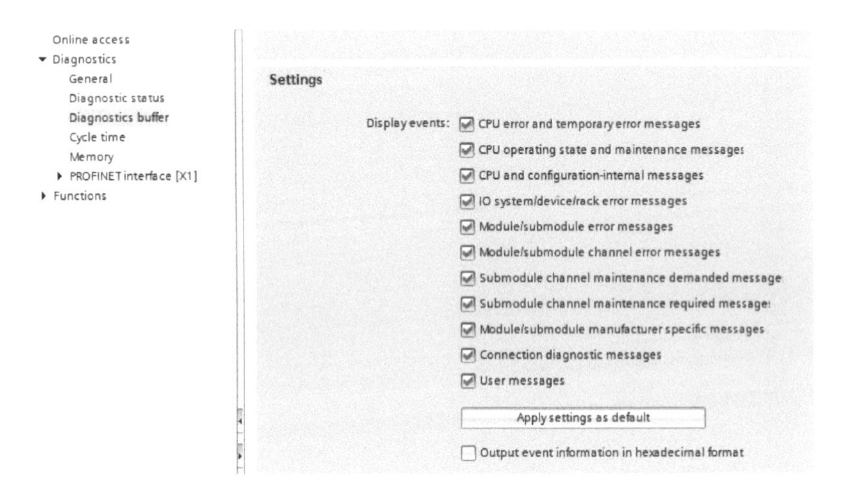

Figure 3-64. Settings that control which Events are shown in the Diagnostic Buffer.

The Diagnostics also include a screen that shows the cycle time of the logic (see Figure 3-65).

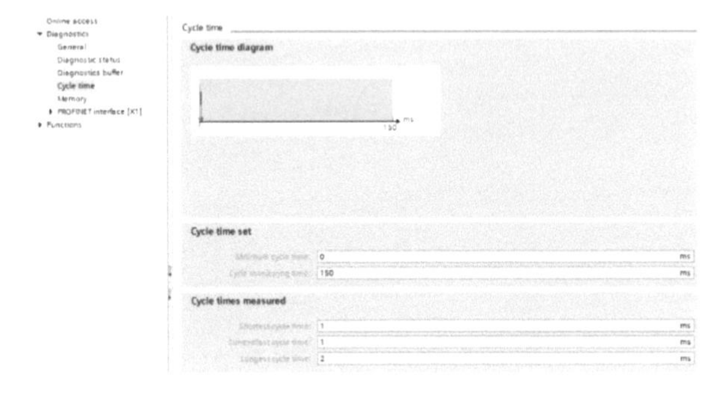

Figure 3-65. Cycle time diagnostics screen.

Diagnostics also has information on memory usage in the PLC. Figure 3-66 shows the memory usage screen.

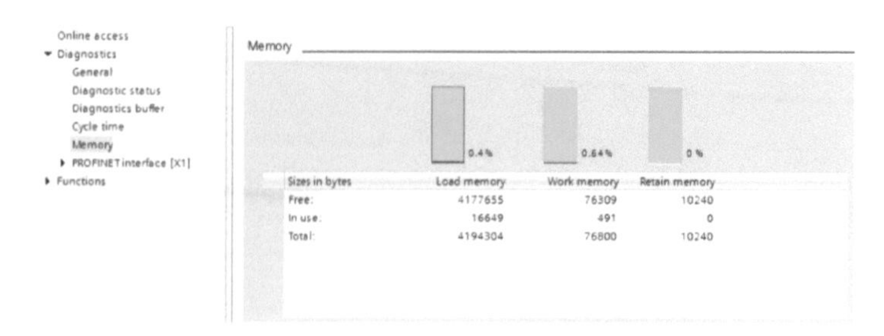

Figure 3-66. Memory usage screen.

The Diagnostics screen can also be used to examine the PROFINET settings. Figure 3-67 shows the PROFINET interface screen. Note that the MAC address is shown in this screen. The Ethernet IP address and subnet are also shown here.

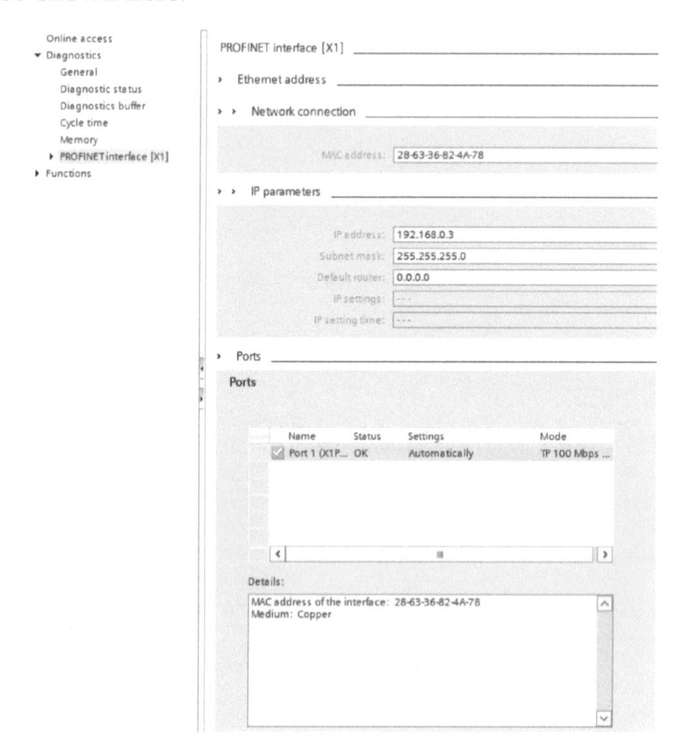

Figure 3-67. The PROFINET settings screen.

Configuration Functions

The IP address and subnet of the PROFINET interface can be set in the Functions portion of Diagnostics (see Figure 3-68).

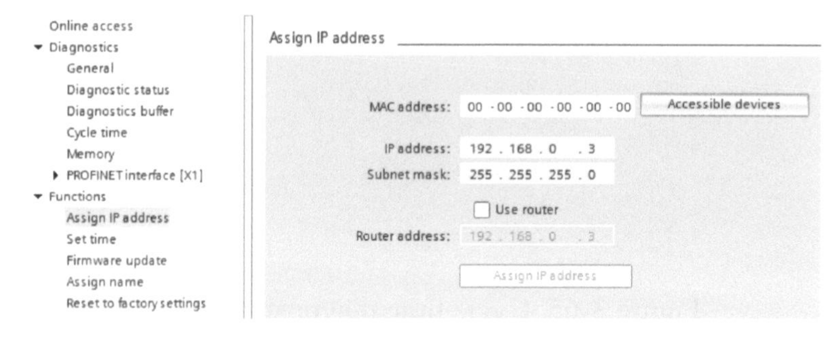

Figure 3-68. The screen where IP address settings can be made.

The time and date settings are configured in this Functions screen (see Figure 3-69).

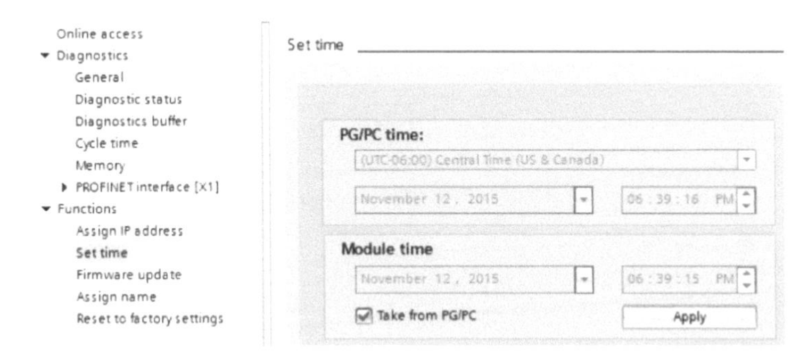

Figure 3-69. The Set time screen in Functions.

The next figure shows the Firmware version and enables the user to upgrade the Firmware.

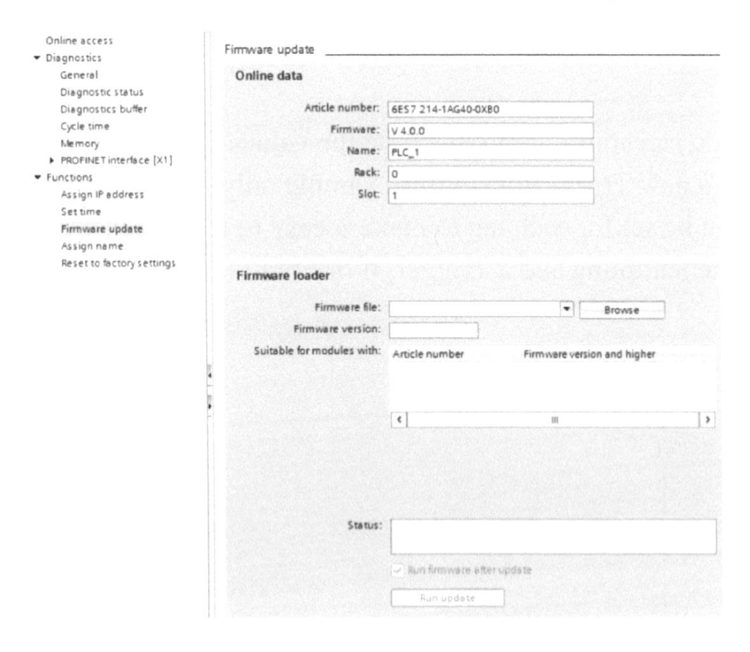

Figure 3-70. Firmware setting and upgrade screen.

Figure 3-71 shows the Assign name screen in Functions. This is where the user can assign a name to the PLC.

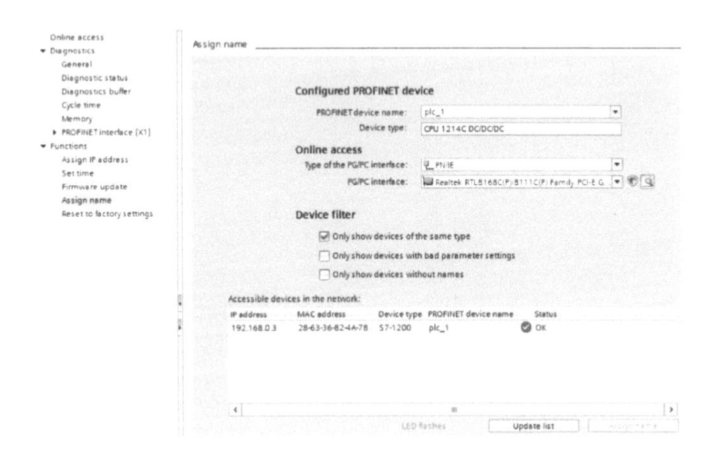

Figure 3-71. The Assign name screen in Functions.

Figure 3-72 shows the screen that can be used to return the PLC to factory settings. Note that the user can choose to retain the IP address or delete the IP address along with the factory reset.

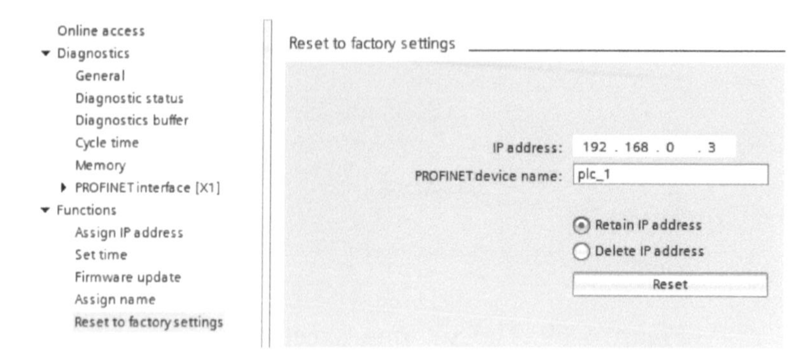

Figure 3-72. Reset to factory settings screen.

Configuring Traces

Traces are graphs that you can configure to illustrate the values of tags over time. Figure 3-73 shows the configuration screen for a new Trace. Note in this example only one tag value will be graphed (Level). The color of the trace can be set for each tag to make it easy to tell which value is which. Note also that the user can configure the Sampling and a Trigger, if desired.

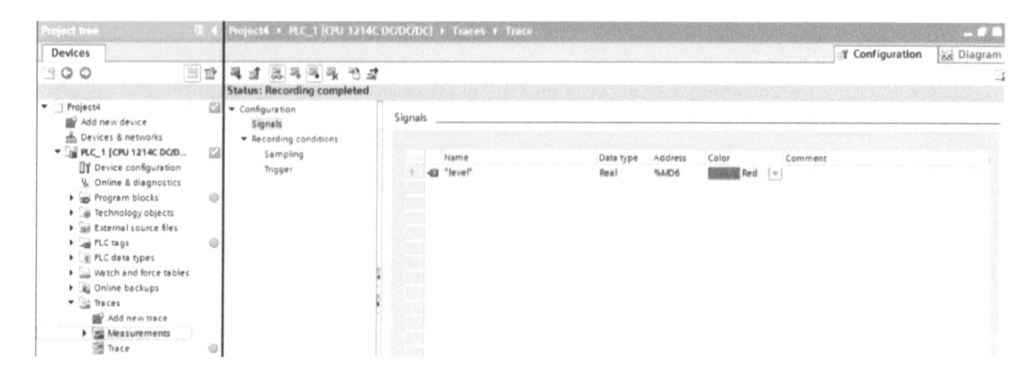

Figure 3-73. Configuring a signal in a trace graph.

Figure 3-74 shows the Trace graphing the value of the tag named “Level”.

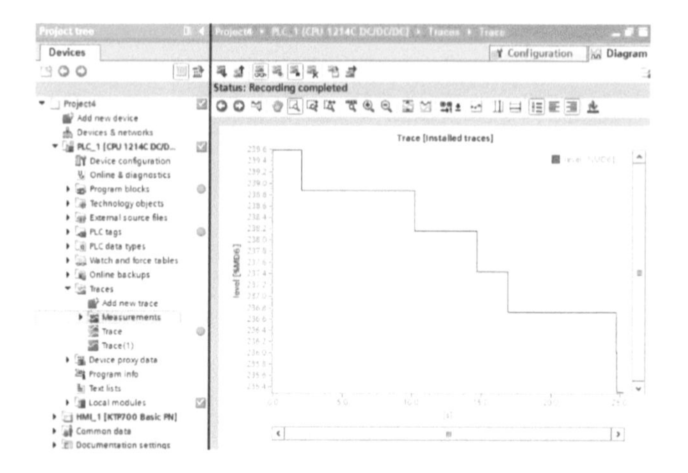

Figure 3-74. Trace of the value of the tag named Level over time.

Chapter Questions

1. What is a contact? A coil?

2. What are transitional contacts used for?

3. Explain the term normally-open.

4. Explain the term normally-closed.

5. What are some uses of normally-open contacts?

6. Explain the terms true and false as they apply to contacts in ladder logic.

7. Design a ladder that shows series input (AND logic). Use I0.5, I0.6, AND NOT I1.1 for the inputs and use Q1.5 for the output.

8. Design a ladder that has parallel input (OR logic). Use I0.2 and XI0.7 for the contacts and Q0.6 for the output.

9. Design a ladder that has three inputs and one output. The input logic should be: I1.1 AND NOT I0.2, OR I0.3. Use Q1.3 for the output.

10. Design a three-input ladder that uses AND logic and OR logic. The input logic should be I0.1 OR I0.3, AND NOT I0.5. Use Q1.2 for the output coil.

11. Design a ladder in which coil Q0.5 will latch itself in. The input contact should be I0.1. The unlatch contact should be I1.4.

12. What is a RET instruction used for?

Examine the rungs below and determine whether the output for each is on or off. The input conditions shown represent the states of real-world inputs.

13.

Real-World	State of Real-World Input
Inp_1	True
Inp_2	True
Inp_3	True
Inp_4	False

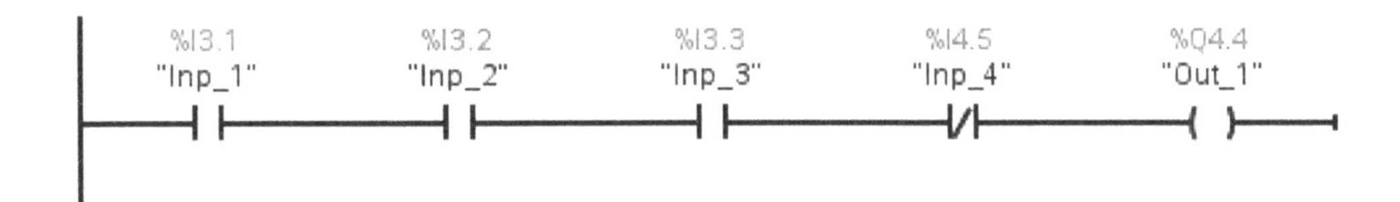

14.

Real-World	State of Real-World Input
Inp_1	True
Inp_2	True
Inp_3	True

%I3.1 "Inp_1" %I3.2 "Inp_2" %I3.3 "Inp_3" %Q4.4 "Out_1"

15.

Real-World	State of Real-World Input
Inp_1	True
Inp_2	True

%I3.1 "Inp_1" %I3.2 "Inp_2" %Q4.4 "Out_1"

16.

Real-World	State of Real-World Input
Inp_1	True
Inp_2	False
Inp_3	True

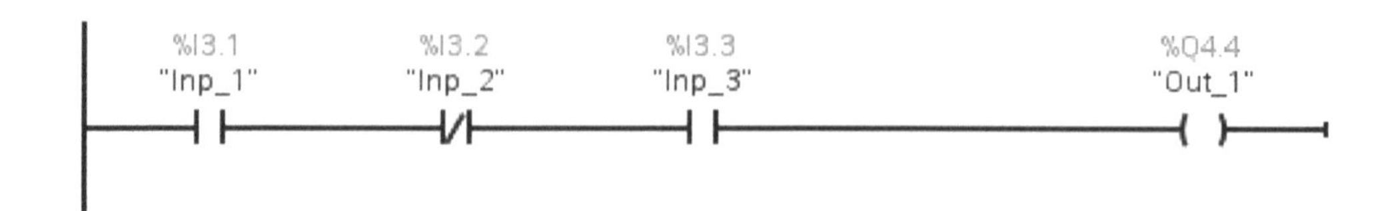

17.

Real-World	State of Real-World Input
Inp_1	True
Inp_2	True

%I3.1
"Inp_1"
%I3.2
"Inp_2"
%I3.2
"Inp_2"
%Q4.4
"Out_1"

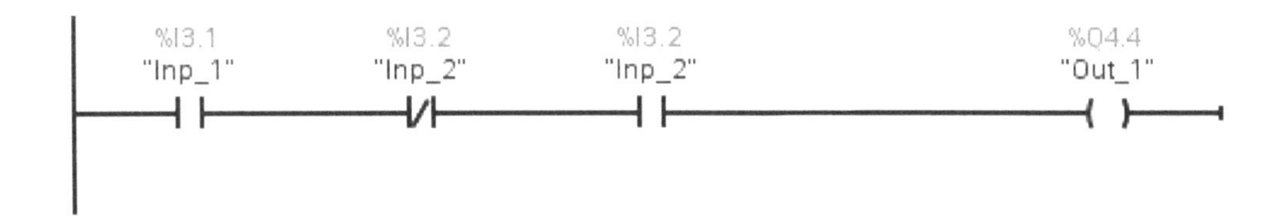

18.

Real-World	State of Real-World Input
Inp_1	False
Inp_2	True
Inp_3	True

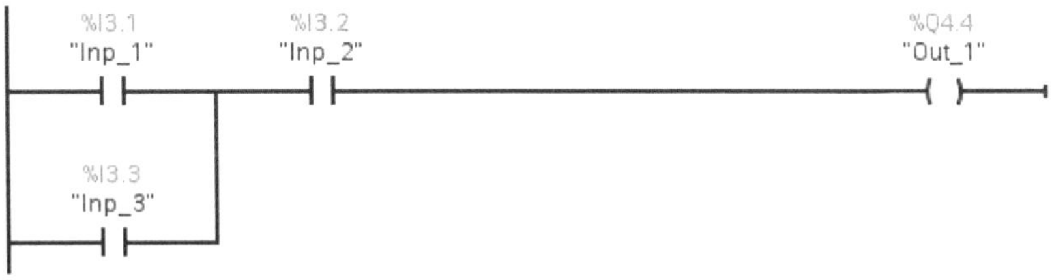

%I3.1
"Inp_1"
%I3.2
"Inp_2"
%Q4.4
"Out_1"
%I3.3
"Inp_3"

19.

Real-World	State of Real-World Input
Inp_1	True
Inp_2	False
Inp_3	True
Inp_4	True

%I3.1
"Inp_1"
%I3.2
"Inp_2"
%Q4.4
"Out_1"
%I3.3
"Inp_3"
%I4.5
"Inp_4"

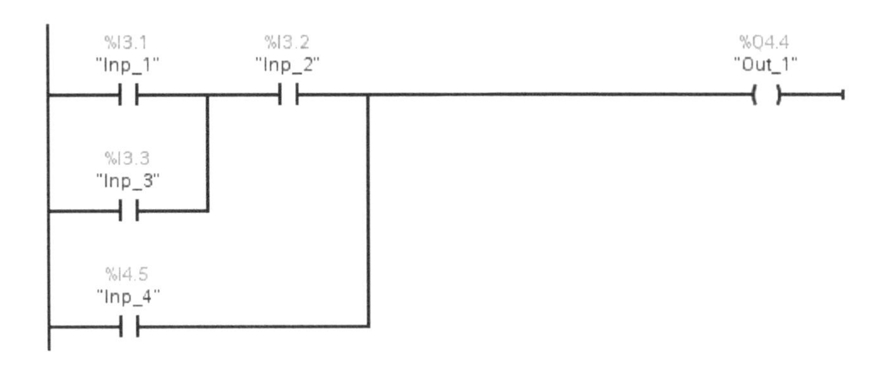

20. Write ladder logic for the application below. Your logic should have a start/stop circuit to start the application. Your logic should assure that the tank does not run empty nor overflow. Use the I/O names from the table for your logic.

I/O Name
Start
Stop
Run
Pump
Hi_Level_Sensor
Low_Level_Sensor

21. Blocks can only be monitored in TIA Portal if:
    a. The offline block is identical to the online block
    b. The PLC is online with TIA Portal
    c. Blocks cannot be monitored
    d. A and B are both correct.

22. Ethernet devices have been assigned a fixed unique address by their manufacturer. The address is called a _________ address.
   a. Ethernet Access Control (EAC) address
   b. Global Access Control (GAC) address
   c. World Access Control (WAC) address
   d. Media Access Control (MAC) address

Chapter 4

Timers and Counters

Objectives

After completing this chapter, the reader will be able to:

Describe the use of timers and counters.

Understand Siemens timer and counters.

Utilize timer and counter tag members in logic.

Define terms such as delay-on, delay-off, preset time, elapsed time, retentive, etc.

Correctly use timers and counters.

Timers

Timing functions are very important in PLC applications. Timers serve many functions in logic. They can be used to control when events occur. They can be used to delay actions in logic. They can also be used to keep track of elapsed time. Timers can also be misused in logic. Whenever possible, real events should be used to control when things happen in logic. Weak programmers sometimes use timers to make bad logic work. Timers should be used for logic that requires timed events.

Timers have some typical entries (see Figure 4-1). Timers have a name. Timers have a preset time (PT) value. The preset time is used to set the total time we would like the timer to accumulate until it turns its output (Q) on. The current amount of time that has accumulated is stored in the elapsed time (ET) tag.

Timers are available in various types to meet the needs of the application. Figure 4-1 shows a typical Siemens timer. The timer in 4-1 is called a TON timer. It is an on-delay timer. The timer's name in Figure 4-1 is "Timer_1". The PT time (PT) is 5s (5 seconds). When the timer is enabled, the elapsed time (ET) starts to increase. The elapsed time will be kept in the ET tag (Total_Time). When the elapsed time is equal to or greater than the preset time (PT) the timer's output (Q) will be true. Note that the tag holding the elapsed time will stop incrementing after it reaches the preset time (PT). So, in this example even if the time reaches more than 5 seconds ET will stop incrementing at 5 seconds.

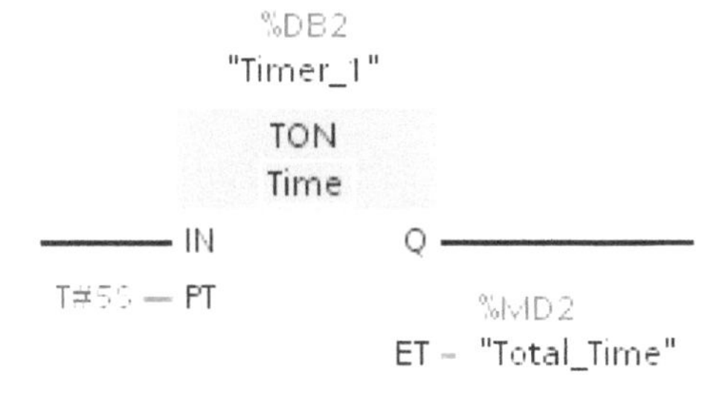

Figure 4-1. A Siemens TON timer.

Timers can be retentive or non-retentive. A retentive timer functions like a stopwatch. Stopwatches can be started and stopped multiple times and still retain their timed value. There is a reset button on a stopwatch to reset the time to zero. Retentive timers do not lose the elapsed time when the rung conditions go False. A retentive timer retains the elapsed time until the timer enable input is true again. When the enable input becomes true again, the retentive timer adds time to the elapsed time again. A reset timer (RT) instruction is used to reset a retentive timer's elapsed time to 0.

Non-retentive timers reset the elapsed time to 0 every time the rung conditions become False. If the timer's enable input (IN) becomes false, the timer elapsed time goes to zero.

On-Delay Timer

The TON instruction is an on-delay timer. A TON timer can be used to turn an output on after a delay. A TON timer begins accumulating time when the rung conditions to the enable input (IN) become true. If the accumulated time in the timer becomes equal to, or greater than the preset time (PT), the timer's output bit (Q) is set to a 1. Figure 4-2 shows an example of a TON timer in a rung of logic. In this example the enable input (Cycle_Start) is true so the timer is accumulating time in the elapsed time output. In this example the current elapsed time is 1S 903 ms (1 second and 903 milliseconds, 1903ms total). The current elapsed time is less than the preset time (PT) of 5 seconds so the timer's output (Q) is false.

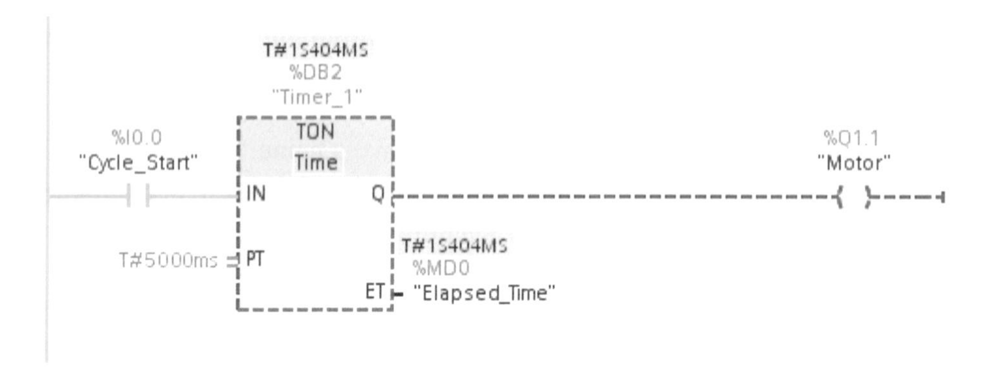

Figure 4-2. Timer example. Elapsed time is less than the preset time so the output is false.

In Figure 4-3 the enable input (Cycle_Start) is true so the timer is accumulating time in the elapsed time output. In this example the current elapsed time is 5S (5 seconds) or more. The current elapsed time is equal to or greater than the preset time (PT) of 5 seconds so the timer's output (Q) is true. The output will remain true until the enable input becomes false or the timer is reset with a reset instruction.

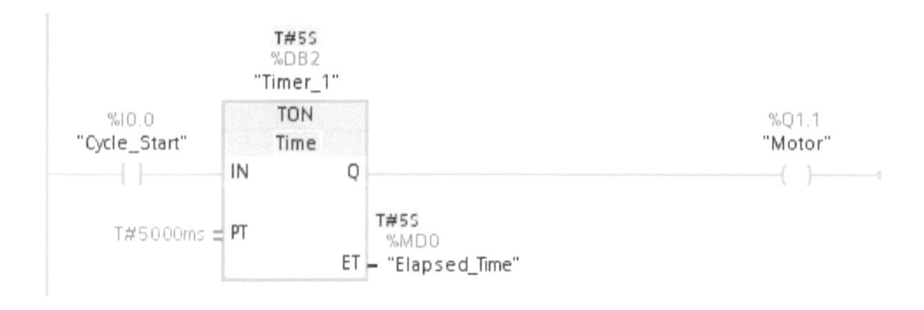

Figure 4-3. Timer example. Elapsed time is equal to or greater than the preset time so the output is true.

Timer Parameters

Timers have parameters that can be set by the user. Figure 4-4 shows a table of input and output parameters and the possible data types for a TON timer.

Parameter	Description	Data Type	Memory Area
IN	Start input	BOOL	I,Q,M,D,L
PT	Preset time	TIME	I,Q,M,D,L, or constant
Q	Output, it is delayed by the time specified in the PT.	BOOL	I,Q,M,D,L
ET	Elapsed time	TIME	I,Q,M,D,L

Figure 4-4. TON timer input and output parameters.

Timers have a PT time that must be set by the programmer. The PT time can be thought of as the number of time increments the timer must count before changing the state of the output. Siemens timers have a time base in days, hours, minutes, seconds, and/or milliseconds. Figure 4-5 shows an example of the preset time (PT) specification using days, hours, minutes, seconds, and milliseconds. Any or all of these may be used to specify the PT.

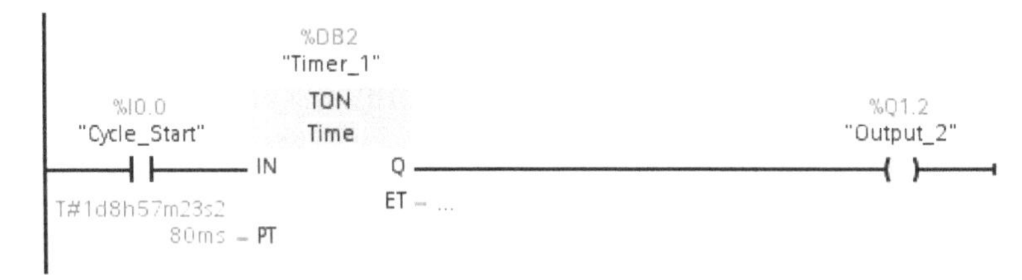

Figure 4-5. TON timer example.

The actual time delay would equal the value in the PT (PT time) multiplied by the time base. For example, if the PT is 35S, this would be a 35 second timer. A PT of 10000ms would be a 10 second time (10000 * 1 ms = 10 seconds). A tag can also be used for the PT value. The tag enables the PT (preset time) to be changed in the ladder logic. Timers have one input that enables the timer (IN). When this input is true (high) the timer will accumulate time in the accumulator.

When a timer is created a data block is also created to hold the parameter data for the timer. Figure 4-6 shows the creation of an instance data block for Timer_1. If the user chooses the Manual option they can enter their own name for the block. If Automatic is chosen the name is generated automatically.

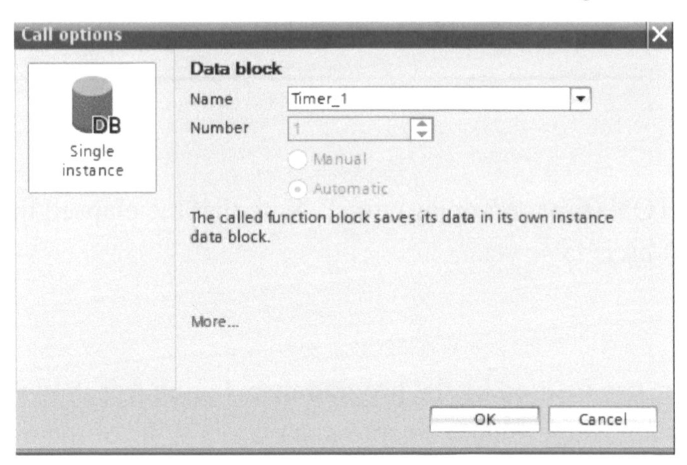

Figure 4-6. Data block for a timer.

Figure 4-7 shows the timer's parameter information in the instance data block for Timer_1.

Timer_1

	Name	Data type	Start value	Retain	Accessible from HMI	Visible in HMI	Setpoint	Comment
1	▼ Static			☐	☐	☐	☐	
2	ST	Time	T#0ms	☐	☑	☑	☐	
3	PT	Time	T#0ms	☐	☑	☑	☐	
4	ET	Time	T#0ms	☐	☑	☑	☐	
5	RU	Bool	false	☐	☐	☐	☐	
6	IN	Bool	false	☐	☑	☑	☐	
7	Q	Bool	false	☐	☑	☑	☐	

Figure 4-7. Timer data block parameter information.

Use of a Timer's Tag Members

The Q bit is the most commonly used timer status bit. In a TON timer the Q bit is false until the elapsed time is equal to or greater than the PT value. The Q bit remains set until the rung goes false or a reset instruction resets the timer. A uses millisecond time base was used in this example. The PT in this example is 10000. This would be a 10 second time delay (10000 X 1 ms = 1 second).

Consider Figure 4-8. The timer is named Timer_1. The PT is 1000 ms. The timer IN (enable) bit becomes true when Cycle_Start is true. The Q bit on the instruction is controlling Output_2 on the rung. The second rung shows the use of the Timer_1.Q bit to control and output named Fan. When the elapsed time (ET) is equal to or greater than the preset time (PT), the Q bit will be true and the outputs will be energized.

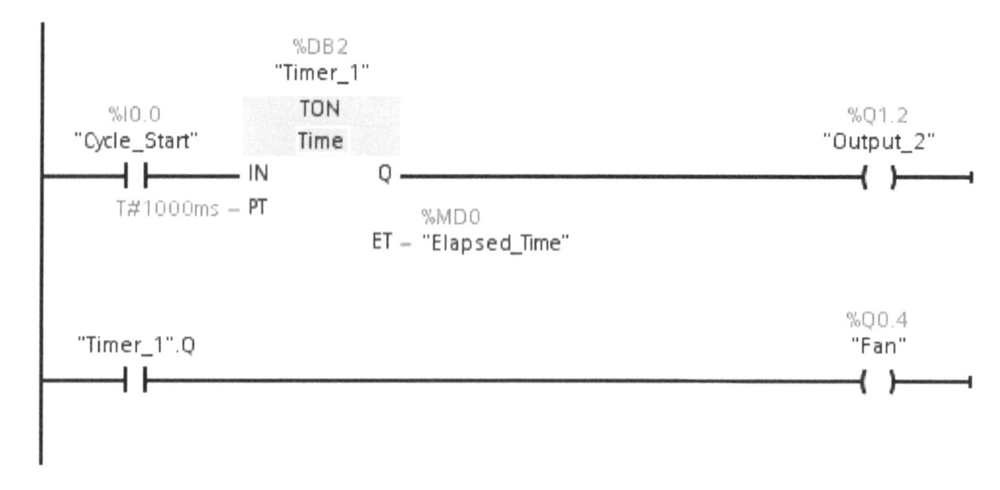

Figure 4-8. Use of a TON (timer on-delay) timer. Note that the elapsed time (ET) has not reached the PT value so the Q bit is false yet.

Elapsed Time Use

The elapsed time (ET) can also be used by the programmer. Figure 4-9 shows the use of comparison instructions. The second rung is to evaluate whether the elapsed time of the timer is greater than or equal to 0 ms and less than or equal to 2500 ms. If this rung is true, Output_3 would be energized.

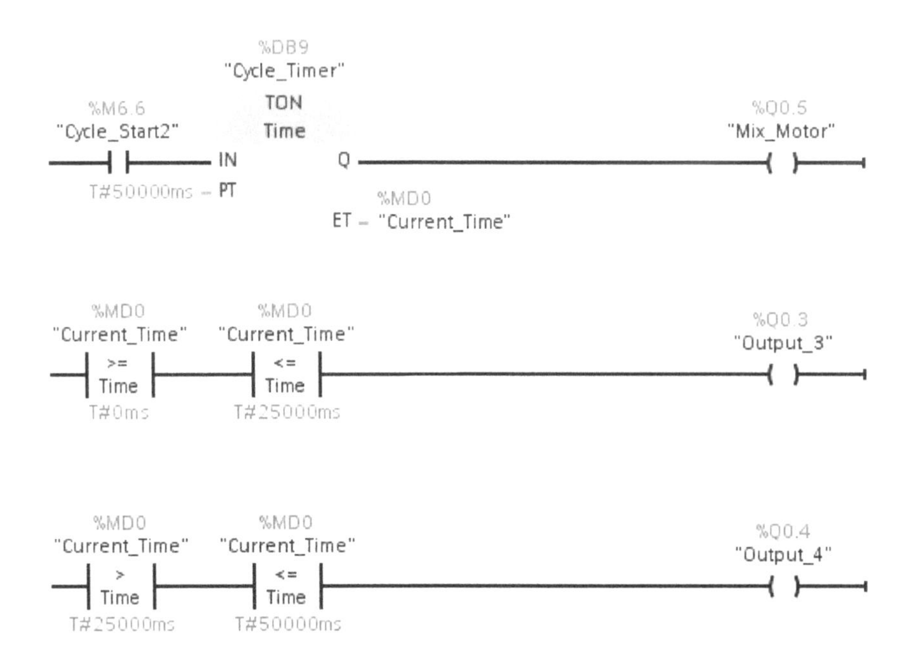

Figure 4-9. Example of the use of the elapsed time in a timer with comparison instructions to control outputs.

The third rung is to evaluate whether the elapsed time of the timer is greater than 2500 ms and less than or equal to 5000 ms. If this rung is true, Output_4 would be energized.

The PT value can also be used in logic. For example, Timer_1.PT would access the PT value of the timer named Timer_1. The PT value would be a double integer (DINT). The value of PT can be modified in logic.

Resetting a Timer

To reset a timer's elapsed time to 0 a reset timer (RT) instruction is used. Figure 4-10 shows a RT instruction. An RT instruction uses the timer's name. In the example in Figure 4-10 the RT would reset the elapsed time to zero in Timer_1.

%DB2
"Timer_1"
RT

Figure 4-10. A reset timer instruction.

Figure 4-11 shows the use of an RT instruction to reset a timer's elapsed time to zero. If contact Inp_1 becomes true, the value of timer Timer_1 ET will be set to zero.

Figure 4-11. Use of a Siemens RES (reset) instruction to reset a timer's elapsed time.

TON timers can also be reset by de-energizing the IN-input of the timer. Figure 7-12 shows an example in which the output bit from the Q-output (Sequence_Complete) is used to reset the timer ET to zero every time the timer elapsed time (ET) reaches the PT. This can be used to make things happen at regular intervals. For example, this could be used to execute an instruction or some logic every 5 seconds. Note that the Q-output was not directly used to reset the timer. An output tag from the Q-output should be used to reset a timer.

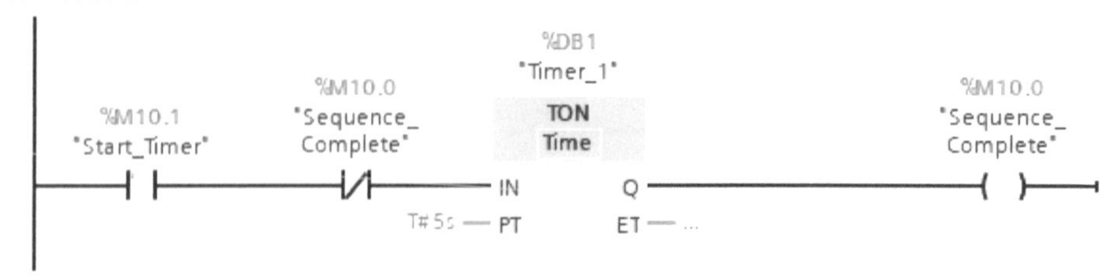

Figure 4-12. Use of an output tag (Sequence_Complete) from the timer Q-output to reset the timer's elapsed time to 0 every time it reaches 5 seconds (the PT value).

Use of MOV (Move) Instructions to Change the Preset Time of a Timer

Figure 4-13 shows an example of using logic to change the PT of a timer based on conditions. In this example, if Product_A bit is set, the PT is set to 10000. If Product_B is set, the PT is set to 20000.

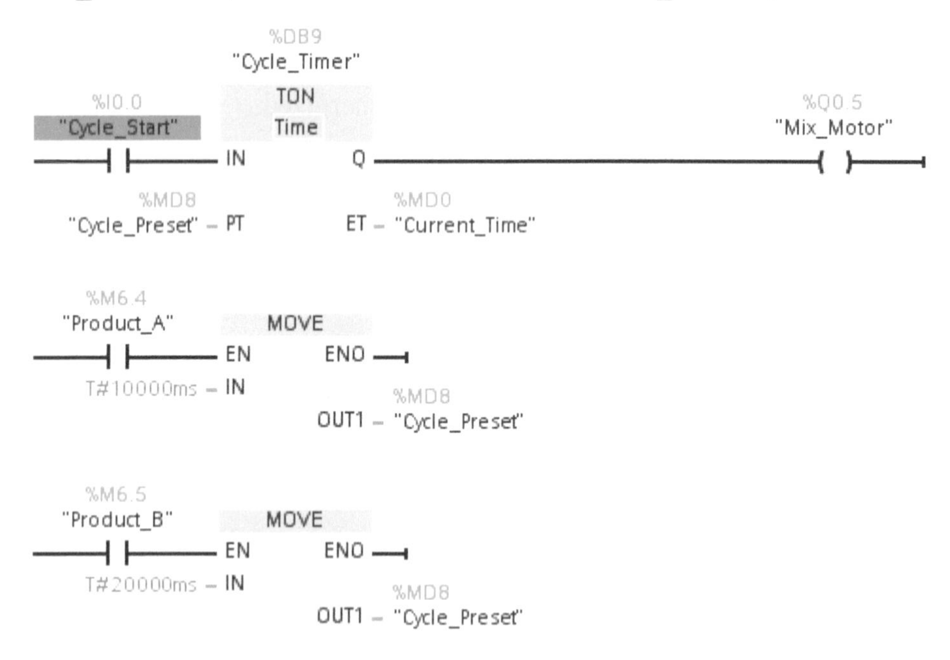

Figure 4-13. Use of move (MOV) instructions to change the PT value in a timer.

There is another style of the TON timer available in STEP 7. Figure 4-14 shows a TON style timer. Too utilize this style of timer you must create a Timer DB first. This is done by using the Add new block choice in the Project Tree. Figure 4-15 shows a part of the project tree. Note that a DB was added and named Timer_1_DB.

After you have created the timer-type DB you can use this style of timer in your logic. Study Figure 4-14. If the input named Tag_1 is true the TON timer will begin to increment the elapsed time in DB Timer_1_DB. The preset was entered as 10S (10 seconds).

A tag could also have been used to set the preset value. If the input to the timer remains true for 10 seconds the Q output of the timer will be true and the second rung will energize and turn on the output named Tag_2. If the input conditions to the timer (Tag_1) become false at any time the elapsed time in Timer_1_DB will be reset to zero. The Q bit is only true if the 10 seconds has elapsed and the timer has not been reset.

Timer Starters

Siemens has some special timer instructions available that are commonly called timer starters. Figure 4-14 shows a TON timer starter. To utilize a timer starter the timer must exist. Figure 4-15 shows part of the project tree showing that a timer DB was created. The timer was named Timer_1_DB. If input Tag_1 becomes true in Figure 4-14 the timer will begin to accumulate time. Note that the TON timer starter instruction has two parameters: name of the timer and a preset value. The preset value can be a constant or a tag. In the second rung of Figure 4-14 the timer's Q bit is used as the input condition. If the timer's elapsed time reaches 10 seconds in this example the output named Tag_2 will be energized.

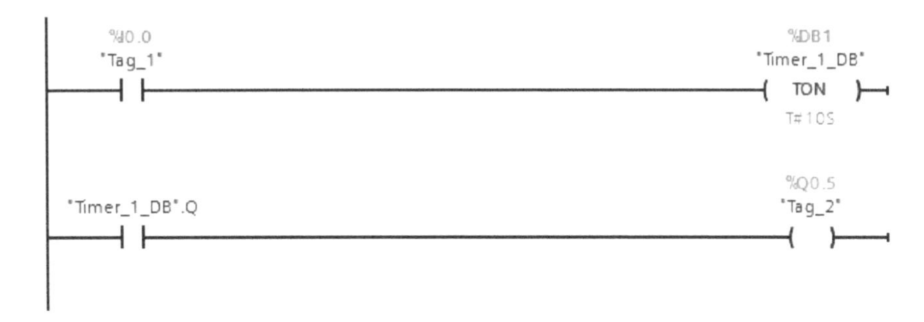

Figure 4-14. TON timer starter example.

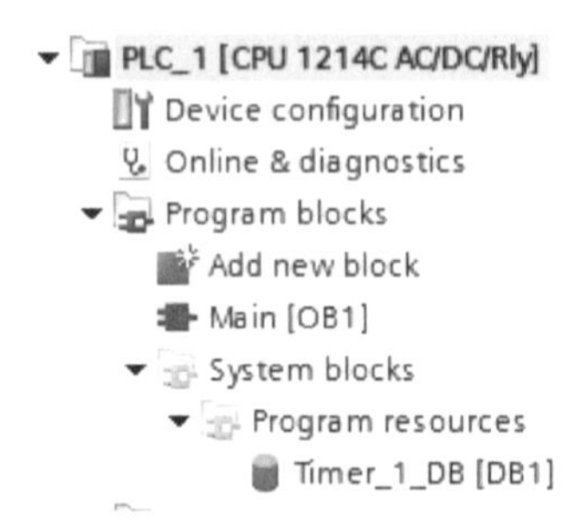

Figure 4-15. Timer DB in the project tree.

Timer Off-Delay

The timer off-delay (TOF) instruction can be used to turn an output coil on or off after the rung has been false for a desired time. Figure 4-16 shows a TOF timer.

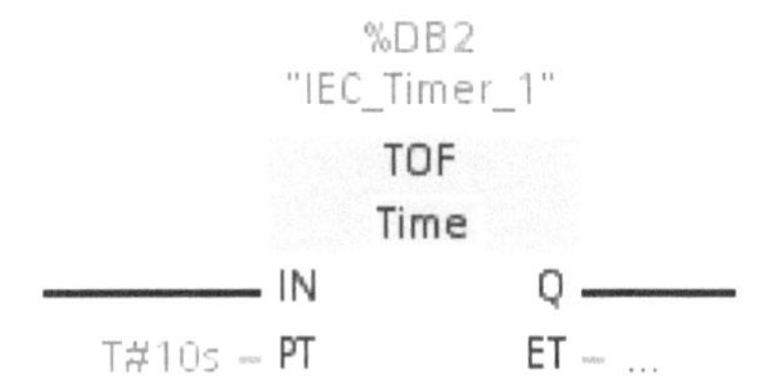

Figure 4-16. A TOF timer.

Figure 4-17 shows the parameters for a TOF timer.

Parameter	Description	Data Type	Memory Area
IN	Start input	BOOL	I,Q,M,D,L
PT	Preset time	TIME	I,Q,M,D,L, or constant
Q	Output, it is delayed by the time specified in the PT.	BOOL	I,Q,M,D,L
ET	Elapsed time	TIME	I,Q,M,D,L

Figure 4-17. TOF timer parameters.

Let's use a non-industrial example to understand the function of a TOF timer. Think of a bathroom fan. It would be nice if we could just push a momentary switch and have the fan turn on immediately and stay on for two minutes and then automatically shut off.

A TOF timer is called a delay-off timer. The timer turns on instantly, counts down time increments and then turns off (delay-off).

Figure 4-18 shows the use of a TOF timer. When the Cycle_Start contact becomes true, the timer Q bit becomes true. Note that Cycle_Time.Q is on, turning the output (Fan) on immediately. The timer's Q bit will stay on forever if the Cycle_Start input remains true. When Cycle_Start becomes false, it starts the timer timing cycle. When the rung goes false the time begins accumulating the elapsed time.

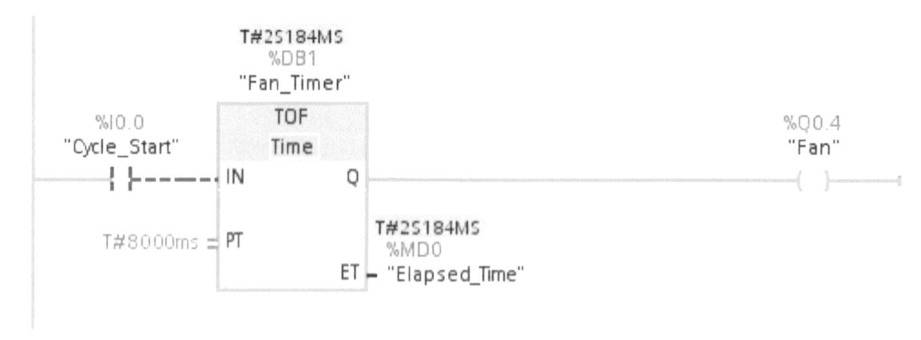

Figure 4-18. Delay-off timing circuit.

When the elapsed reaches 8 seconds the Q bit will become false. Figure 4-19 shows that when the elapsed time reaches 8 seconds, the Fan output is false.

Remember that a TOF timer's output will be true when the IN input becomes true. A TOF timer starts to accumulate time when the rung makes a transition from true to false. TOF timers the accumulate time until the elapsed time (ET) equals the PT value or the timer enable (IN) becomes true again (see Figure 4-19).

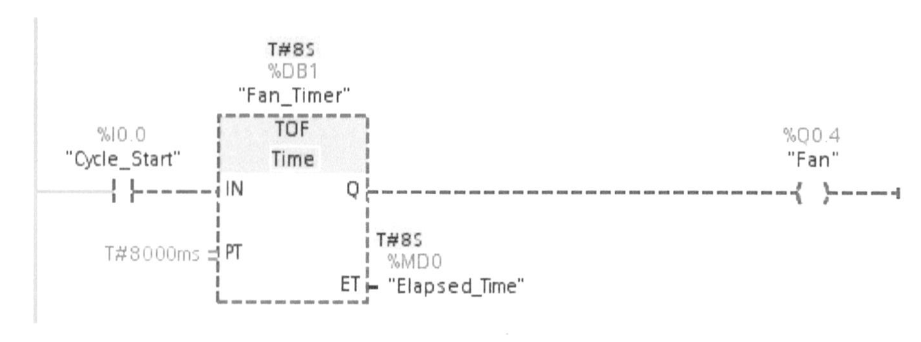

Figure 4-19. Use of a TOF timer.

TOF Timer Starter

Figure 4-20 shows a TOF timer starter. To utilize a timer starter the timer must exist. Note that the TOF timer starter instruction has two parameters: name of the timer and a preset value. The preset value can be a constant or a tag. In the second rung of Figure 4-14 the timer's Q bit is used as the input condition. If input Tag_1 becomes true in Figure 4-14 the output bit (Q) will become true immediately which will turn on the output named Tag_2 in the second rung timer. When Tag_1 becomes false (true-to-false transition) the timer will begin to accumulate time until the elapsed time reaches 10 seconds. When the elapsed time reaches 10 seconds, the Q output bit will be reset to false. And the output on the second rung will be reset to false.

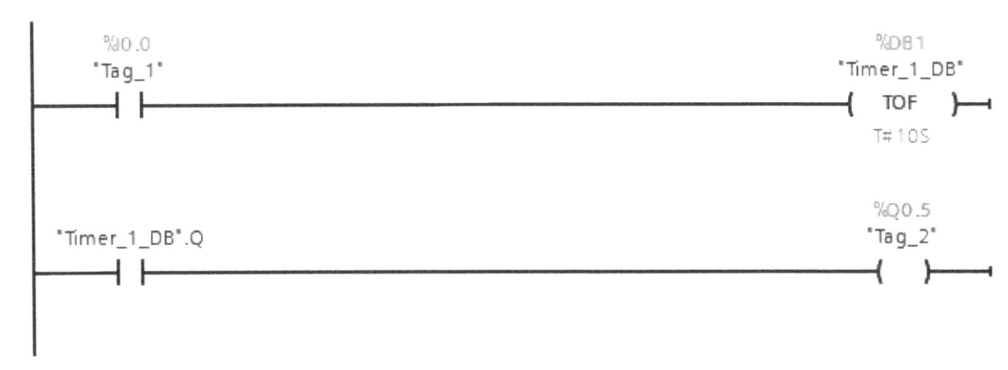

Figure 4-20. TOF timer example.

Retentive Timer On

The TONR (retentive timer-on instruction) is used to turn an output on after a PT time has accumulated (see Figure 4-21). The TONR timer is an accumulating timer. It retains the elapsed time (ET) when the enable input of the timer (EN) goes false. The TONR retains the accumulated time even if power is lost, or you switch modes, or the enable input (IN) becomes false.

Remember that a retentive timer functions like a stopwatch. Stopwatches can be started and stopped multiple times and still retain the elapsed time. There is a reset button on a stopwatch to reset the time to zero. Retentive timers do not lose the elapsed time when the rung conditions go False. A retentive timer retains the elapsed time until the timer enable input is true again. When the enable input becomes true again, the retentive timer adds time to the elapsed time again. A retentive timer's elapsed time can be reset by the reset input to the instruction. In this example an input named TMR_Reset can be used to reset the elapsed time. A reset timer (RT) instruction can also be used to reset a retentive timer's elapsed time to 0.

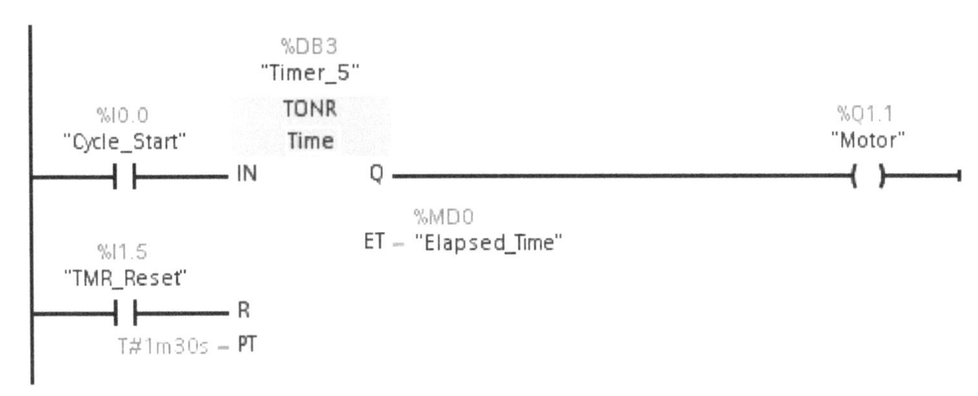

Figure 4-21. Use of a TONR timer.

Figure 4-22 shows the parameters for a TONR timer.

Parameter	Description	Data Type	Memory Area
IN	Start input	BOOL	I,Q,M,D,L
R	Reset input	BOOL	I,Q,M,D,L
PT	Preset time	TIME	I,Q,M,D,L, or
Q	Output, it is set when the PT time expires.	BOOL	I,Q,M,D,L
ET	Elapsed time.	TIME	I,Q,M,D,L

Figure 4-22. Parameters for a TONR timer.

TONR Timer Starter

Figure 4-23 shows a TONR timer starter. To utilize a timer starter the timer must exist. Note that the TONR timer starter instruction has two parameters: name of the timer and a preset value. The preset value can be a constant or a tag. When Tag_1 becomes true, the timer will begin to accumulate time until the elapsed time reaches 10 seconds. If Tag_1 becomes false before the timer's elapsed time reaches 10 seconds the timer's elapsed time is not reset to zero, it is retained and will increase when the input conditions become true again. When the elapsed time reaches 10 seconds, the timer's Q bit will be set to true. In the second rung of Figure 4-22 the timer's Q bit is used as the input condition to control the output (Tag_2). A RT (reset timer) instruction must be used to reset the elapsed time in a retentive timer to zero.

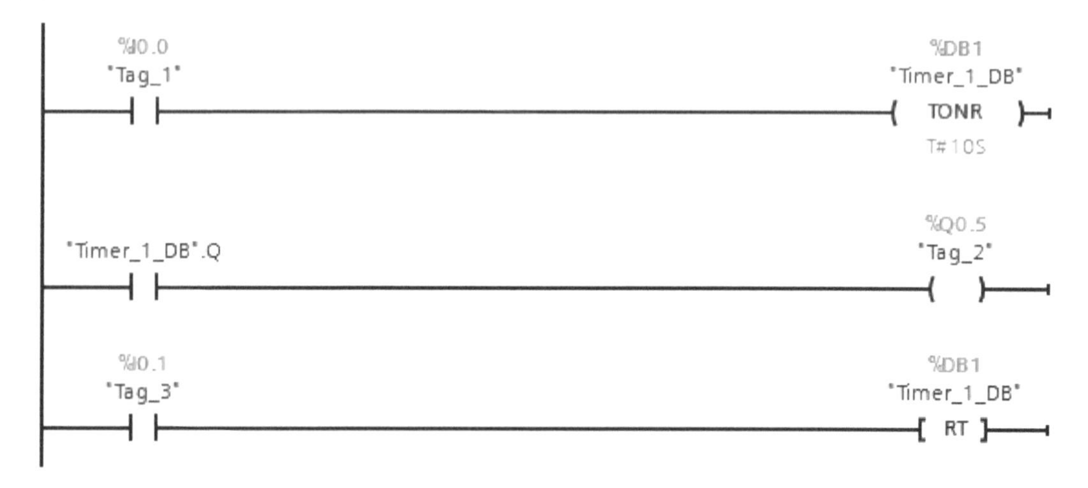

Figure 4-23. TONR timer example.

Pulse Generation (TP) Instruction

A TP instruction can be used to turn on an output for a set period of time. When the enable input (IN) becomes true the output Q will be turned on for the amount of time that the preset time (PT) specifies. In the example shown in Figure 4-24 when Part_Sensor becomes true output Q will be on for 500 ms. This means that the output named Motor will be turned on for 500 ms. A tag may be specified to hold the elapsed time at the ET output of the timer.

Figure 4-24. Use of a TP instruction.

The parameters for a TP instruction are shown in Figure 4-25.

Parameter	Description	Data Type	Memory Area
IN	Start input	BOOL	I,Q,M,D,L
PT	Duration of the pulse- Must be a positive value.	TIME	I,Q,M,D,L, or constant
Q	Output to be pulsed.	BOOL	I,Q,M,D,L
ET	Elapsed time.	TIME	I,Q,M,D,L

Figure 4-25. Parameters for a TP instruction.

TP Timer Starter

Figure 4-26 shows a TP timer starter. To utilize a timer starter the timer must exist. Note that the TP timer starter instruction has two parameters: name of the timer and a preset value. The preset value can be a constant or a tag.

A TP instruction can be used to turn on an output for a set period of time. When the input condition to the TP instruction (Tag_1 in this example) becomes true, the timer's Q output will be turned on for the amount of time that the preset time specifies (5 seconds in this example). In the example shown in Figure 4-26 when Tag_1 becomes true, output Q will be on for 5 seconds. This means that the output named Tag_2 will be turned on for 5 seconds.

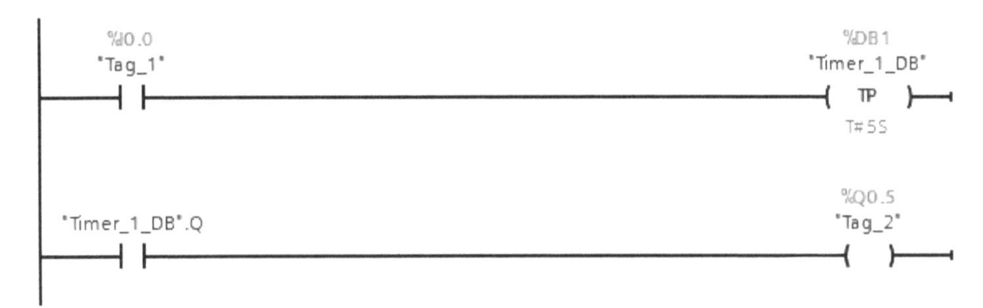

Figure 4-26. TP timer starter example.

Summary of Timers

Figure 4-27 shows the types of timer instructions.

Timers	Description
TON	On delay timer
TOF	Off delay timer
TONR	Retentive timer
TP	Pulse generation instruction

Figure 4-27. Timer instructions

Counters

Counting is a very common function in industrial applications. Actions must often be based on product counts. In case packing for example, there may be 4 rows of 6 cans making up one case of product. In this simple example we may need to count to 6 for the 6 cans in each row and then 4 for the number of rows in a completed case. Actions would be based on each count. We would also need another counter to count the number of cases that had been produced. We might need another down-counter to show how many more need to be produced to complete the order.

Up counters, down counters, and up/down counters are available. For example, if we are counting the number of filled and capped bottles leaving a bottling line, and we are tracking how many parts remain in a storage system, we might use a up/down counter.

Down-counters cause a count to decrease by one every time there is a pulse. Up/Down counters can be used to increase or decrease the count depending on inputs.

Counters have a counter name, a preset value (PV), a current count value (CV) value, and several other tag members. A data block is created for each counter (see Figure 4-28). Figure 4-29 shows the parameters for a counter data block.

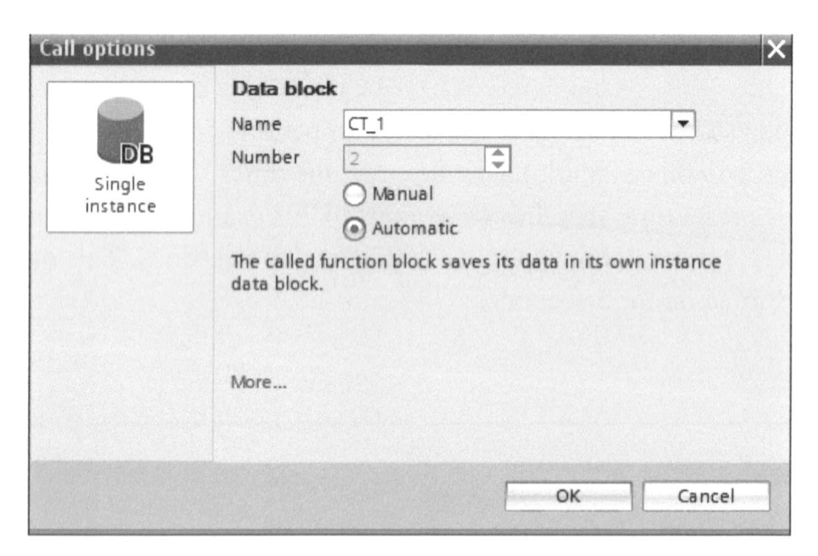

Figure 4-28. Instance block creation for a counter.

CT_1

	Name	Data type	Start value	Retain	Accessible from HMI	Visible in HMI	Setpoint	Comment
1	▼ Static			☐	☐	☐	☐	
2	CU	Bool	false	☑	☑	☑	☐	
3	CD	Bool	false	☑	☑	☑	☐	
4	R	Bool	false	☑	☑	☑	☐	
5	LD	Bool	false	☑	☑	☑	☐	
6	QU	Bool	false	☑	☑	☑	☐	
7	QD	Bool	false	☑	☑	☑	☐	
8	PV	Int	0	☑	☑	☑	☐	
9	CV	Int	0	☑	☑	☑	☐	

Figure 4-29. A counter data block.

Figure 4-30 shows the use of a CTU counter. Each time contact Cycle_Start makes a transition, 1 is added to the current value (CV- tagname is Current_Count) of the counter. When the CV is equal to or greater than the value of PV (12) the output will be true.

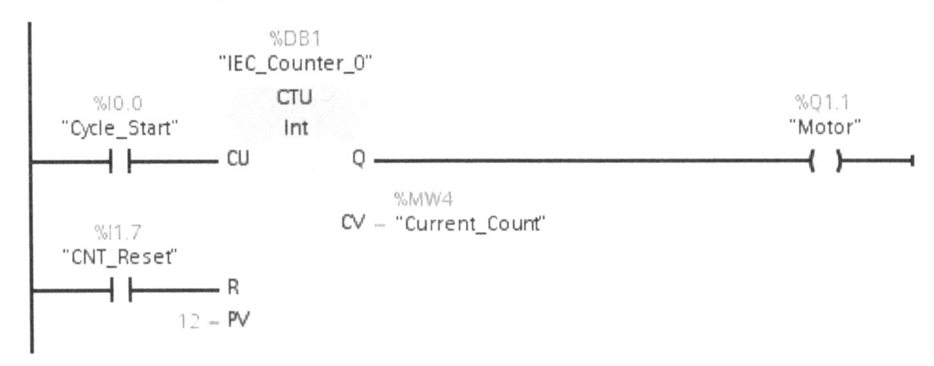

Figure 4-30. A CTU instruction.

Count-Up (CTU) Counter

Figure 4-31 shows the use of a CTU's CV in a ladder diagram. Each time input Part_Sensor makes a false-to-true transition the counter's CV is incremented by one. The CV of the counter is being used in the equal math instruction in the second rung to turn on an output when the CV reaches 6. Note that a RES (reset) instruction can be used to reset the value of a counter's CV.

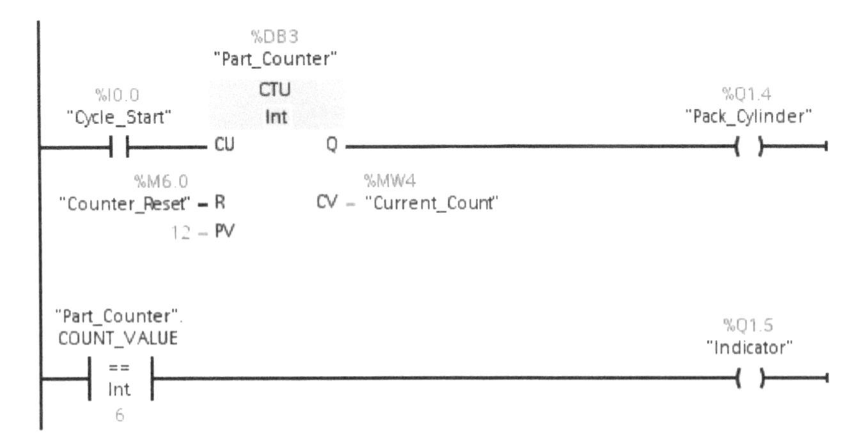

Figure 4-31. Use of an equal instruction and the counter's CV.

Count-Down (CTD) Counters

A CTD counter can be used to count down from a PV number. Figure 4-32 shows an example. Although it seems illogical, when the CV equals or is greater than the PV the Q-bit will be energized, just like a CTU counter. The CTD is almost always used in conjunction with a CTU as an up/down counter by assigning the same tag name to them both.

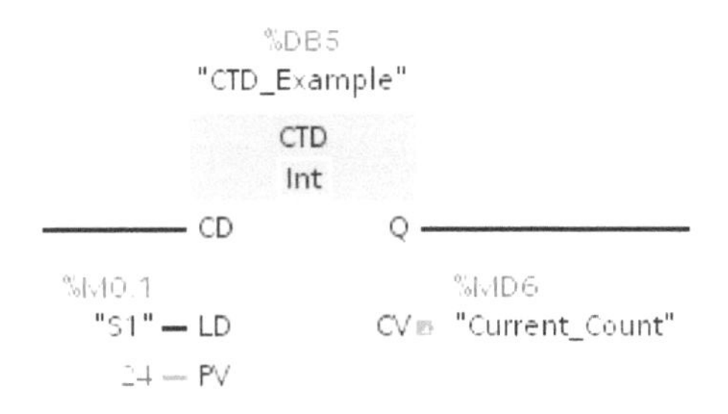

Figure 4-32. A CTD instruction.

Figure 4-33 shows a rung of logic using a CTD counter. In this example the PV is 24. This counter will be used to count down from 24 to 0.

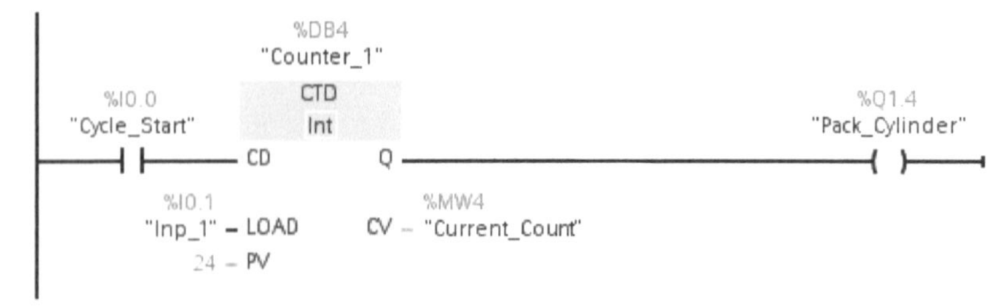

Figure 4-33. A CTD counter.

Count Up/Down (CTUD) Counter

A counter that can count up or down is also available in two Siemens programming languages. The CTUD counter is used to count up and count down. Figure 4-34 shows the use of a CTUD counter. A transition at the CU input will increment the counter's CV. A transition at the CD input will decrement the counter's CV. One possible use for this would be in an area where a manufacturer likes to keep a buffer of accumulated parts for processing. The sensor for adding to the count would be at the entrance to the accumulator.

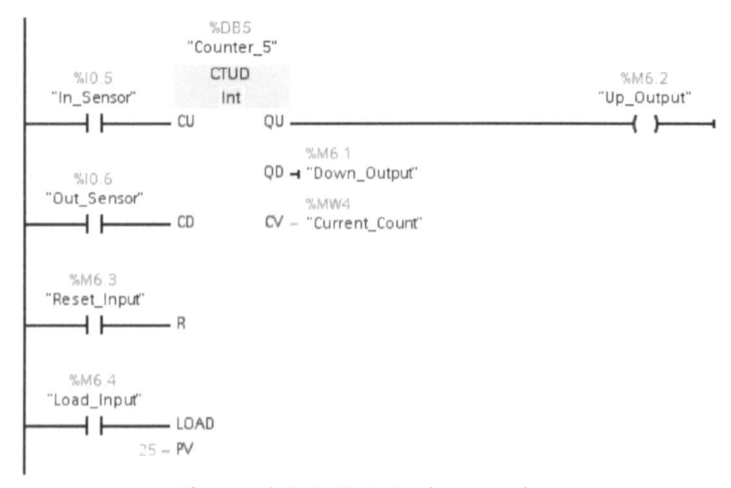

Figure 4-34. CTUD instruction.

Control High Speed Counters (CTRL_HSC)

High speed counters are able to count pulses independent of a control programs cycle time. Counting frequency of up to 200 kHz is possible. The counting range of a high-speed counter uses a DINT so the range of counts is -2,147,483,648 to +2,147,483,647. High speed counters must be configured and activated prior to use. The CPU model determines how many are available. A 1211 CPU has 3, a 1212 has 4, and a 1214 has 6 available. High speed counters can be used as a single-phase or 2-phase counter. With a high-speed counter, specific input channels are permanently assigned to that counter. The data that is required for a high-speed counter is saved in a data block. The inputs are either CPU integrated inputs or on a signal board. If a counter uses the assigned inputs they cannot be used elsewhere in logic.

High speed counters can utilize specific inputs that are integrated into the CPU or on a signal board. They are assigned to a high-speed counter. If a counter uses the input it cannot be used elsewhere. Figure 4-35 shows how CPU inputs can be assigned to counters. High speed counters can be used in three modes: counting, as a frequency meter, or for motion. The first two columns are counting modes and the third column is A/B quadrature mode.

Modes of a High-Speed Counter

Single-Phase Counter:
This type of counter is controlled by a single pulse train. The counting direction is specified in the CTRL_HSC instruction or externally by an input. Each rising edge of the input increases or decreases the count depending on the counting direction.

Two-Phase Counter
A two-phase counter is controlled by two pulse trains. One pulse train is used for counting up and the other for counting down. Each rising edge increases or decreases the count. Note that with the first pulse of the other input there is a change in the counting direction. An associated process interrupt can be generated if it is activated.

A/B Quadrature with Single Speed
This counter is controlled by two pulse trains that are offset by 90 degrees (phase A and phase B). This mode is used to take encoder feedback for velocity and position control. Note that not all counters can be used in in every mode, as some of the modes use multiple inputs for one counter. If pulse train B has a signal state of 0 between the pulses, counting is enabled and each rising edge of the pulses of phase A increases the count value and each falling edge a decreases the count.

A/B Quadrature with Quadruple Speed

This counter is controlled by two pulse trains offset by 90 degrees. Each edge of the A and each edge of the B is counted.

CPU Input	HSC #	Single Phase Counter Mode	2-Phase Counter Mode	A/B Quadrature Mode
I0.0	HSC 1	Clock Input	Clock Input Up	Clock Phase A
I0.1	HSC 1	Direction	Clock Input Down	Clock Phase B
	HSC 2	Reset	Reset	Clock Phase Z
I0.2	HSC 2	Clock Input	Clock Input Up	Clock Phase A
I0.3	HSC 1	Reset	Reset	Clock Phase Z
	HSC 2	Direction	Clock Input Down	Clock Phase B
I0.4	HSC 3	Clock Input	Clock Input Up	Clock Phase A
I0.5	HSC 3	Direction	Clock Input Down	Clock Phase B
	HSC 4	Reset	Reset	Clock Phase Z
I0.6	HSC 4	Clock Input	Clock Input Up	Clock Phase A
I0.7	HSC 3	Reset	Reset	Clock Phase Z
	HSC 4	Direction	Clock Input Down	Clock Phase B
I1.0	HSC 5	Clock Input	Clock Input Up	Clock Phase A
I1.1	HSC 5	Direction	Clock Input Down	Clock Phase B
I1.2	HSC 5	Reset	Reset	Clock Phase Z
I1.3	HSC 6	Clock Input	Clock Input Up	Clock Phase A
I1.4	HSC 6	Direction	Clock Input Down	Clock Phase B
I1.5	HSC 6	Reset	Reset	Clock Phase Z

Figure 4-35. Assignment of CPU inputs to counters.

A CTRL_HSC instruction can be used to make parameter settings and control high speed counters by loading new values into the counter. Figure 4-36 shows a CTRL_HSC counter. To use a CTRL_HSC instruction you must enable the high-speed counter in the hardware configuration. Only one CTRL_HSC instruction may be used per high-speed counter. The programmer must enter the hardware identifier of the high-speed (HW-ID) whose values are to be assigned at the high-speed counter (HSC) input.

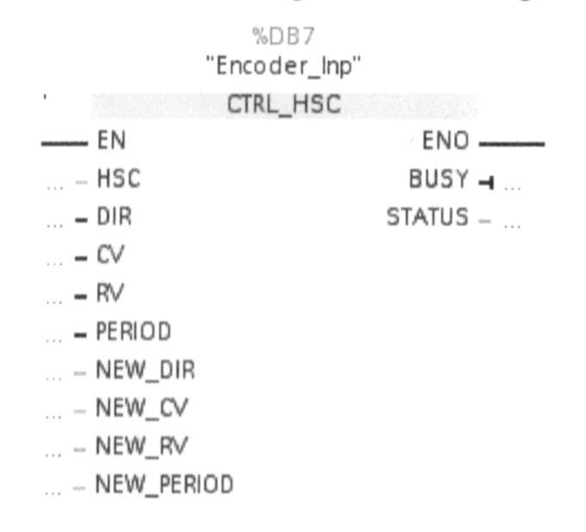

Figure 4-36. CTRL_HSC instruction.

Only tags of the data type HW_HSC can be used for the HSC input. There are inputs that are integrated into the CPU module that can be used for high-speed counters. A Signal Board can also be added to the CPU module with additional inputs that can be used for high-speed counters. Figure 4-37 shows a Signal Board that has high-speed inputs on a Siemens S7- 1215 CPU. Figure 4-38 shows the device overview showing the addressing for the 4 high-speed inputs on the 1214 CPU Signal board.

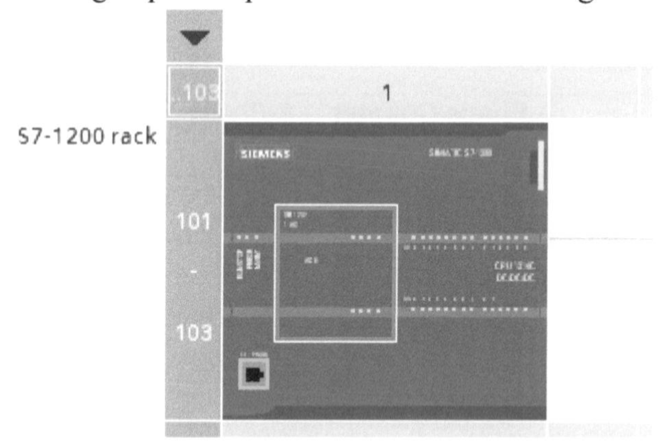

Figure 4-37. A 1215 Siemens CPU showing the Signal board that houses the high-speed inputs. Courtesy Siemens.

Device overview

Module	Slot	I address	Q addre...	Type	Order no.
AO1 x 12bit..	1.3		80...81	AO1 signal board	6ES7 232-4H..
HSC_1	1.16			High speed counter..	
HSC_2	1.17			High speed counter..	
HSC_3	1.18			High speed counter..	
HSC_4	1.19			High speed counter..	

Figure 4-38. The device overview showing the addressing for the 4 high-speed inputs.

The parameters for a CTRL_HSC instruction are shown in Figure 4-39.

Parameter	Description	Data Type	Memory Area
EN	Enable input	BOOL	I,Q,M,L,D
ENO	Enable output	BOOL	I,Q,M,L,D
HSC	Hardware address of the high-speed counter.	HW_HSC	L,D, or Constant
DIR	Enables the new count direction.	BOOL	I,Q,M,L,D
CV	Enables the new count value.	BOOL	I,Q,M,L,D
RV	Enables the new reference value.	BOOL	I,Q,M,L,D
PERIOD	Enables the new period of a frequency measurement.	BOOL	I,Q,M,L,D
NEW_DIR	Count direction loaded when DIR = TRUE.	INT	I,Q,M,L,D
NEW_CV	Count value loaded when CV = TRUE.	DINT	I,Q,M,L,D
NEW_RV	Reference value loaded when CV = TRUE.	DINT	I,Q,M,L,D
NEW_PERIOD	Period of the frequency measurement loaded when PERIOD = TRUE.	INT	I,Q,M,L,D
BUSY	Processing status.	BOOL	I,Q,M,L,D
STATUS	Status of the operation.	WORD	I,Q,M,L,D

Figure 4-39. CTRL_HSC instruction parameters.

The following parameters can be loaded into a high-speed counter using a CTRL_HSC instruction.

NEW_DIR - This is the count direction. The count direction determines whether the instruction counts up or down. A 1 in the input would define the direction as up. A -1 would define the direction as down. The direction may only be changed if direction control is set in the parameters. The count direction specified at the NEW_DIR input is loaded into a high-speed counter when the bit at the DIR input is set.

NEW_CV - This is the count value. The count value is the initial value at which the counter starts counting. The count value can be between -2147483648 to 2147483647. The count value that is specified in NEW_CV is loaded into a high-speed counter when the bit at the RV input is set.

NEW_PERIOD - This is the period of the frequency measurement. This value is specified as 10 = 0.01s, 100 = 0.1s, 1000 = 1s in the NEW_PERIOD input. The period may be updated when the frequency measurement function is set in the high-speed counter parameters. The period that is specified by the NEW_PERIOD input is loaded into a high-speed counter if the bit at the PERIOD input is set.

The CTRL_HSC instruction is only executed if the EN input is true. The BUSY output of the instruction will be set while the instruction is executing. When the instruction is done executing, the BUSY output will be reset. The enable output (ENO) output is only set when the EN input has a 1 and no errors occurred during the execution of the operation. An instance data block is created for the instruction when a CTRL_HSC instruction is programmed.

A status output is provided to provide information on whether errors occurred during the execution of the instruction. Figure 4-40 shows the error codes for a CTRL_HSC counter.

Error Code (Hex)	Description
0	No Error.
80A1	Hardware identifier of the HSC is not valid.
80B1	Count direction (NEW_DIR) is not valid.
80B2	Count value (NEW_CV) is not valid.
80B3	Reference value (NEW_RV) is not valid.
80B4	Period of frequency measurement (NEW_PERIOD) is not valid.
80C0	Multiple access to the high-speed counter.

Figure 4-40. Error codes for CTRL_HSC STATUS tag.

There are a variety of counters available for Siemens PLCs. Figure 4-41 shows a table of the counter instructions and their use.

Counter	Description
CTU	Up counter
CTD	Down counter
CTUD	Up/down counter
CTRL_HSC	Control high-speed counter

Figure 4-41. Available counter types.

Chapter Questions

1. What is a TON timer?

2. What is a TOF timer?

3. What is the time base for a Siemens PLC?

4. What is the PT used for in a timer?

5. What is the ET used for in a timer?

6. Describe two methods of resetting the elapsed time of a TON timer to 0.

6. What does the term retentive mean?

9. Give an example of how the Q bit for a TOF timer could be used.

10. What is a timer starter?

11. In what way are counters and timers very similar?

12. What is a CTU instruction?

13. What is a CTD instruction?

14. What is a CTUD instruction?

15. What languages are CTUD instructions available in?

16. How can the accumulated count be reset in a counter?

17. Complete the descriptions in the tables below.

Siemens Timers	Description of Address
PV	
ET	
Q	
T_1IN	

Siemens Counters	Description of Address
CU	
PV	
CV	

18. Write ladder logic for the following application.

This is a simple heat treat machine application. The operator places a part in a fixture then pushed the start switch. An inductive heating coil heats the part rapidly to 1500 degrees Fahrenheit. When the temperature reaches 1500 a discrete sensor's output becomes true. The coil turns off and a valve is opened which sprays water on the part to complete the heat treatment (quench). The operator then removes the part and the sequence can begin again. Note there must be a part present or the sequence should not start.

I/O	**Type**	**Description**
Part_Presence_Sensor	Discrete	Sensor used to sense a part in the fixture.
Temp_Sensor	Discrete	For simplicity assume this sensor's output becomes true when the temperature reaches 1500 degrees Fahrenheit.
Start_Switch	Discrete	Momentary normally-open switch.
Heating_Coil	Discrete	Discrete output that turns coil on.
Quench_Valve	Discrete	Discrete output that turns quench valve on.

Chapter 5

Input/Output Modules and Wiring

Objectives

Upon completion of this chapter, the reader will be able to:

Define terms such as: discrete, digital, analog, resolution, producer, consumer, and so on.

Describe types of digital I/O modules.

Describe types of analog I/O modules.

Find the resolution for an analog module.

Describe how analog modules are calibrated.

Wire digital and analog modules.

I/O Modules

There are a wide variety of modules available for Siemens systems. Modules are available for digital and analog I/O, communications, motion, and many other purposes. Modules can be located in the same chassis as the controller or remotely.

Digital Modules

Digital modules are also called discrete modules. Discrete means that each input or output has two states: true (on) or false (off). Most industrial automation devices are discrete.

Digital Input Modules

Digital input modules are used to take input from the real world. Inputs to a discrete module are provided by devices such as switches or sensors that are either on or off.

Input and output modules must be able to protect the CPU from the real world. Assume an input voltage of 110 VAC. The input module must change the 110 VAC level to a low-level DC logic level for the CPU. This is accomplished through opto-isolation. An opto-isolator uses a photo transistor. The LED in the photo transistor is controlled by the input signal. The light from the LED falls on the base of the photo transistor and turns the transistor output on (allows collector/emitter current flow). Figure 5-1 shows how this is done for AC input modules. The light totally separates the real-world electrical signals from the PLC internal electrical system.

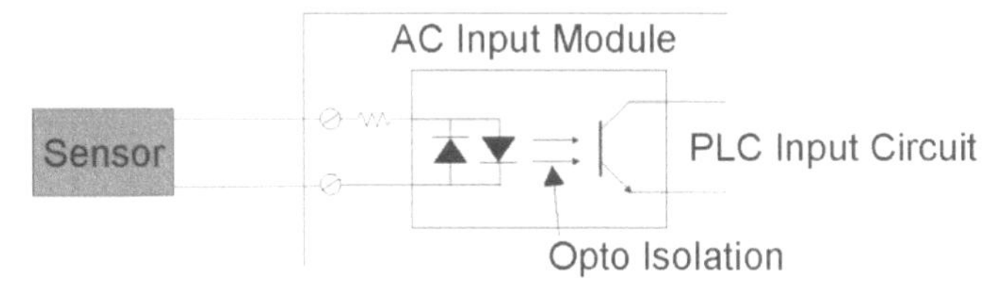

Figure 5-1. Opto-isolation for an AC input module. Note in the figure that the AC input signal is transmitted via an LED to a photo transistor. The output of the photo-sensor is a low-level DC signal for the CPU.

DC Input and output modules must also be able to protect the CPU from the real world. Assume a 24 VDC input module. The input module must change the 24 VDC level to a low-level DC logic level for the CPU. This is accomplished through opto-isolation. An opto-isolator uses a photo transistor. The LED in the photo transistor is controlled by the input signal. The light from the LED falls on the base of the photo transistor and turns the transistor output on (allows collector/emitter current flow). Figure 5-2 shows how this is done for DC input modules. The light totally separates the real-world electrical signals from the PLC internal electrical system.

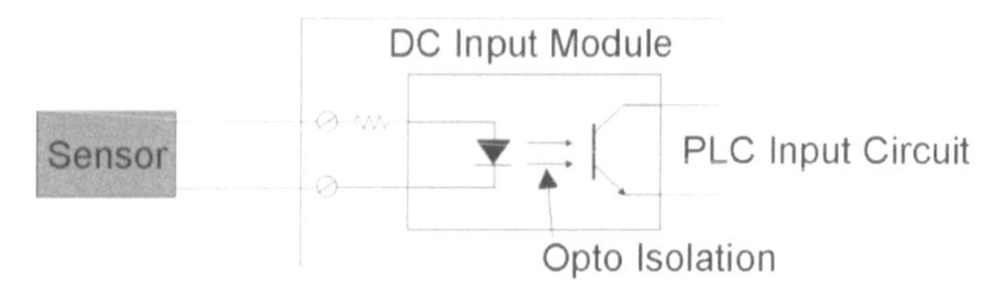

Figure 5-2. Opto-isolation for a DC input module.

Figure 5-3 shows an example of a signal board being installed in a S71200 PLC CPU. Signal boards can be purchased in various configurations of I/O. This one is a signal board (SB) that can be plugged into a S7-1200 CPU. Note the wiring terminals on the bottom of the SB.

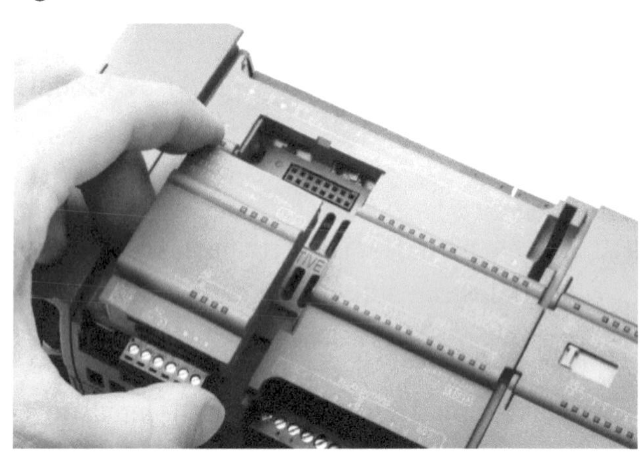

Figure 5-3. A signal board being plugged into a S7-1200 CPU. Courtesy Siemens.

SBs are available with 4-24 VDC inputs, or with 4-5 VDC inputs, or with 4-24 VDC outputs or with 4-5 VDC outputs. SBs are also available as combination input/output modules. They are available with 2 inputs and 2 outputs in either 24 VDC or 5 VDC. The SB connector accepts wire sizes from 16 AWG to 22 AWG.

Analog SBs are also available. SBs are available with I analog input. Analog SBs are also available with 1 analog output. These modules can be configured to work with current or voltage. SBs are also available to take thermocouple or RTD input.

PLC Signal Modules

Modules are chosen to meet the needs of an application. They are available to take inputs, control outputs, handle communications, and many other special functions. This chapter will take a look at input modules first. Figure 5-4 shows a typical input signal module. Siemens calls these signal modules (SMs).

Figure 5-4. A typical I/O signal module. Courtesy Siemens.

There are a wide variety of input modules available for Siemens PLCs. They are available in multiple DC and AC voltages. They are available with various numbers of inputs from 8 to 64. There are also combination modules available that have inputs and outputs on the same module.

Study Figure 5-5. This figure explains how a signal module can be installed. You must make sure that the PLC and all associated equipment has been disconnected from power before installing or removing modules.

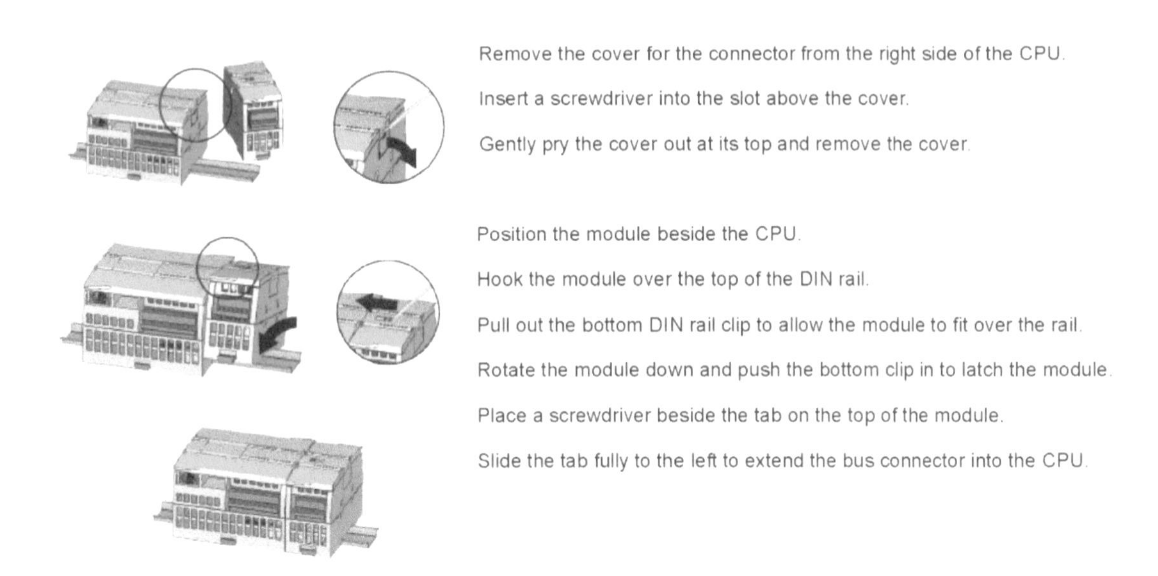

Figure 5-5. Installing a signal module on a S7-1200 PLC. Courtesy Siemens.

LED Status Information

Input modules have light-emitting diodes (LEDs) for monitoring the inputs. If the input is true, the LED is energized. Modules also have a diagnostic LED (DIAG) the DIAG LED shows the status of the module.

Removable Terminal Connectors

To make installation and maintenance quick and easy modules are provided with removable terminal connectors. The connectors can be removed while the wires are connected if it makes installation easier. The biggest advantage is if there is a problem during operation and the module needs to be replaced. The wires can be left connected to the wiring connector. The whole connector can be removed, a new module installed and the connector replaced. The S7-1200 CPU also has removable terminal connectors for the integrated I/O. Study Figures 5-6 and 5-7. These figures show how to remove a wiring terminal block from a S7-1200 CPU. You must be sure that the PLC and all associated equipment has been disconnected from power before installing or removing a wiring terminal block.

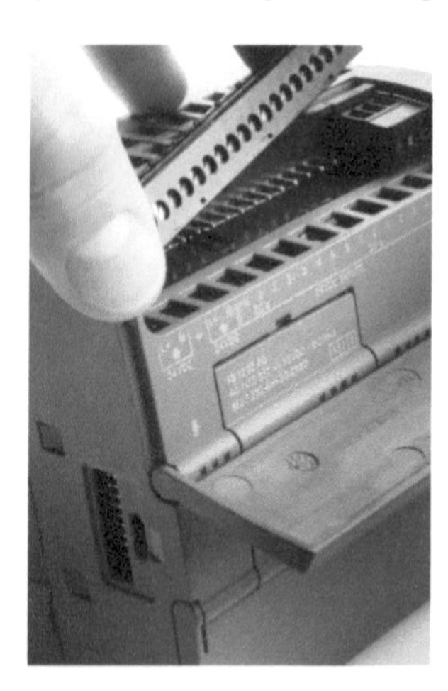

Figure 5-6. Removable a wiring terminal block. Courtesy Siemens.

Open the cover above the connector.

Inspect the top of the connector and locate the slot for the tip of the screwdriver.

Insert a screwdriver into the slot.

Gently pry the top of the connector away from the CPU.

The connector will release with a snap.

Grasp the connector and remove it from the CPU.

Figure 5-7. How to remove a wiring terminal block. Courtesy Siemens.

Configuring Digital Input Filters

Figure 5-8 shows how filters can be configured for digital inputs. Input filters are used to set the input delay time. This is used to improve the immunity to high-frequency noise. The longer the delay time that is chosen, the greater the immunity to noise. This also increases the detection period until a change in the input signal is recognized by the CPU.

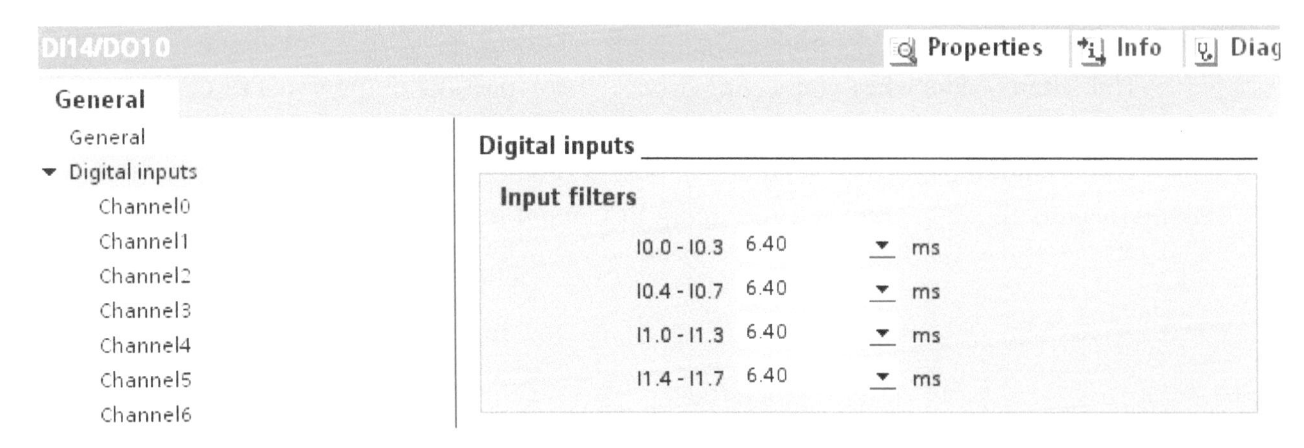

Figure 5-8. Configuring digital inputs.

Wiring Input Modules

This chapter will only cover a few of the available modules. If you understand the wiring of these it will be easy to look up a wiring diagram for a specific PLC module and understand it.

Sinking Versus Sourcing Modules

Inputs to a PLC input module can be sinking or sourcing. Sensors, for example, are available with positive or negative output. Devices such as robots, vision systems, and other devices may have positive or negative outputs available. They may also require positive or maybe negative inputs. A device with a positive output is called sourcing. A device with negative output is called sinking. When connected these devices to a PLC module we must be careful to wire them correctly. A sensor that has a positive output must be connected to a module input that requires a positve input. The sensor would be a sourcing device and the module input would be a sinking input.

The easiest way to understand this is to remember that a sinking PLC input module would require a positive input signal. For example, if we have a sensor that has a positive output signal (sourcing), we would need a sinking PLC input module. Figure 5-9 shows an example of a sinking PLC input module. Remember that an output from a device becomes an input to the PLC.

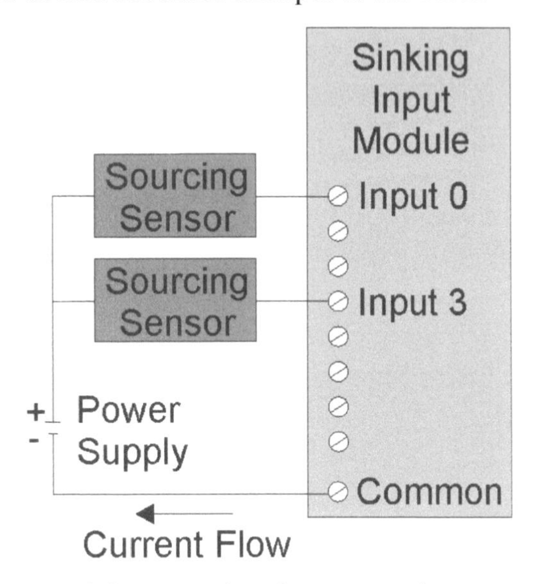

Figure 5-9. Sinking input module. Note that the sensors have sourcing (positive) outputs.

If we have a sensor or device with negative output, we would need a sourcing PLC input module (see Figure 5-10). This often occurs when connecting drives, vision or robot outputs to a PLC. You must be careful to check the output polarity required and wire it correctly. Remember – opposites attract. If you have a sensor with a sourcing (positive) output you need a sinking input module. If you have a sensor with a sinking (negative) output you need a sourcing PLC input module.

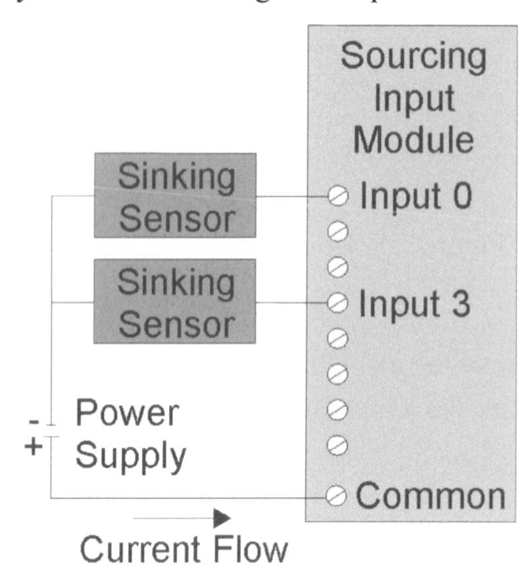

Figure 5-10. Sourcing input module. Note the sensors have sinking (negative) outputs.

It is a good idea to first determine what the module requires for power before we worry about wiring the actual input devices to the module. In other words, study the module to see what connections it requires directly from a power supply and then look at how the device is connected.

Some brands of PLCs have separate modules available for sinking or sourcing. With Siemens modules you can wire a group of inputs or outputs to be sinking or sourcing. The whole group must be the same however. A module may have 1 group, 2, 4 or more groups.

Figure 5-11 shows the wiring of inputs and outputs for an S7-1214 CPU. This CPU can act as a sinking or sourcing input module. In this example the CPU is wired as a sinking input module. This means that the actual input devices are sourcing devices. Note that that the CPU can also be used as a sourcing input module by connecting the + side of the power supply to M instead of the -.

Fourteen inputs are connected to the inputs of the CPU. The input devices are normally-open switches. One side of each switch is "commoned" together connected to the + side of the power supply. The bottom of each switch is connected directly to the desired input terminal. Note that in this example the negative side of the power supply is connected to the common labeled M. The first 8 input terminals are labeled .0 - .7 (this represents the first byte of memory for this module) and the last 6 inputs are labeled .0 - .5 (this represents the first 6 bits in the second byte of memory for this module). The CPU connector accepts wire sizes from 14 AWG to 22 AWG.

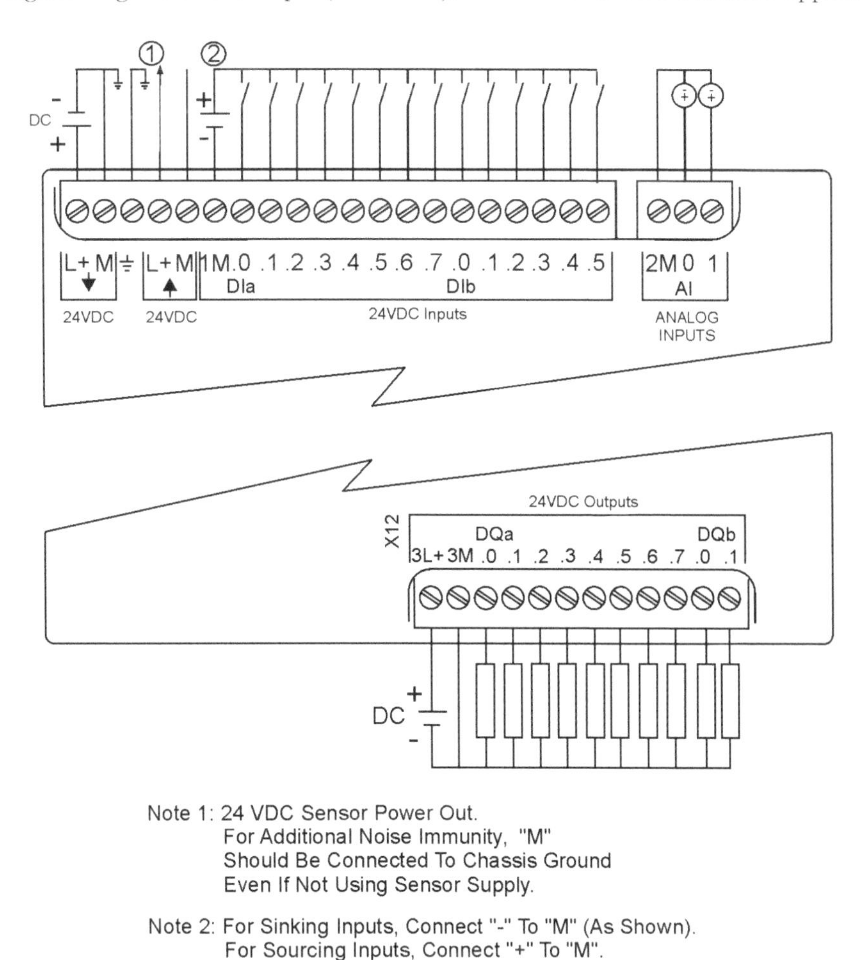

Figure 5-11. S7-1200 CPU wiring.

Figure 5-12 shows the wiring of inputs and outputs for an S7-1214 signal module (SM 1221). This module has 16 inputs available. This module can act as a sinking or sourcing module. In this example the CPU is wired as a sinking input module. This means that the actual input devices are sourcing devices. Note that that the module can also be used as a souring input module by connecting the positive side of the power supply to M instead of the negative. Note also that the 16 inputs are organized in groups of 4. This means that some groups could be connected as sinking and some sourcing.

The positive side of the power supply is not directly connected to the module shown in Figure 5-12. There is a direct connection from the negative side of the power supply to the module. The negative is connected to the grounds (common). They are labeled 1M, 2M, 3M and 4M. The SM connector accepts wire sizes from 14 AWG to 22 AWG.

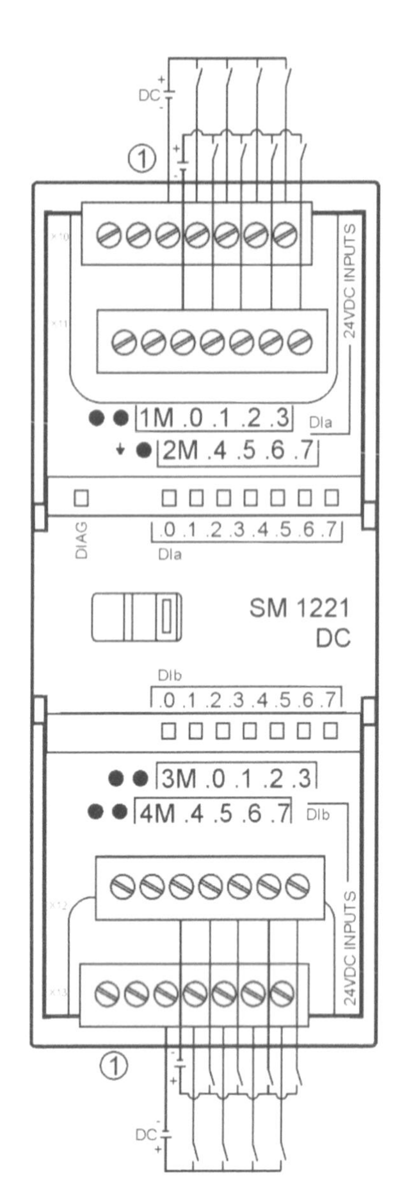

Note 1: For Sinking Inputs, Connect "-" to M.
For Sourcing Inputs, Connect "+" to M.

Figure 5-12. S7-1200 SM 1221 signal module input wiring.

Figure 5-13 shows an example of the wiring of a S7-300 input signal module. In this example the module has been wired as a sinking module. The input devices are sourcing devices. Four inputs are shown on the right of the module. The switches are normally-open switches. The negative side of the power supply is connected to terminal 20. The right side of the switch is connected to the positive side of the power supply. The left sides of the switches are connected directly to the desired input terminals. Note that the diagram shows terminal numbers, not input numbers. Note that terminal number 2 is used for input 0 on the module's first byte. Terminal 9 is used for input number 7 on the first byte of the module memory. Terminal 12 is used for input 0 in the second byte of module memory and terminal 19 is used for input 7 on the second byte of module memory.

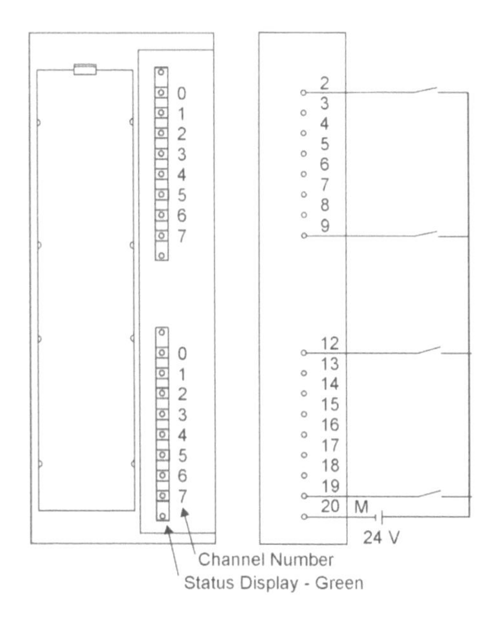

Figure 5-13. Example of S7-300 DC digital input module wiring.

Figure 5-14 shows an example of the wiring of a S7-300 input signal module. In this example the module has been wired as a sourcing module. The input devices are sinking devices. Four inputs are shown on the right of the module. The switches are normally-open switches. The positive side of the power supply is connected to terminal 1. The right side of the switch is connected to the negative side of the power supply.

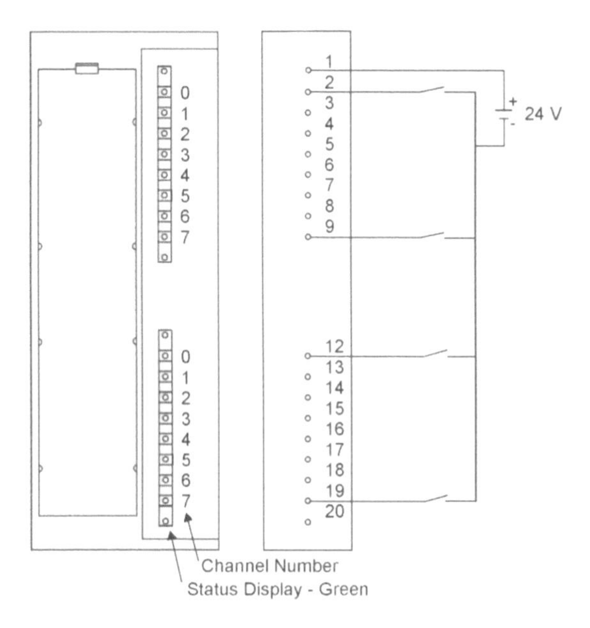

Figure 5-14. Wiring inputs to a S7-300 PLC signal module.

The left sides of the switches are connected directly to the desired input terminals. Note that the diagram shows terminal numbers, not input numbers. Note that terminal number 2 is used for input 0 on the module's first byte. Terminal 9 is used for input number 7 on the first byte of the module memory. Terminal 12 is used for input 0 in the second byte of module memory and terminal 19 is used for input 7 on the second byte of module memory.

AC Input Modules

Figure 5-15 shows an S7-300 AC output module. Note that this module has 16 inputs that are electrically isolated in 4 groups of 4. This module can utilize 120/230 AC volts.

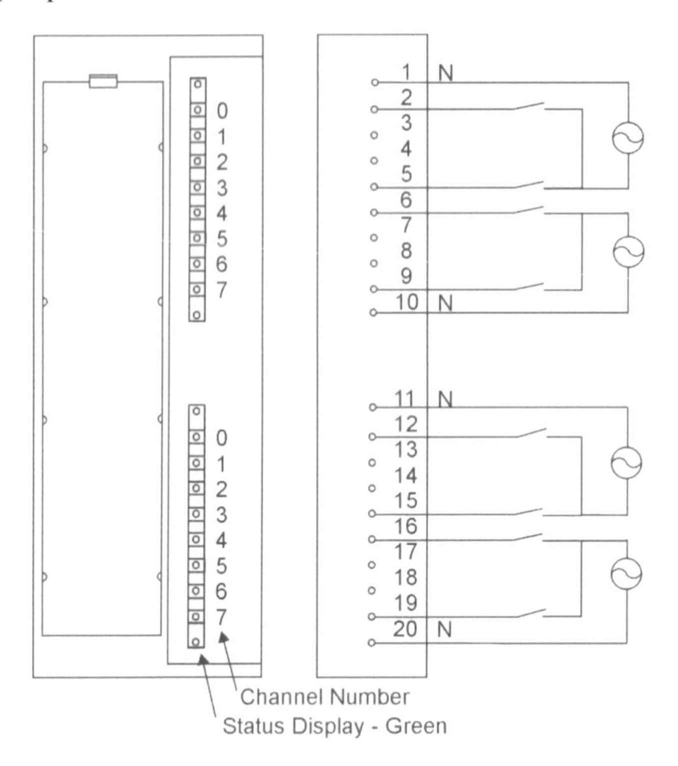

Figure 5-15. S7-300 SM-321 AC input signal module wiring.

Discrete Output Modules

Discrete output modules are used to turn real-world output devices off or on. Output modules are available for AC and DC devices. They are also available in various voltage ranges and current capabilities. The actual output device used for each output includes transistors, triac output, or relay output. The transistor output would be used for DC outputs. Triac outputs are used for AC devices. TTL (transistor-transistor logic) output modules are also available. Figure 5-16 shows an example of output modules.

Figure 5-16. AC digital Modules. Courtesy Siemens.

Sinking Versus Sourcing DC Outputs.

Just like with input modules there are two choices: sinking output or sourcing output. This is often an issue when connecting PLC outputs to other devices such as vision systems, robots, etc. The devices might specifically require a positive or a negative signal for their input. Remember that an output from the PLC becomes an input to another device.

A sourcing output module supplies a positive signal to an output. Figure 5-17 shows that that each output for a sourcing output module is connected to the output terminal and the other side of the output is tied to the negative side of a supply.

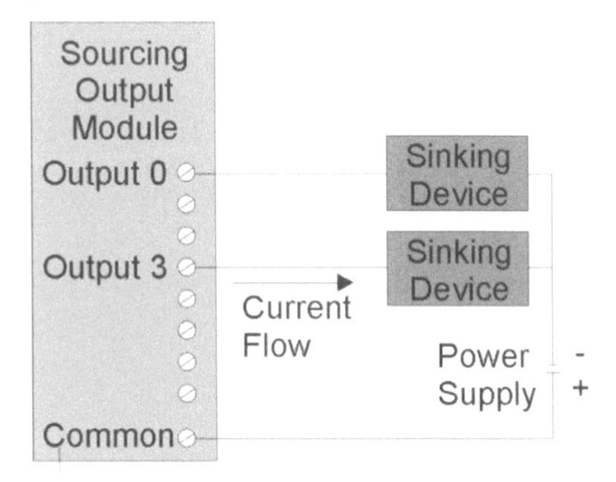

Figure 5-17. Wiring a sourcing output module.

Figure 5-18 shows a sinking output module. Note that the negative side of the power supply, in this example, is directly connected to the common terminal and the power supply positive is not directly connected to the module. One side of the output is connected to an output terminal and the other side of the output is tied to positive.

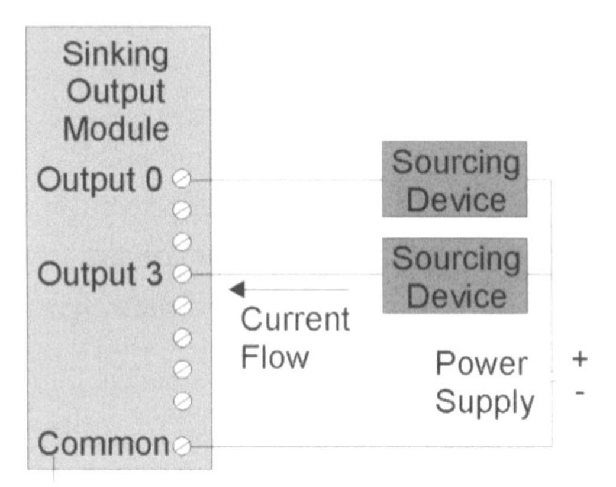

Figure 5-18. Wiring a sinking output module.

Output Current Limit

The current limit specifications for an output module are normally specified for each individual output and as a group current limit. Remember that I/O is divided into groups on Siemens modules. A group may include all of the I/O on the module or a subset. The total current for the group must not exceed the total that the group can handle. Normally each PLC output will not each draw the maximum current, nor will they all be on at the same time. Consider the worst case when choosing an appropriate output module.

Let's take a look at the outputs for a S71200 CPU (see Figure 5-19). The output current limit for the SM 1222 module's DC outputs is .5 A per output. The maximum current for the whole module's 16 DC outputs is 8 amps. If the relay output module is used the current limit for each relay output is 2 amps. The modules 16 relay outputs are divided into 4 isolation groups of 4 outputs each. The current limit per isolation group is 10 amps.

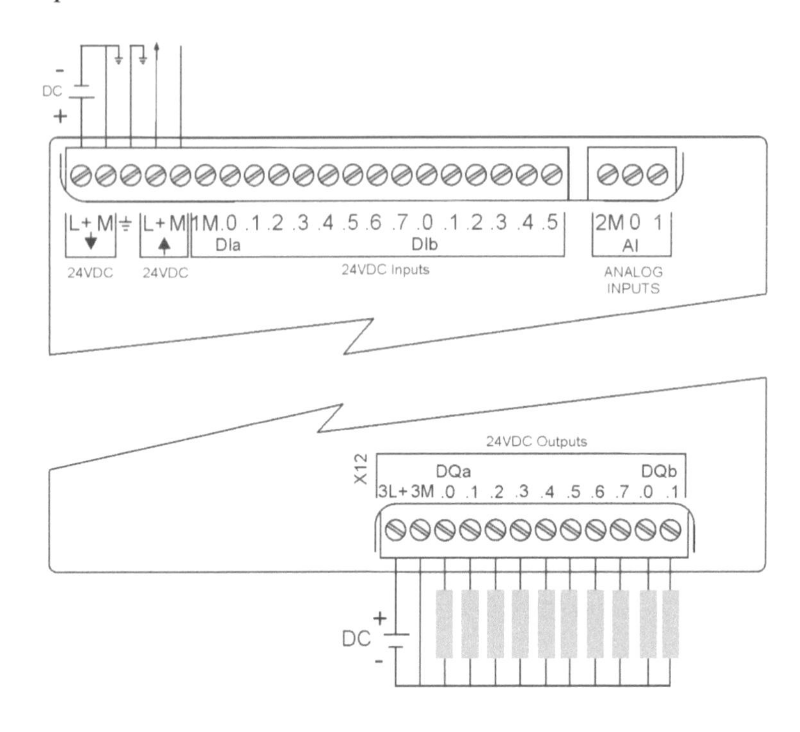

Figure 5-19. Wiring the DC outputs on a S7-1200 CPU.

Output Modules

There are a wide variety of output modules available for Siemens PLCs. They are available in multiple DC and AC voltages. They are available with various numbers of outputs from 8 to 64. There are also combination modules available that have inputs and outputs on the same module. This chapter will only cover a few of the modules. If you understand the wiring of these it will be easy to look up a wiring diagram for a specific module and understand it.

Figure 5-20 shows the wiring for a S7-1200 DC output signal module. One of the first considerations when you are figuring out the wiring for a module is: does the module require direct connections to a power supply? In Figure 5-13 we see that the positive side of the power supply must be connected to terminal L+. The negative side of the supply must be connected to terminal M.

Now that we have the module powered correctly, we can examine the wiring of the outputs. The output devices are connected to an output terminal and then tied to the negative side of the power supply. Note that there is only one common (M) used in this example meaning that there is only one group of 16 outputs.

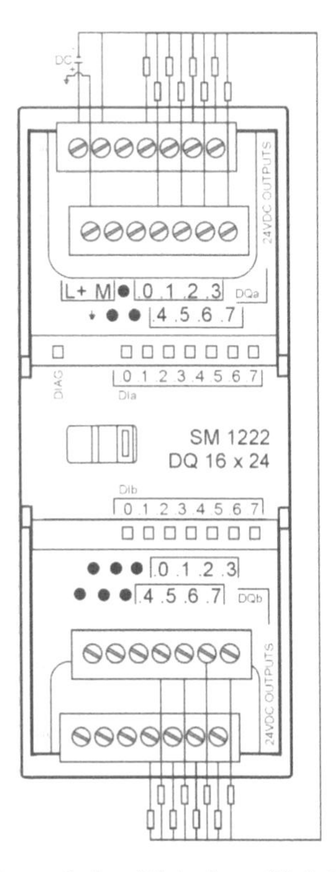

Figure 5-20. S7-1214C signal module. This is a SM 1222 module with DC outputs.

Figure 5-21 shows the wiring of outputs for a S7-300 SM 322 DC 16-output module. Note that this module is electrically isolated into two groups of 8 outputs each. The output current limit for this module is .5 amp per output. In this module the positive side of the power supply is connected to 1L+ and 2 L+ on the module. The negative side of the supply is connected to 1M and 2 M.

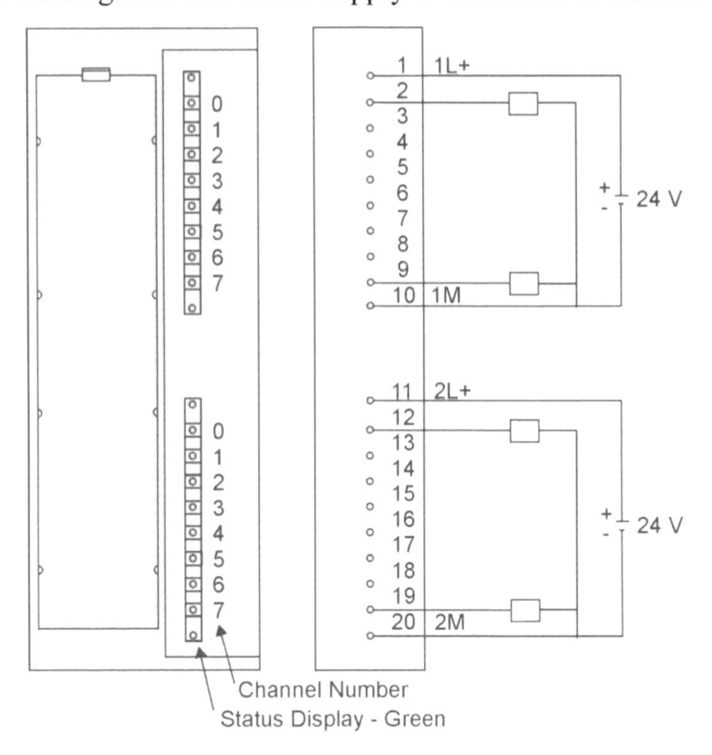

Figure 5-21. S7-300 SM 322 16 output, DC 24 VDC output module wiring.

Figure 5-22 shows a S7-1223 AC combination input and output module.

This particular model has 8 AC digital inputs and 8 digital AC relay outputs. Each output can handle up to 2 amps. The 8 outputs are electrically isolated into two groups of 4.

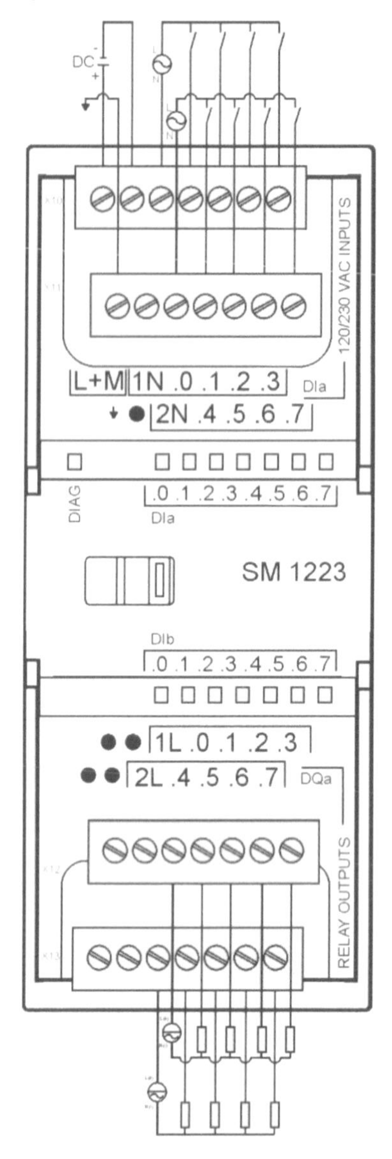

Figure 5-22. A S7-1223 AC combination input and output module.

Figure 5-23 shows a S7-300 SM 322 AC output module.

This particular model has 16 digital AC outputs. Each output can handle up to 1 amp. The 16 outputs are electrically isolated into two groups of 8. Note that the AC neutral is connected to N and the hot is connected to L1

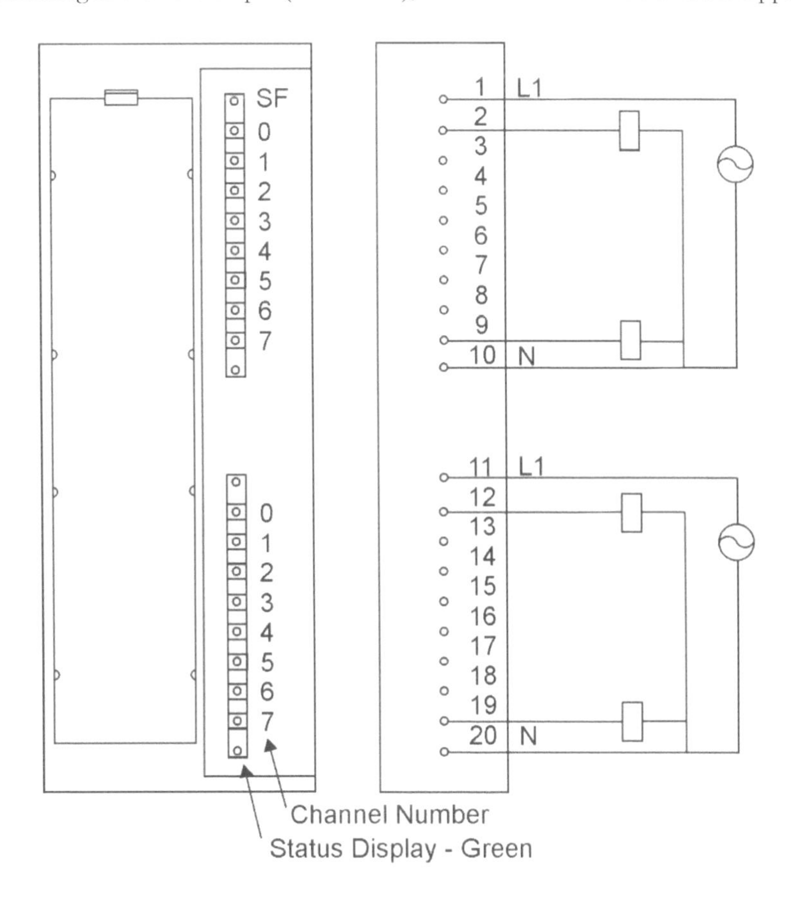

Figure 5-23. Wiring diagram for an S7-300 SM 322 AC output module.

High-Density I/O Modules

The most common I/O modules have 16 inputs or outputs. A high-density module may have up to 64 inputs or outputs. The advantage is that there are a limited number of slots in a PLC rack. Each module uses a slot. With the high-density module, it is possible to install up to 64 inputs or outputs in one slot. The only disadvantage is that the high-density output modules typically cannot handle as much current per output.

Analog Input Modules

Analog input modules are designed to take analog information from devices and convert the analog signal to digital information. Digital modules are discrete devices they only have two states: 1 or 0. A discrete input cannot tell us an actual temperature for example. Analog devices must be used to get a range of numbers, such as temperature, velocity, flow rate, pressure, etc. Analog modules are available to handle analog signals.

The two most common types are: current sensing and voltage sensing. These cards will take an analog current or voltage and change it to digital data for the PLC.

Analog input or output modules can be unipolar or bipolar. Analog input modules are commonly available in 0 to 10 V (unipolar) and -10 to +10 V (bipolar). Current sensing analog modules are also available. 0-20 mA and 4 to 20 MA are the most common input range, although other ranges are available.

Many analog modules will accept voltage or current input. Current or voltage may be individually selected for each channel group on Siemens analog SMs. Figure 5-24 shows the analog input connections on a S7-1200 PLC.

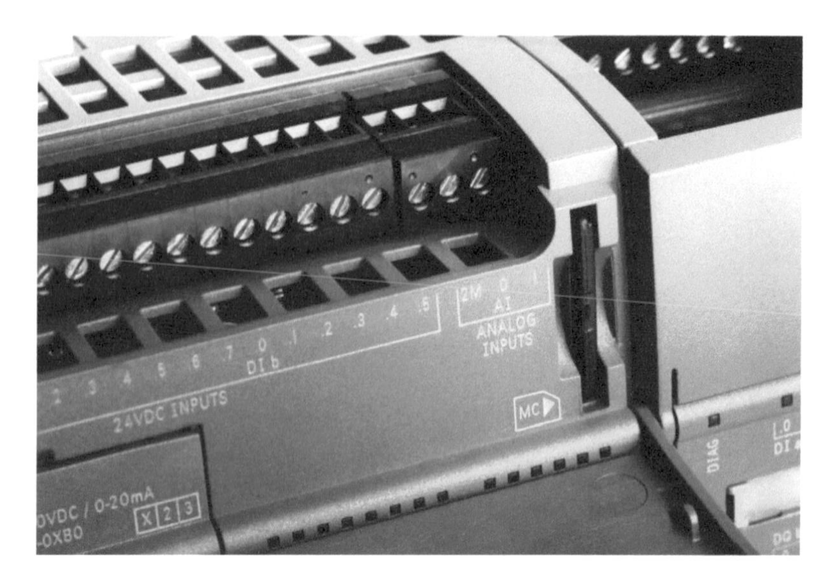

Figure 5-24. S7-1200 CPU analog input terminals. Courtesy Siemens.

Analog Resolution

Resolution has to do with how closely something can be measured. Imagine a ruler. If the only graduations on the ruler were inches, the resolution would be 1 in. If the graduations were every 1/8 in., the resolution would be a 1/8 in. The closest we could measure any object would be l/8 inch. The CPU in a PLC only works with digital information. The analog-to-digital (A/D) card changes the analog source into discrete steps. The higher the resolution, the finer the measurement. Another way to think of resolution is in terms of a pie. If you have people over for thanksgiving the pie will be divided based on the number of people. The pie represents what we are measuring (maybe 0-10 volts) the number of people represents the size of each piece of pie. So, the higher the number of people (bits of resolution) the smaller each piece is.

Resolution is the smallest amount of change that a module can detect. Analog modules are available in different resolutions. Output modules are typically available in 8 to16-bit resolution. A 16-bit module would have 65,536 counts. This can be calculated by raising 2 to the number of bits the module has. For example, a 16-bit module would be 2^{16} or 65,536.

Figure 5-25 shows a table of binary values.

Bit	15	14	13	12	11	10	9	8	7	6	5	4	3	2	1	0
Bit Value	2^{15}	2^{14}	2^{13}	2^{12}	2^{11}	2^{10}	2^{9}	2^{8}	2^{7}	2^{6}	2^{5}	2^{4}	2^{3}	2^{2}	2^{1}	2^{0}

Figure 5-25. Bit values for binary.

Bit 15 (the 16th bit) is used to keep track of the sign of the number positive or negative. The sign of the analog value is always set at bit 15. If bit 15 is a 1 the number is negative. If bit 15 is 0 the number is positive.

Resolution less than 16 bits.

Study Figure 5-26. On analog modules with a resolution of less than 16 bits, the analog value is stored left-aligned in Siemens PLCs. In the figure you will notice that for an 8-bit resolution the sign bit and all 7 bits in the high byte are used and only the left-most bit (most significant) is used in the low byte. In the figure the unused right-most (least significant) bits are shown as Xs. The unused least significant bit positions are actually filled with zeros in PLC memory. The one in the right most position would represent the finest resolution.

Resolution in Bits	**Analog Value**	
	High Byte	Low Byte
8 + Sign Bit	Sign Bit 0000000	1XXXXXXX
9 + Sign Bit	Sign Bit 0000000	01XXXXXX
10 + Sign Bit	Sign Bit 0000000	001XXXXX
11 + Sign Bit	Sign Bit 0000000	0001XXXX
12 + Sign Bit	Sign Bit 0000000	00001XXX
13 + Sign Bit	Sign Bit 0000000	000001XX
14 + Sign Bit	Sign Bit 0000000	0000001X
15 + Sign Bit	Sign Bit 0000000	00000001

Figure 5-26. This figure shows that analog values are stored left-aligned in PLC memory.

Figure 5-27 shows a theoretical example of resolution. In this example the module's input range is 0-21 mA. The module is 13-bit which means that there are 8192 (0-8191) counts. If we divide the 21 mA by 8192, we get the measurement resolution. In this example the resolution is 0.00256 mA. 21 mA input would result in a bit count of 8191 in memory. If the input was 10 mA the bit count in the input would be 3900. If the input was 20 mA the input in this example would be 7801. Note that this is just a theoretical example. One must look up the resolution and range for the particular module you are working with.

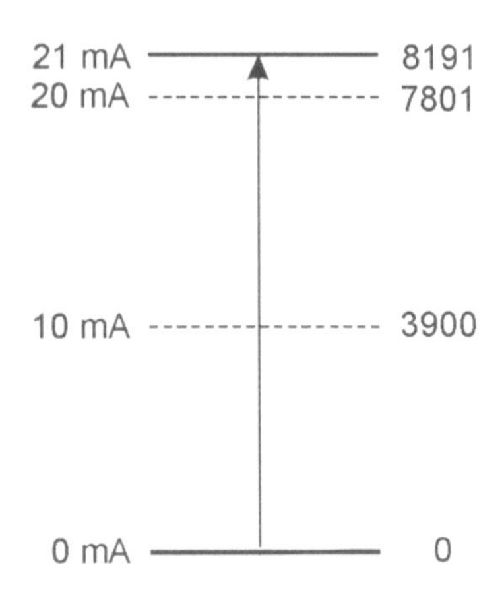

Figure 5-27. Resolution example for a 13-bit module.

Figure 5-28 shows a simple tank application. An analog sensor is being used to measure the height of fluid in a tank. The sensor outputs 0 mA at 0 feet and 20 mA at 20 feet (240 inches). The tank is only 10 feet tall (120 inches). The setpoint for the application is normally set at 6 feet (72 inches). Note that the sensor is not perfectly matched to the application. The sensor's range is 20 feet and the application is only 10 feet. This is often the case in an application. In this example we will be wasting half of the sensor range. This means that our resolution will be less than it would be if the sensor's output range was 0 – 10 feet. We are essentially throwing away half of the sensor's range in this example.

If the fluid level is 10 feet in this tank the counts in the analog input would be 3900, because 10 feet would represent half of the sensor's output (10 mA). If we wanted to calculate how closely we could measure the tank level in inches we could divide the sensor output range by the counts for that value. For example, 240 inches / 7801 would equal 0.03076 inches per count. This means that we could measure to within 0.03076 inches. A count of 0 would be 0 inches and a count of 1 would be 0.03076 inches.

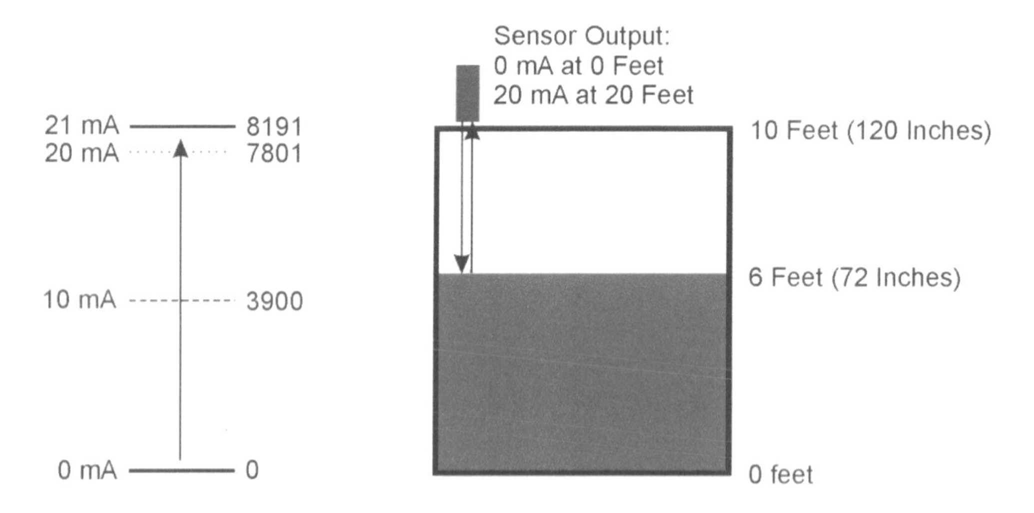

Figure 5-28. Tank resolution example.

Figure 5-29 shows an example of analog input voltage resolution for a 16-bit module. Note that bipolar and unipolar ranges are shown for various input voltage choices. The first voltage column for example, shows the use of +- 10 VDC. For this input range a zero in memory would represent 0 volts. A value of 27648 in memory would represent + 10 volts. A value of -27648 would represent – 10 volts in memory. Note that a value larger than 27648 or smaller that -27648 would represent an overshoot or undershoot in the rated range. A value of 1 in memory would represent 361.7 microvolts. This would be the finest resolution for this range. Bipolar ranges are also shown for 5 volts, 2.5 volts and 1.25 volts. The last voltage range represents a unipolar input range for 0-10 VDC.

Decimal Value	Input Voltage Measurement Range						
	+- 10V	+- 5V	+- 2.5V	+- 1.25V	Range	0-10V Range	
32767	11.85V	5.926V	2.963V	1.481V	Overflow	11.85	Overflow
32512							
32511	11.76V	5.879V	2.940V	1.470V	Overshoot Range	11.759	Overshoot Range
27649							
27648	10V	5V	2.5V	.938V	Rated Range	10V	Rated Range
20736	7.5V	3.75V	1.875V	0.75V	Rated Range	7.5V	
1	361.7uV	180.8uV	90.4uV	45.2uV	Rated Range	361.7uV	
0	0V	0V	0V	0V	Rated Range	0V	
-1					Rated Range	Negative Values Are Not Supported.	
-20736	-7.5V	-3.75V	-1.875V	-0.938V	Rated Range		
-27648	-10V	-5V	-2.5V	-1.25V	Rated Range		
-27649					Undershoot Range		
-32512	-11.76V	-5.88V	-2.94V	-1.47V			
-32513					Underflow		
-32768	-11.85V	-5.926V	-2.963V	-1.481V			

Figure 5-29. S7-1200 analog input voltage resolution.

Figure 5-30 shows an example of configuring an analog module. The choices for input signal are plus or minus 2.5 VDC, plus or minus 5 VDC, plus or minus 10 VDC or current (0-20 mA. Note that voltage was chosen. The range chosen was +- 10 volts (bipolar). Also note that the module has been configured to enable overflow and underflow diagnostics.

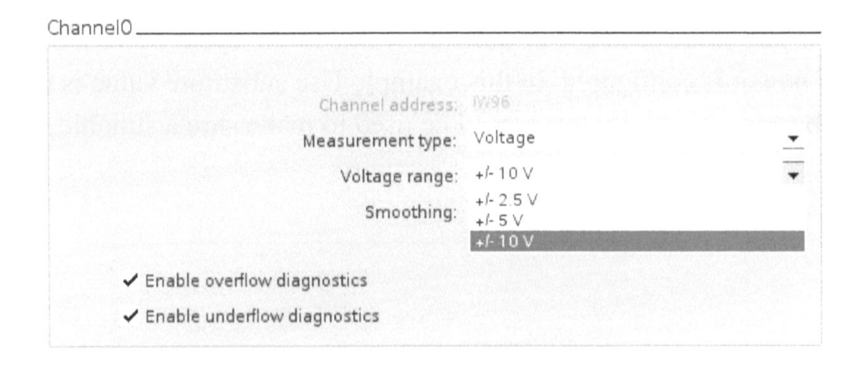

Figure 5-30. Configuring an analog module for voltage range in a S7-1200 PLC.

Figure 5-31 shows an example of analog input voltage resolution for a 16-bit module. The voltage column shows the use of +- 10 VDC range. For this input range a zero in memory would represent 0 volts. A value of 27648 in memory would represent + 10 volts. A value of -27648 would represent – 10 volts in memory. A value of 1 in memory would represent 361.7 microvolts. This would be the finest resolution for this range.

Decimal Value	Voltage Output Range	
	+- 10 V	
32767	Note 1	Overflow
32512	Note 1	
32511	11.76 V	Overshoot Range
27649		
27648	10V	Rated Range
20736	7.5 V	Rated Range
1	361.7 uV	Rated Range
0	0 V	Rated Range
-1	-361.7 uV	Rated Range
-20736	-7.5 V	Rated Range
-27648	-10 V	Rated Range
-27649		Undershoot Range
-32512	-11.76 V	
-32513	Note 1	Underflow
-32768	Note 1	

Note 1: In an overflow or underflow condition, analog outputs will behave according to the device configuration properties set for the analog signal module. In the Reaction to CPU STOP parameter, select either: Use substitute value or Keep last value.

Figure 5-31. S-71200 output voltage resolution.

Analog modules can be configured on how to respond to a CPU Stop condition. The user can configure the module to keep the last value or to use a substitute value in the event of a CPU Stop. Figure 5-32 shows an example of how it is configured. In this example Use substitute value is being chosen and the substitute value will be zero. Normally this would be used to make sure a suitable, safe value is used if a CPU stop would occur.

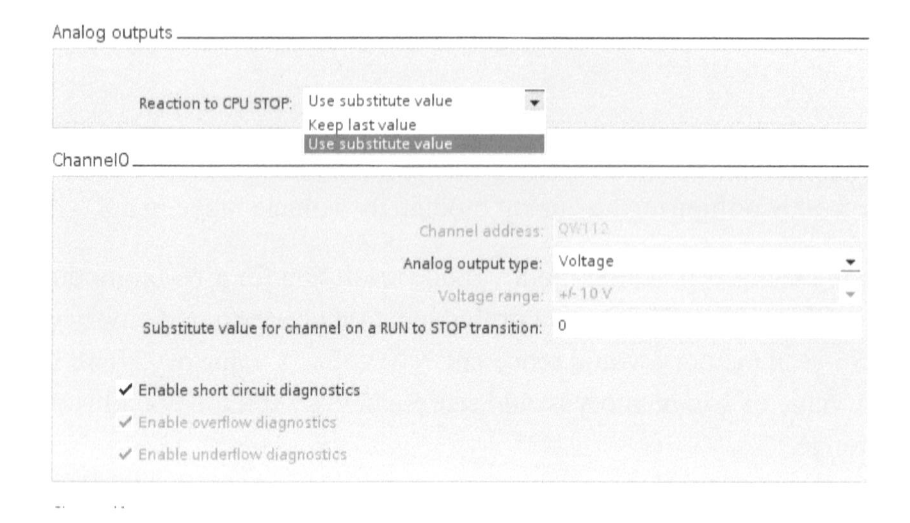

Figure 5-32. Configuring the analog module to keep the last value or to use substitute value as a reaction to a CPU stop.

Noise Reduction

Analog signals are low level signals and are susceptible to noise from the line frequency. Noise reduction can be configured and used to help reduce the effects of noise on an analog signal. A time should be chosen that is different than the line frequency. This helps to reduce the effect of crosstalk from the line power. Figure 5-33 shows how noise reduction is configured. In this example 50 Hz was chosen. Smoothing can also be configured in this screen. The choices are None – 1 cycle, Weak – 4 cycle, Medium – 16 cycle, or Strong – 32 cycle. These are the number of cycles the signal will be smoothed over.

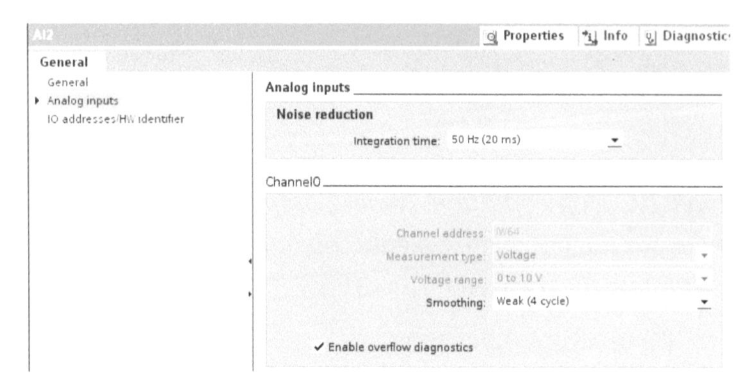

Figure 5-33. Configuring noise reduction.

Figure 5-34 shows an example of analog input current resolution for a 16-bit module. Note that unipolar ranges are shown for 2 different input current choices. The first Current column for example, shows the range of 0-20 mA. For this input range a zero in memory would represent 0 mA. A value of 27648 would represent 20 mA. Note that a value larger than 27648 or smaller that 0 would represent an overshoot or undershoot in the rated range. A value of 1 in memory would represent 723.4 nanoamps. This would be the finest resolution for this range. The values are also shown for a 4-20 mA range.

Decimal	Range		
Value	0 - 20 mA	4 – 20mA	
32767	23.70 mA	22.96mA	Overflow
32512			
32511	23.52 mA	22.81 mA	Overshoot Range
27649			
27648	20 mA	20 mA	Rated Range
20736	15 mA	16 mA	
1	723.4 nA	4 ma +578.7 nA	
0	0 mA	4 mA	
-1			Undershoot Range
-4864	-3.52	1.185 mA	
-4864			Underflow
-32768			

Figure 5-34. S7-1200 Current input resolution.

Figure 5-35 shows an example of analog current voltage resolution for a 16-bit module. The module is configured to use the 0-20 mA range. For this input range a zero in memory would represent 0 mA. A value of 27648 in memory would represent 20 mA. A value of 1 in memory would represent 723.4 nanoamps. This would be the finest resolution for this range.

This figure also illustrates that an analog module can be configured on how to respond to a CPU stop condition. The user can configure the module to keep the last value or to use a substitute value if an overflow condition exists.

Decimal Value	Range	
	Note 1	
32767	Note 1	Overflow
32512		
32511	23.52 mA	Overshoot Range
27649		
27648	20 mA	Rated Range
20736	15 mA	Rated Range
1	723.4 nA	Rated Range
0	0 mA	Rated Range

Note 1: A reaction to a CPU stop condition, analog outputs will behave according to the device configuration properties set for the analog signal module. In the Reaction to CPU STOP parameter, select either: Use substitute value or Keep last value.

Figure 5-35. S7-1200 Current output resolution.

Figure 5-36 shows an example of analog input current resolution for a 16-bit bipolar module. Note that bipolar ranges are shown for 3 different input current configurations. The first current column for example, shows the range of -20 mA to +20 mA. For this input range a zero in memory would represent 0 mA. A value of 27648 would represent 20 mA. A value of -27648 in memory would represent -10 mA. Note that a value larger than 27648 or smaller than -27648 would represent an overshoot or undershoot in the rated range. A value of 1 in memory would represent 723.4 nanoamps. This would be the finest resolution for this range. The values are also shown for a -10 to +10 mA range and for a -3.2 mA to a +3.2 mA range. Note that negative analog values are stored as two's complement numbers in memory. This is a common way to represent negative numbers in computer memory.

Decimal Value	Input Current Measurement Range			Range
	+- 20 mA	+- 10 mA	+- 3.2 mA	
32767	23.70 mA	11.85 mA	3.79 mA	Overflow
32512				
32511	23.52 mA	11.76 mA	3.76 mA	Overshoot Range
27649				
27648	20 mA	10 mA	3.2 mA	Rated Range
20736	15 mA	7.5 mA	2.4 mA	Rated Range
1	723.4 nA	361.7 nA	115.7 nA	Rated Range
0	0 mA	0 mA	0 mA	Rated Range
-1				Rated Range
-20736	-15 mA	-7.5 mA	-2.4 mA	Rated Range
-27648	-20 mA	-10 mA	-3.2 mA	Rated Range
-27649				Undershoot Range
-32512	-23.52 mA	-11.76 mA	-3.76 mA	
-32513				Underflow
-32768	-23.70 mA	-11.85 mA	-3.79 mA	

Figure 5-36. Bipolar resolution.

Figure 5-37 shows an example of ladder logic that could be used to scale analog input resolution to match an application's specific range (engineering units). It is not important at this point to understand the logic completely. These instructions will be covered in detail in a later chapter. The analog input module is being used to input a 4-20 mA signal. The NORM_X instruction is used to convert the 4-20 mA signal to a real number between 0 and 1. The module will have 5530 for a value if the input is 4 mA. If the input value is 20 mA the number will be 27648. If the input is 12 mA it would be half way between 4 and 20 mA. The value in the input would be 16589. The instruction would then output .5 to the Norm_Value tag. This would mean that the input is .5 of the input range. The Norm_Value is then used as the input value to the VALUE input of the SCALE_X instruction. The MIN value of the instruction is 0.0 and the MAX value is 850.0. In this example the 0-850 represents the RPM of a motor based on a 4-20 mA value at the analog module input. The output of the SCALE_X instruction would be 425.0, because .5 is half way between the 0 MIN and 850 MAX.

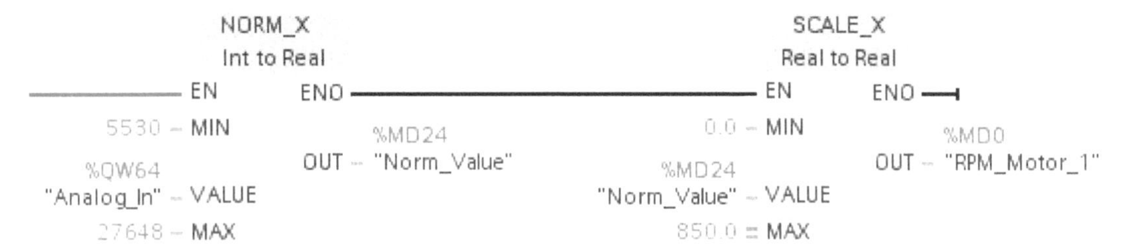

Figure 5-37. Converting an analog input signal to an RPM.

Wiring Analog Inputs

Figure 5-38 shows an example of how analog inputs can be connected to a S7-1200 CPU. The power supply positive is connected to L+ and the negative is connected to M. The negative side of both analog inputs are connected to 2M and the positive side of the inputs are connected to analog input 0 and 1.

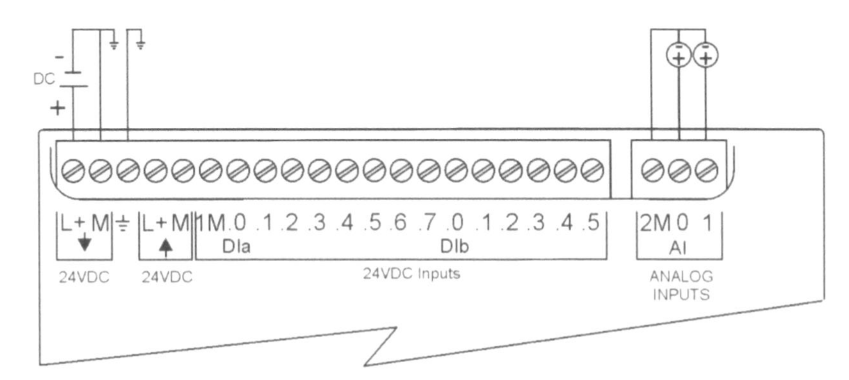

Figure 5-38. Wiring the analog inputs on a S7-1200 CPU.

Figure 5-39 shows how analog inputs are connected to a S7-1200 signal module. Note that the power supply is connected to L+ and M. The analog inputs are connected to 0+ and 0-, 1+ and 1-, and so on.

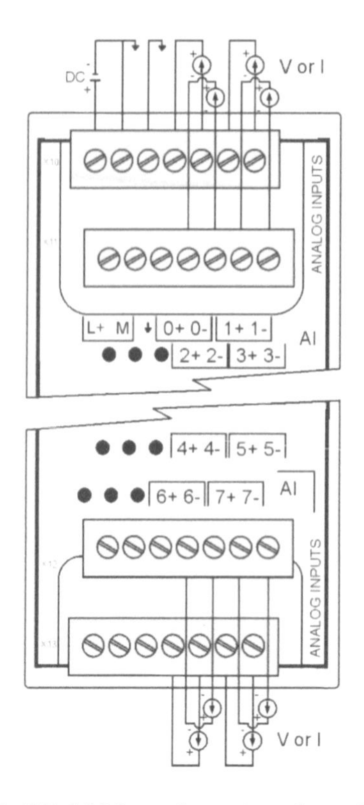

Figure 5-39. S7-1200 analog signal module wiring.

Figure 5-40 shows how 3 and a 4-wire analog device can be connected to the inputs of an analog signal module.

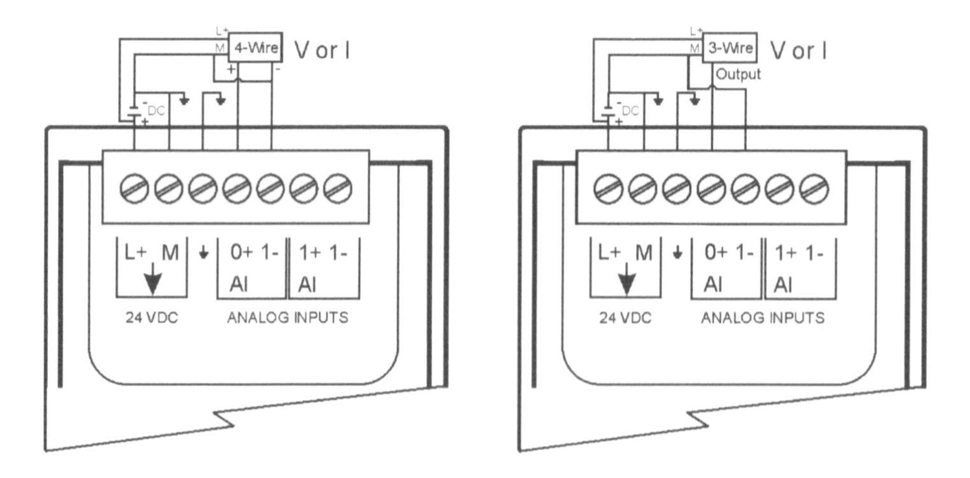

Figure 5-40. Wiring 3 and 4 wire analog inputs.

Analog input voltage wiring for a S7-300 analog input module is shown in Figure 5-41. Note the shield around the signal wires. The shield helps make the wiring more noise immune. Note that it is only grounded at one end. It is normally grounded to the chassis at the PLC.

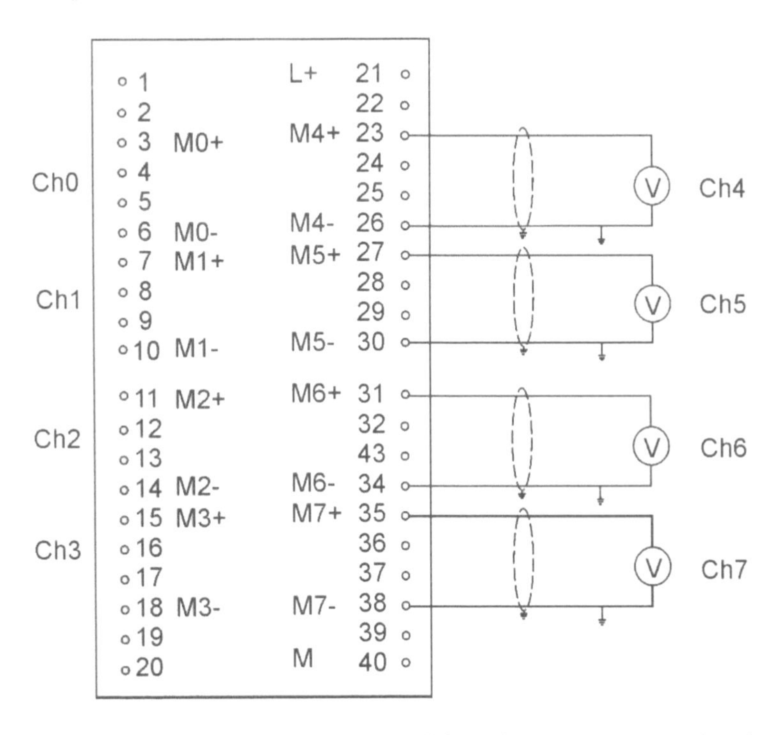

Figure 5-41. Analog input voltage wiring for a S7-300 analog input.

Analog input wiring for 2 and 4- wire transducers on a S7-300 analog input signal module is shown in Figure 5-42.

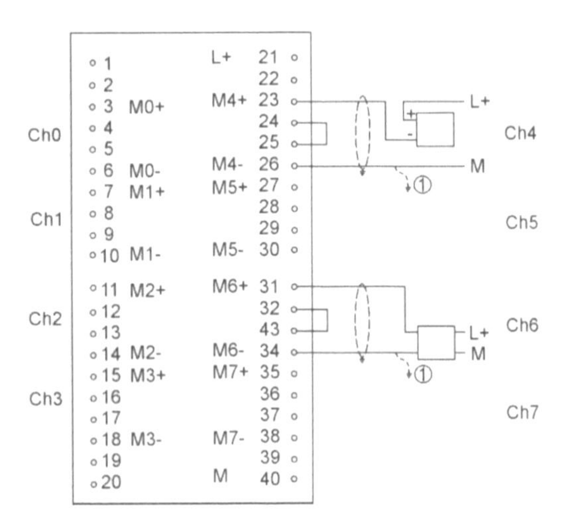

Note 1: Equipotential bonding.

Figure 5-42. Wiring 2 and 4 wire analog inputs to a S7-300 signal module.

Specialty modules are also available for Siemens PLCs. Figure 5-43 shows how thermocouples are connected to a S7-1200 thermocouple signal module.

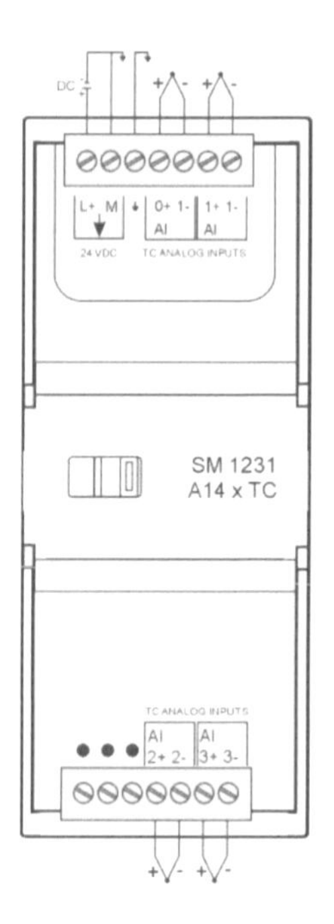

Figure 5-43. Wiring a thermocouple signal module.

Analog Output Modules

Analog output modules are used to convert digital values to analog output signals. Analog output modules are available with voltage or current output. Typical outputs are 0 to 10 volts, -10 to +10 volts and 4 to 20 milliamps.

Output Resolution

Most modules are capable of 16-bit resolution. 16 bits represents 65,536 counts (0-65535).

Output Scaling

Scaling is used to change a quantity from one notation to another. Scaling is only available with the floating-point data format in CL modules. When a channel is scaled, two points along the module's operating range are chosen and low and high values are applied to the points.

STEP 7 Function Calls for Scaling Analog Values

You can use function call FC105 SCALE (scale values) and function call FC106 UNSCALE (unscale values) blocks to read and output analog values in STEP 7. These function calls are available in the STEP 7 standard library, in the TI-S7-Converting Blocks subfolder.

Analog Output Module Wiring

Analog output modules can be used as current or voltage sources for analog outputs. Shielded, twisted-pair cables should be used to wire analog signals. Two twisted pairs of the QV and S+, and M and S- signals should be used to reduce interference. Any potential difference between the cable ends may cause an equipotential current on the shield and disturb analog signals. This can be avoided by grounding only one end of the shield.

Figure 5-44 shows the wiring for a typical S7-1200 analog output module. Note that this example is for a current output. Note the shield around the signal wires. Note also that this module is also capable of outputing voltage.

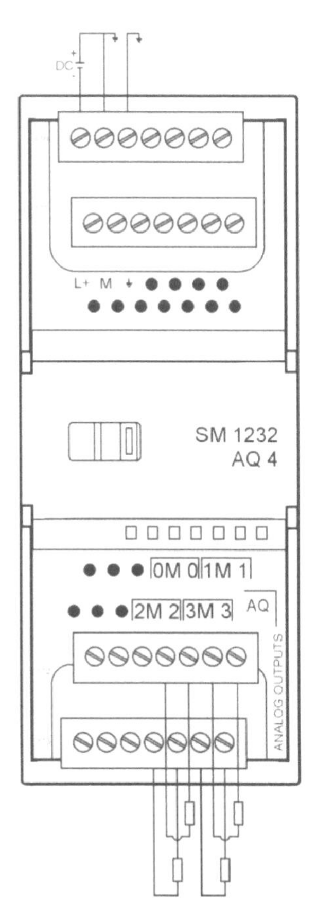

Figure 5-44. Analog wiring for a S7-1200 analog output signal module.

Wiring Loads to Analog Voltage Outputs

Analog voltage outputs support the wiring and connecting of 2-wire and 4-wire loads. Not all analog output modules support both types of wiring. The 4-wire load circuit provides high precision through the wiring and connecting the S- and S+ sense lines directly to the load. This allows direct measurement and correction of the load voltage.

Interference or voltage dips may lead to potential differences between the sense line S- and the reference loop of analog circuit M_{ANA}. This potential difference may not exceed set limits.

Any potential difference above limits has a negative impact on analog signal precision.

When wiring 2-wire loads to the voltage output of non-isolated module, wire the loads to the QV terminals and to the reference point of measuring circuit M_{ANA}. Interconnect terminal S+ to QV with terminal S to M_{ANA} in the front connector. Two-wire circuits do not provide for compensation of line impedance.

Figure 5-45 shows the wiring of 2 and 4-wire analog devices.

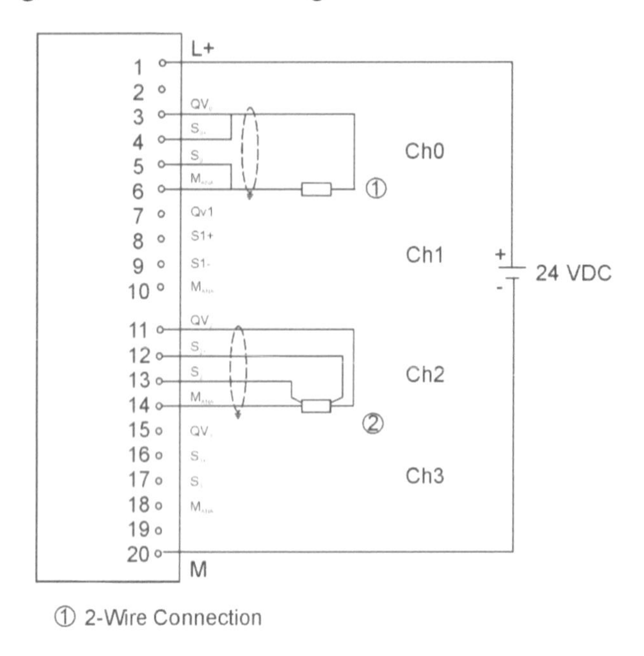

Figure 5-45. Analog output wiring for a S7-300 analog signal module.

Figure 5-46 shows the wiring of two outputs and also shows equipotential bonding for the output on channel 2 (see note 1 in the figure). Equipotential simply means equalizing electrical potential. Note that the output common was tied to functional ground.

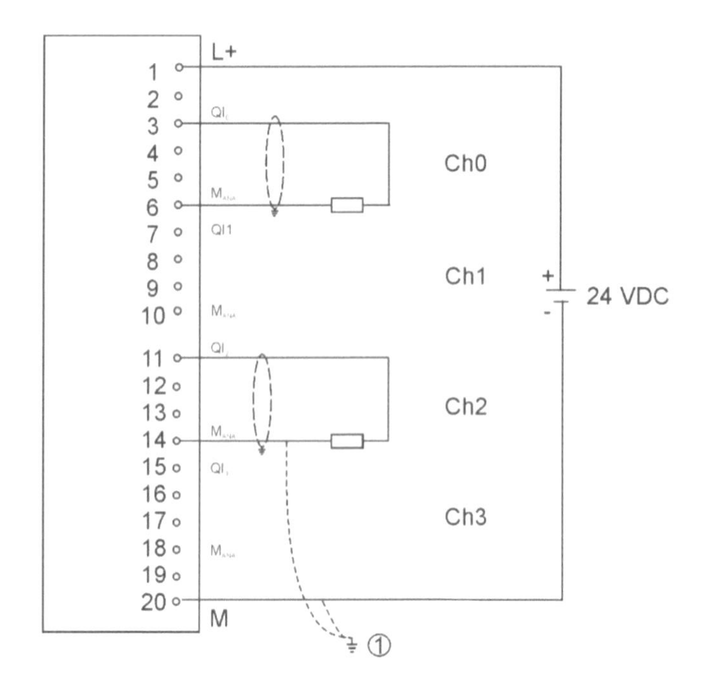

Figure 5-46. Typical analog output wiring.

Electrically Isolated Analog Output Modules

Electrically isolated analog output modules do not have a connection between the reference point of the measuring circuit common (MANA) and the CPU's common (M) terminal. Electrically isolated analog input modules should be used if there is any risk of a difference in potential developing between the reference point of measuring circuit Common (MANA) and the common (M) terminal of the CPU.

Non-isolated Analog Output Modules

When using on-isolated analog output modules, always interconnect the reference point common (MANA) of the measuring circuit with terminal common (M) of the CPU. Wire the MANA terminal to the M terminal of the CPU. Any potential difference between MANA and the M terminal of the CPU could otherwise corrupt the analog signal.

Overall Wiring Considerations

A single disconnect switch should be provided for a PLC system that can remove power from the CPU power supply and from all input and output circuits simultaneously.

Fuses or circuit breakers should be provided to limit fault currents on the supply wiring. Over-current protection should also be considered for output devices. This can be accomplished by fusing. Surge suppression devices should be used with any wiring that could be subject to surges due to lightning.

Low-voltage signal wires and communications cables should not be routed in the same wiring tray as high-energy, rapidly switched DC wires or AC wiring. Wire should always be routed in pairs, with the neutral or common wire paired with the hot or signal-carrying wire.

The shortest wire possible should be used and the wire must be properly sized to carry the required current. Where electrical noise may be an issue shielded wire should be used. The shield should typically be grounded at the PLC. When wiring input circuits that are powered by an external power supply, include an overcurrent protection device in that circuit.

Inductive output loads should be protected with suppression circuits to limit voltage rise when the output is turned off. The use of suppression devices in output circuits can protect your outputs from failure due to high voltages associated with turning off inductive loads. Suppression devices can also limit the electrical noise generated when switching inductive loads. Suppression devices are most effective when they are located close to the load.

Output Configuration

Analog outputs can be configured on how to respond to a Stop to Run change in the CPU. Outputs can be configured to use a configured value, keep the previous value, or switch off.

Chapter Questions

1. What is opto-isolation?

2. What is the difference between a signal board and a signal module?

3. Look up the manual for a Siemens SM 1221 DI 8 x 24 VDC (S7-1200) signal module or a SM 321 DI 8 x AC (S7-300) signal module on the Siemens website and draw a wiring diagram for 6 inputs.

4. Look up the manual for a Siemens SM 1222 DQ 8 x 24 VDC (S7-1200) signal module or a SM 322; DO 16 x DC 24 V (S7-300) signal module on the Siemens website and draw a wiring diagram for 6 outputs.

5. Look up the manual for a Siemens SM 1223 DI 8 x 24 VDC, DQ 8 x Relay (S7-1200) signal module or a SM 322; DO 16 x AC 120/230 V (S7-300) signal module on the Siemens website and draw a wiring diagram for 5 AC outputs.

6. Explain the term resolution as it applies to analog modules.

7. If we have a 16-bit analog current module that (see Figure 5-32) that is being used to take the input from a sensor is measuring the height of fluid in a tank. The sensor outputs 0 ma at 5 feet and 20 mA at 12 feet, what is the resolution in inches?

8. A 16-bit analog current input module (see Figure 5-32) is used to measure the level in a tank. The tank can hold between 0 and 8 feet of fluid. The sensor outputs 0 mA at 1 foot and 6 mA at 8 feet. What is the resolution in inches in this application?

9. The user can configure the module as to which value to use in the event an underflow or overflow condition exists. What are the two choices? Give an example when each might be used.

10. What is scaling?

11. An S7-1200 SM 1234 AI 4 x 13 bit / AQ 2 x 14-bit signal module (or you may use an S7-300 SM 332; AO 4 x 16-bit signal module for this question) will be used to output a 4-20 mA output to control a valve. Look up the module to find its resolution. Calculate the counts that would be used to output 4 mA and 20 mA. Calculate the resolution in mA/count.

12. An S7-1200 SM 1234 AI 4 x 13 bit / AQ 2 x 14 bit (or you may use an S7-300 SM 332; AO 4 x 12 signal module for this question) will be used to output a 0-10 VDC output to control a valve. Look up the module to find its resolution. Calculate the counts that would be used to output 0 VDC and +10 VDC. Calculate the resolution in volts/count.

13. If you need to generate a diagnostic interrupt in the event that a measured value exceeds the range of the channel, you would:
   a. Check the Enable Alarm checkbox and enter an upper value
   b. Check the Enable Overflow Diagnostics checkbox
   c. Add the logic to a FB
   d. Add the logic to an OB

14. When the CPU is switched from Stop to Run, an analog output will:
  a. be turned off
  b. Keep the last value it had
  c. Use a configured value
  d. One of the above based on how it was configured
  e. Be forced to zero

15. Negative numbers are represented in CPU memory as:
  a. Binary
  b. Hexadecimal
  c. Two's complement
  d. Negated binary

16. If the resolution of a Siemens analog module is less than 16 bits, the analog value is ____________ in memory.
  a. Left-justified
  b. Right-justified
  c. compressed
  d. interpolated

17. If an anlog module is used to measure a voltage of -10V to + 10V in the range of -27648 to +27648, which instruction would be used to convert this to a linear scale between the values of 0 and 1?
  a. Scale
  b. Linearize
  c. normalize
  d. None of the above

18. Once a value has been converted to a linear scale between 0 and 100, which instruction would be used to linearly map it to a specified range (enginerring units)?
  a. Scale
  b. Normalize
  c. Linearize
  d. None of the above

Chapter 6

Math Instructions

Objectives

Upon completion of this chapter, the reader will be able to:

Utilize math instructions in programs.

Explain terms such as comparison instructions, precedence, logical instruction, and so on.

Explain how to determine which instructions are available in each programming language.

Introduction

Arithmetic instructions are very useful in programming industrial applications. Many types of instructions are available. This chapter will over many of the more commonly used instructions. Siemens PLCs have a wide variety of instructions that can ease application development. Once you become familiar with some of the more common instructions it will be easy to learn new ones.

Math instructions are very useful in developing automated systems. For example, many times a bit count in memory must be changed to a more understandable value for display on an operator screen. Bit counts may not make much sense to an operator, but the actual RPM does. Input values must usually be modified with math instructions before a value is sent to an analog output module. This chapter will examine some of the Siemens math instructions that are available. The instruction help file in the programming software is very useful in explaining each instruction. Figure 6-1 shows a few of the math instructions that are available.

Instruction	Description
ADD	Add instructions are used to add two values.
SUB	Subtract instructions are used to subtract two values.
MUL	Multiply instructions are used to multiply two values.
DIV	Divide instructions are used to divide two values.
MOD	A modulo instruction is used to find the remainder of a division.
NEG	A negate instruction is used to change the sign of a value.
SQRT	A square root instruction is used to find the square root of a value.
SQR	A square instruction is used to square a value.
CALC	The calculate instruction can be used to execute a mathematical expression for calculations.

Figure 6-1. A few of the common math instructions.

The use of math instructions can help simplify the task of programming a complex system. We will look at few simple examples of the use of math instructions in applications as we go through the chapter. The first is a stoplight application. Developing a stoplight application in ladder logic can be cumbersome and a little difficult to follow. The use of math instructions can make it an easy and straightforward task.

By using math instructions, we would only need one timer to control the timing of the 3 east-west lights and the 3 north-south lights.

For example, we might want the east-west green light to be on from 0 seconds to 20 seconds. The yellow from 20 to 30 and the red on from 30 to 60. Figure 9-2 has a simple ladder to control the east-west lights. IN_RANGE math instructions were used to compare the timer elapsed time to the desired time limits for each light. As you can see, the use of the IN_RANGE math instructions made it possible for the logic to be very short and also quite understandable. You do not have to understand the IN_RANGE instruction at this point. They will be covered in more detail later in the chapter. Note that a memory bit output was used to reset the timer, not the Q-bit directly.

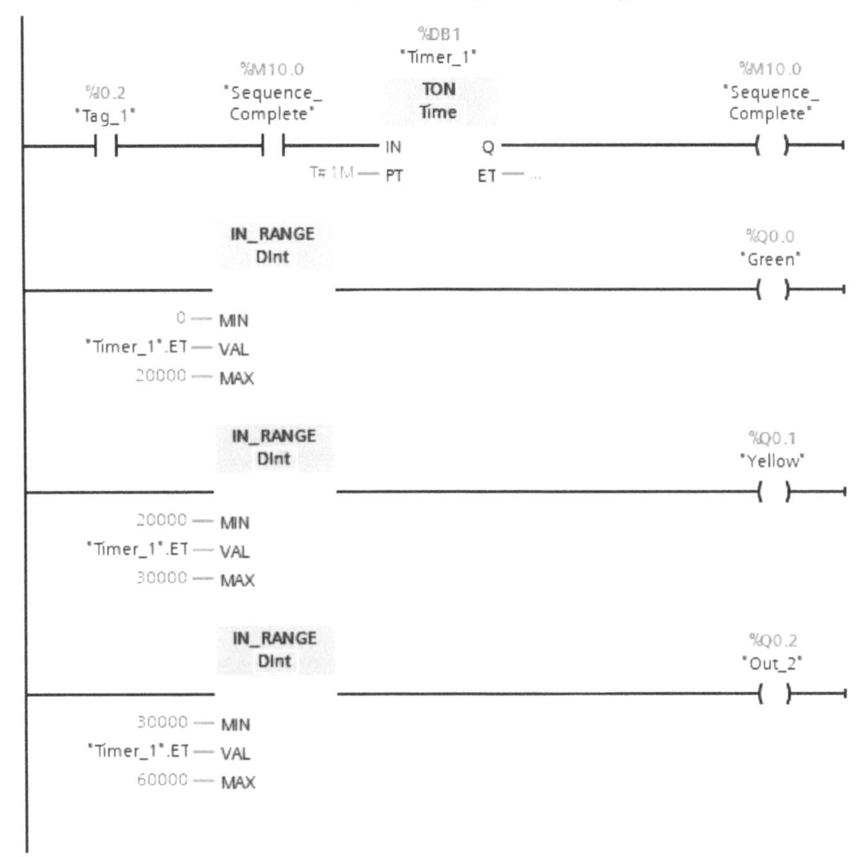

Figure 6-2. Stop light program using math instructions.

Add (ADD) Instruction

The ADD instruction can be used to add two numbers. The instruction adds the values from IN1 and IN2. The source can be a constant value or a tag. The result of the ADD is put in the OUT tag. An example of an ADD instruction is shown in Figure 6-3. If the enable input (EN) for the instruction is true, the add instruction will add the value from IN1 (Num_1) and the value from IN2 (Num_2).

The result will be stored in the OUT tag (Answer). Note that the instruction was declared as a Real-type instruction. The data type for math instructions is chosen from a drop-down box (see Figure 6-4). Right under the ADD in the instruction Real was chosen in this example. The programmer could have chosen any of the available types such as Dint, Real, LReal, Etc.

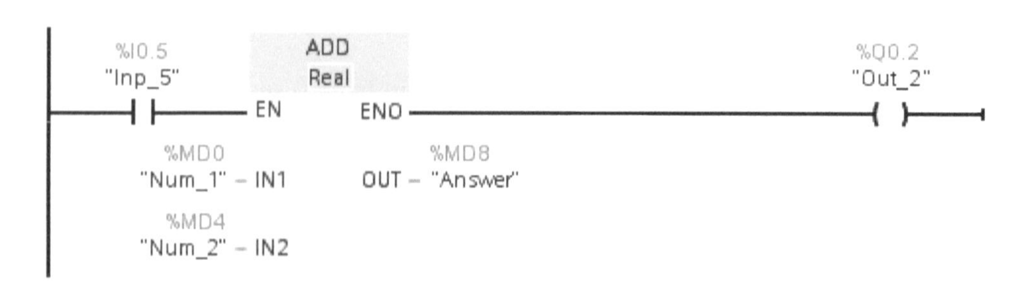

Figure 6-3. Use of an ADD instruction.

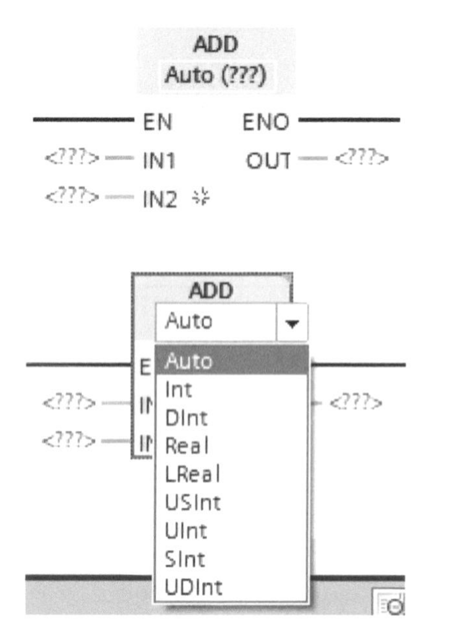

Figure 6-4. Data-type drop-down menu.

Subtract (SUB) Instruction

SUB (subtract) instructions are used to subtract two numbers. It might be nice if the operator knew how many parts they still had to make. We could subtract the number that is ordered from the number that are complete at the time. The result would be the number of parts that still need to be produced.

The source of the numbers can be constants or tags (addresses). A SUB instruction subtracts input 2 (IN2) from input 1 (IN1). The result is stored in the OUT address (tag).

The use of an SUB instruction is shown in Figure 6-5. If the EN input is true, the SUB instruction is executed. IN2 (Num_2) is subtracted from IN1 (Num_1); the result is stored in the OUT tag named Answer in this example. Note that Real was chosen for the data type of this instruction.

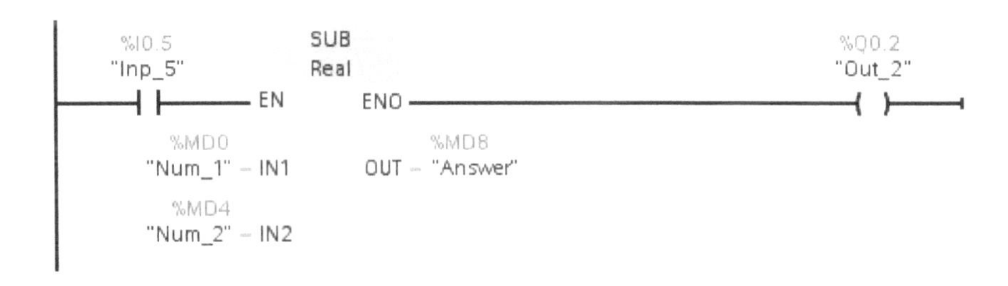

Figure 6-5. This figure shows the use of a SUB instruction.

Multiply (MUL) Instruction

A multiply instruction (MUL) is used to multiply two numbers. When we looked at a use for the subtract instruction, we calculated how many products still had to be made by using a subtract instruction. We could use a multiply instruction to calculate how much time it will take to complete the remaining parts. If there are 15 that still need to be made and our cycle time to make one is 37 seconds the result would be 555 seconds (15*37).

Figure 6-6 shows an example of the use of a MUL instruction. When a MUL instruction executes, the value at input (IN1) is multiplied by value at input (IN2). The result of the multiplication is stored in the OUT tag. IN1 (Value_1) and IN2 (Value_2) can be numbers, tags or addresses. Figure 6-5 shows the use of a multiply instruction. If the EN (enable) input is true, IN1 (the value in tag Num_5) is multiplied by IN2 (Num_6) and the result is stored in OUT tag Result. Note that this instruction was declared to be a Dint type.

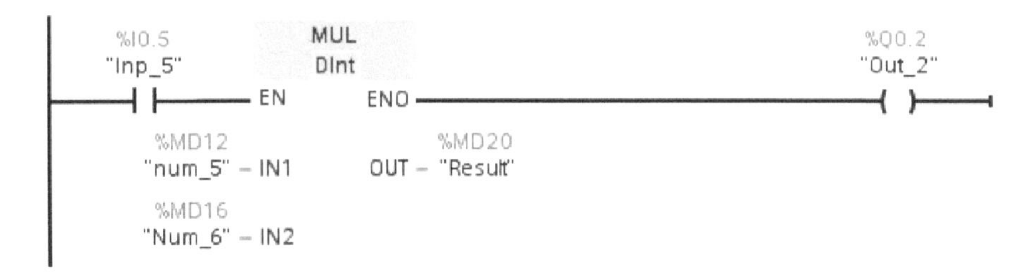

Figure 6-6. Use of a Multiply (MUL) instruction.

Divide (DIV) Instruction

The DIV instruction can be used to divide two numbers. When we looked at the multiply instruction, we calculated that it would take 555 seconds to complete the required parts. We could use a divide instruction to calculate how many minutes that would be. If we used a divide instruction and divided 555 seconds by 60 seconds the result would be 9.25 minutes.

The source of the two numbers can be constants or tags. When a DIV instruction is executed, IN1 is divided by IN2 and the result is placed in the OUT tag (see Figure 6-7).

Figure 6-7 shows an example of the use of a division instruction. The data type that was chosen for this example is Real. If the EN (enable) input is true, the divide instruction will divide the number from IN1 by the number from IN2. The result is stored in the OUT tag (Answer).

In this example the result is a decimal (Real) number. Note that a tag (Num_1) was used for IN1 and a tag (Num_2) was used for IN2. Tags, addresses, or constants could have been used for IN1 and/or IN2.

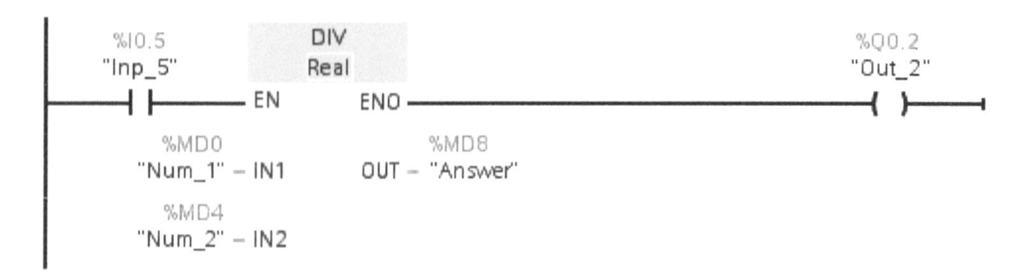

Figure 6-7. DIV instruction example.

Figure 6-8 shows what happens to the remainder when a Dint-type DIV instruction divides two integers. The answer is actually 17.666666666666666666666666667, but the DIV instruction returns 17 as the answer because the Dint type was declared for the instruction.

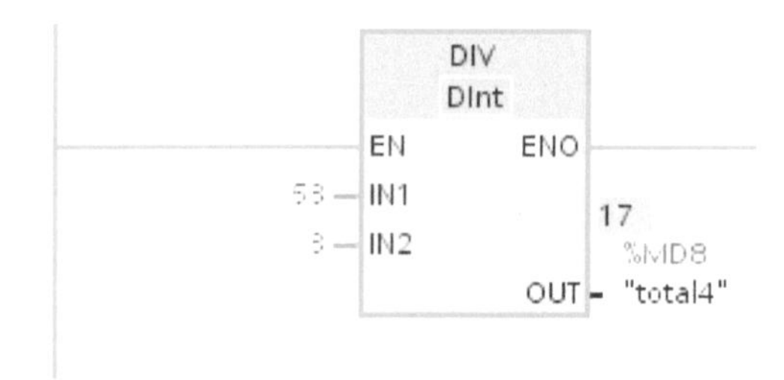

Figure 6-8. Use of a Dint-type DIV instruction.

Modulo (MOD) Instruction

The Modulo instruction is a very interesting and useful instruction. A MOD instruction returns the remainder of a division. For example, if we divide 13 by 5, we would get 2 fives and a remainder of 3. A Modulo instruction would return 3 as the answer. One example that shows the power of modulo instructions would be a machine that has to make change for a given amount of money. Imagine we would like to know how many quarters, dimes, nickels and pennies are in a dollar amount. Let's use the amount $1.43. First, we would divide by the size of the largest coin (quarter). If we use a MOD instruction there would be 5 quarters in $1.43 with a remainder (modulo) of 18. Next, we would divide by the next largest coin (dime). There would be 1 with a remainder (modulo) of 8. Next there would be one nickel with a remainder (modulo) of 3. Then there would be three pennies with no remainder.

Study Figure 6-9. The MOD instruction divides the value at IN1 (Num_5) by the value at IN2 (Num_6) and places the remainder in the OUT-output tag named Result. In this example 13 (the value in Num_5) was divided by 5 (Value in Num_6). In this example the MOD instruction would divide 12 by 5 for an integer result of 2 with a remainder of 3. The remainder (3) is put in the OUT tag. The OUT-tag name is named Result in this example. Note that this instruction was declared as a Dint type.

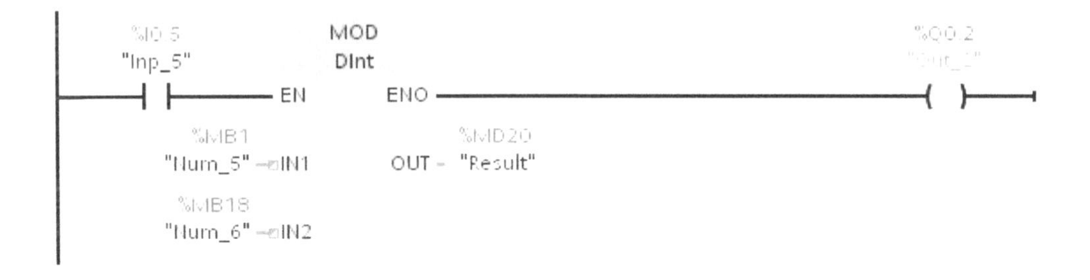

Figure 6-9. Example of a Modulo instruction.

Negate (NEG) Instruction

The negate instruction (NEG) is used to change the sign of a value. If it is used on a positive number, it makes it a negative number. If it is used on a negative number, it will change it to a positive number. Remember that this instruction will execute every time the EN input to the instruction is true. Use transitional contacts if needed. The use of a negate instruction is shown in Figure 6-10.

If contact Inp_5 is true, the value in input IN (num_5) will be given the opposite sign and stored in destination tag (Result). In this example if num_5 contained 5, the Negate would change it to -5 and store it in the tag named Result.

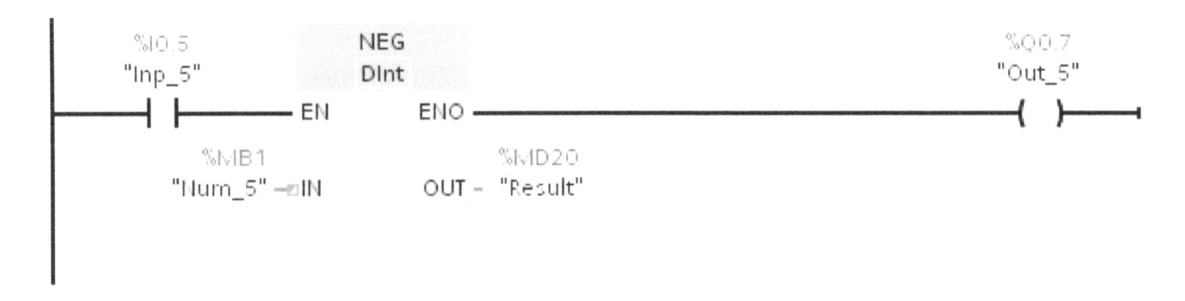

Figure 6-10. Use of a Negate (NEG) instruction.

Square Root (SQRT) Instruction

The square root instruction (SQRT) is used to find the square root of a number. The result is stored in an OUT tag. The source can be a value, tag, or address. Figure 6-11 shows the use of a square root instruction. If the EN input is true, the SQRT instruction will find the square root of the IN-input value (Num_1). The result will be stored in the OUT-output tag named Answer in this example. Note that this instruction was declared as a Real type.

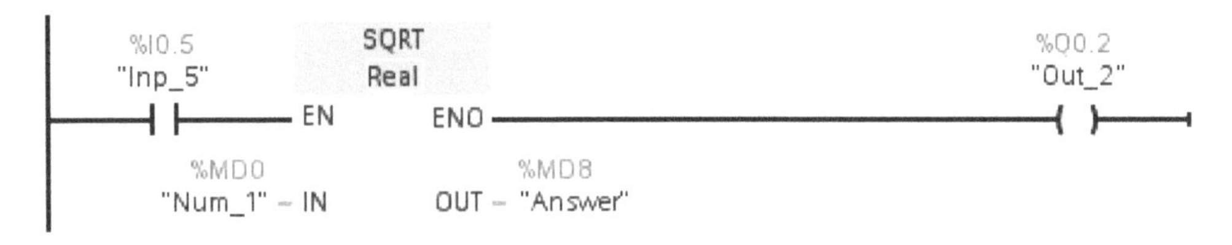

Figure 6-11. Use of a Square Root (SQRT) instruction.

Square (SQR) Instruction

The SQR instruction is used to find the square of a number. The result is stored in an OUT tag. The source can be a value, tag, or address. Figure 6-12 shows the use of a SQR instruction. If the EN input is true, the SQR instruction will find the square of the IN-input value (Num_1). The result will be stored in the OUT-output tag named Answer in this example. Note that this instruction was declared as a Real type.

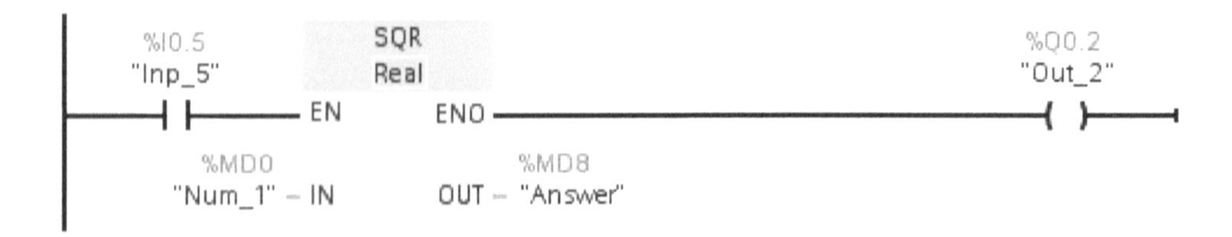

Figure 6-12. Use of a Square (SQR) instruction.

Calculate (CALC) Instruction

You can use the calculate instruction to define and execute an expression for calculating mathematical operations or complex logic combinations, depending on the selected data type.

You can select the data type of the instruction from the drop-down list of the instruction box. Depending on the data type selected, you can combine the functions of certain instructions to perform a complex calculation. The information for the expression to be calculated is entered in a dialog, which you can open with the icon at the upper right edge of the instruction box. The expression can contain names of input parameters and the syntax of the instructions. Operand names and operand addresses cannot be specified.

The table in Figure 6-13 shows the instructions that can be executed together in the expression of the calculate instruction, depending on the selected data type.

In its initial state, the instruction box contains at least 2 inputs (IN1 and IN2). The number of inputs can be extended. The inserted inputs are numbered in ascending order in the box.

The values of the input parameters are used to execute the specified expression. Not all of the defined input parameters have to be used in the expression. The result of the instruction is transferred to the output OUT.

If you use inputs in the expression that are not available in the box, they are inserted automatically. This requires that there are no gaps in the numbering of the inputs to be newly defined in the expression. For example, you cannot use the IN4 input in the expression unless the IN3 input has been defined.

Data Type	Instruction	Example
Bit Strings	AND	IN1 OR IN2 AND IN3
	OR	
	XOR	
	INV	
	SWAP	
Integers	ADD	(IN1 – IN2) / IN3
	SUB (Subtract)	
	MUL (Multiply)	((SQRT(IN1)) * ((SQR (IN2))
	DIV (Divide)	
	MOD (Modulo)	
	INV (Inverse)	
	NEG (Negate)	
	ABS (Absolute Value)	
Floating Point Numbers	ADD	((TAN(IN1) * COS(IN2) – ((TAN(IN3) * IN4))
	SUB (Subtract)	
	MUL (Multiply)	
	DIV (Divide)	
	NEG (Negate)	
	ABS (Absolute Value)	
	SQR (Square)	
	SQRT (Square Root)	
	LN (Logarithm)	
	EXP (Exponentiate)	
	FRAC (Fraction)	
	SIN (Sine)	
	COS (Cosine)	
	TAN (Tangent)	
	ASIN (Arcsine)	
	ACOS (Arccosine)	
	ATAN (Arctangent)	
	NEG (Twos complement)	
	TRUNC (Truncate)	
	ROUND	
	CEIL (Form Next Higher Integer)	
	FLOOR (Form Next Lower Integer)	

Figure 6-13. Math instructions.

The calculate instruction is only executed if the signal state at the enable input EN is 1. When all the individual instructions of the specified expression are executed without errors, the ENO enable output also has the signal state 1.

The ENO enable output has the signal state 0 if one of the following conditions is fulfilled:

- Enable input EN has the signal state 0.
- The result of the Calculate instruction is outside the range permitted for the data type specified at the OUT output.
- A floating-point number has an invalid value.
- An error occurred during execution of one of the instructions in the expression.

In the example in Figure 6-14 there are 4 inputs. In the equation IN1 would first be divided by IN2, then the result of the division would be multiplied by IN3. The result of the multiplication would then be divided by IN4. The end result will be stored in the OUT-tag Result.

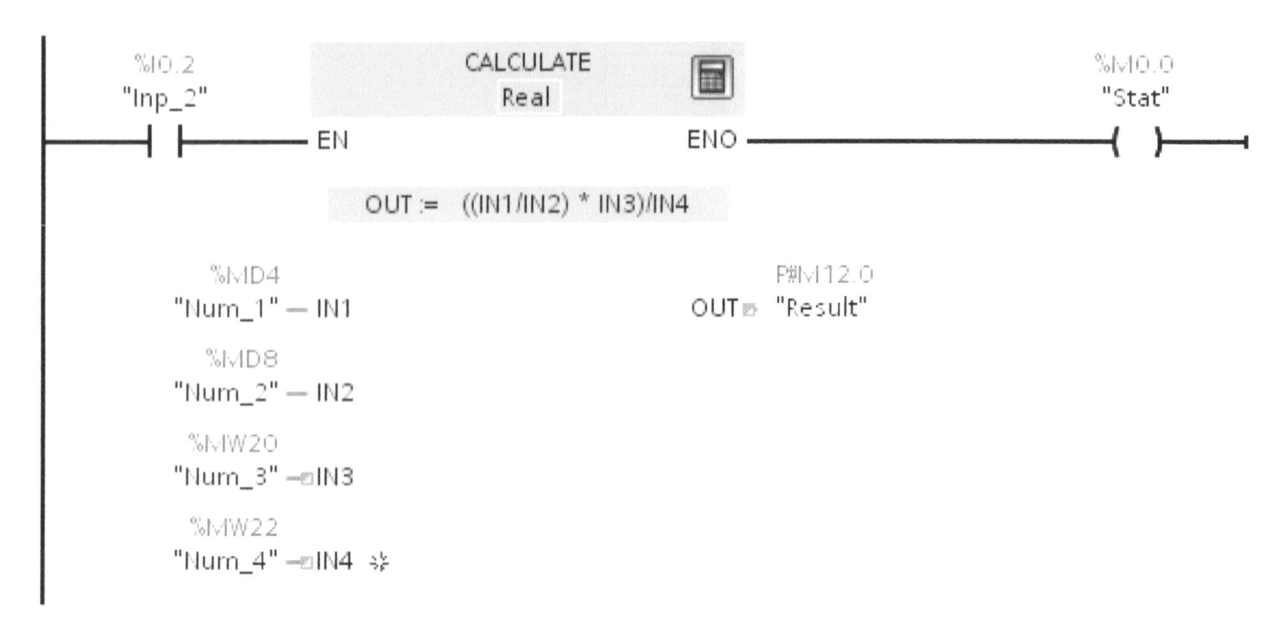

Figure 6-14. Use of a CALC instruction.

Relational Instructions

Relational instructions have many uses in programming industrial applications. Most relational instructions are used to compare two values. They can be used to see if values are equal, if one is larger, if one is between two other values etc. Figure 6-15 shows some of the available relational instructions.

CMP	**Compare values based on an expression**
EQU	Test whether 2 values are equal.
GEQ	Test whether one value is greater than or equal to another value.
GRT	Test to see if one value is greater than a second value.
LEQ	Test to see if one value is less than or equal to a second value.
LES	Test whether one value is less than a second value.
LIM	Test whether one value is between two other values.
MEQ	Pass two values through a mask and test if they are equal.
NEQ	Test whether one value is not equal to a second value.
IN_RANGE	You can use the Value Within Range instruction to determine if the value at the VAL input is within a specific value range.
OUT_RANGE	You can use the Value Outside Range instruction to determine if the value at the VAL input is outside a specific value range.
OK	The Check Validity instruction checks if the value of an operand is a valid floating-point number.
NOT_OK	The Check Invalidity instruction checks if the value of an operand is an invalid floating-point number.

Figure 6-15. Relational operators.

Equal to Instruction

The equal to instruction is used to test if two values are equal. The values tested can be actual values or tags that contain values. An example is shown in Figure 6-16. The first value (Value_5) is compared to the second value (Value_6) to test to see if they are equal. In this example Value_5 is equal to Value_6 so the instruction is true and the output is true.

Figure 6-16. Example of an Equal to instruction.

Greater Than or Equal to Instruction

The greater than or equal to instruction (GEQ) is used to test two sources to determine whether or not one value is greater than or equal to the second value. For example, in a process we might want the heating coil to be on until the temperature is greater than or equal to 200 degrees. The use of a GEQ instruction is shown in Figure 6-17. The first value in this example is a tag named "CT_1".COUNT_VALUE. The second value is 6 in this example. If "CT_1".COUNT_VALUE is greater-than- or equal to 6, the instruction will be true and the output coil will be turned on.

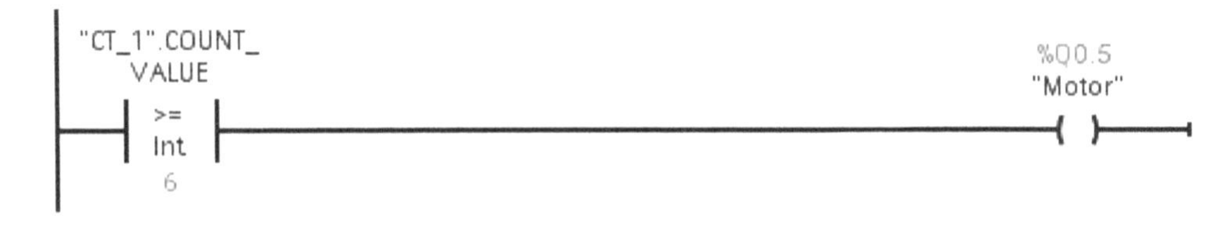

Figure 6-17. Use of a Greater Than or Equal to instruction.

Greater Than Instruction

The greater than instruction is used to see if a value from one source is greater than the value from a second source. An example of the instruction is shown in Figure 6-18. If the first value ("CT_1".COUNT_VALUE) is greater than the second value (10), the rung will be true and the output coil named Motor will be energized.

"CT_1".COUNT_
VALUE
>
Int
10
%Q0.5
"Motor"

Figure 6-18. Use of a Greater Than instruction

Less Than Instruction

Figure 6-19 shows an example of a less than instruction. A less than instruction can be used to check if a value from one source is less than the value from a second source. If the value of "CT_1".COUNT_VALUE is less than 24 in this example, the rung will be true and the output named Motor will be energized.

Figure 6-19. Use of a Less Than instruction.

Not Equal To (NEQ) Instruction

The not equal to instruction (NEQ) is used to test two values for inequality. The values tested can be constants or addresses that contain values. An example is shown in Figure 6-20. If "CT_1".COUNT_VALUE is not equal to (6), the instruction is true and output named Motor is energized.

"CT_1".COUNT_
VALUE
<>
Int
6
%Q0.5
"Motor"

Figure 6-20. Use of a Not Equal to (NEQ) instruction.

Figure 6-21 shows an example of two comparison instructions in series. In this example Num_1 must be greater than or equal to Num_2 AND Num_2 must be less than 150.5. If both conditions are true, the rung will be true and the output named Out_2 will be energized.

%MD0
"Num_1"
>=
Real
%MD4
"Num_2"
%MD4
"Num_2"
<
Real
150.5
%Q0.2
"Out_2"

Figure 6-21. Use of a Greater Than or Equal to (GEQ) in series with a less than instruction.

LIMIT Instruction

A LIMIT instruction is used to limit the output value of the instruction to a value greater-than-or equal to the MIN value AND less-than-or equal to the MAX value (See Figure 9-23).

The programmer must provide three pieces of data to the LIMIT instruction when programming. The programmer must provide a MIN value. The MIN value can be a constant or an address that contains the desired value. The address can be an integer or floating-point value.

The programmer must also provide an input value at the IN input. This is a constant or the address of a value. The third value the programmer must provide is the MAX value. The MAX can be a constant or the address of a value. If the input value is between the MIN and MAX values, the input value is sent to the OUT-output (Result in this example). If the input value is less that the MIN value, the MIN value is sent to the OUT-output. If the input value is greater than the MAX value, the MAX value is sent to the OUT-output. So, the smallest output from the instruction will be the MIN value. The largest output from the instruction will be the MAX value. If the instruction is executed without errors, the "Out_2" output is set.

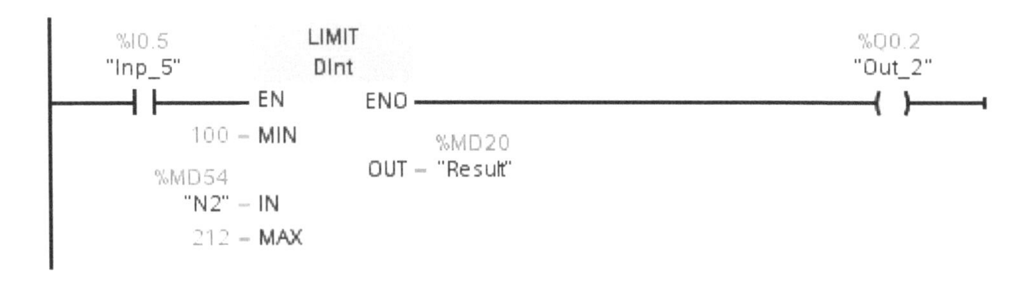

Figure 6-22. Use of a LIMIT instruction.

IN_RANGE Instruction

You can use the IN_RANGE instruction to determine if the value at the VAL input is within a specific value range. Figure 6-23 shows the use of an IN_RANGE instruction. The limits of the value range are specified with the MIN and MAX parameters. When the instruction is processed it compares the value at the VAL input to the values of the MIN and MAX parameters and sends the result to the box output. If the value at the VAL input is greater than or equal to the MIN and less than or equal to the Max, the output will have a signal state of 1. If the comparison is not true, the signal state of the output will be 0.

The enable input (EN) must be true for the instruction to execute.

The instruction requires that the values to be compared are of the same data type and the output is connected in logic.

Figure 6-23. Use of an IN_RANGE instruction.

OUT_RANGE Instruction

The value outside range (OUT_RANGE) instruction can be used to determine if the value at the VAL input is outside of a specific range. Range limits are specified by the MIN and MAX input parameters. When the instruction executes, the instruction compares the value at the VAL input to the MIN and MAX parameters and sends the result to the instruction's output. If the value at the VAL input is less than MIN or it is greater than MAX, the instruction output will have a signal state of 1. If the comparison is not true, the output will be 0. The values to be compared must be the same data type and the output must be connected to use the instruction.

In the example shown in Figure 6-24 if Num_6 (VAL input) is less than 0 or Num_6 is greater than 150, the instruction will be true and the output will be true.

Figure 6-24. OUT_RANGE Instruction

Check Validity (OK) Instructions

The check validity instruction (OK) can be used to check if the value of a tag is a valid floating-point number. The instruction is executed in each program cycle when the signal state at the input of the instruction is true. If the tag value is a valid floating-point number, the instruction is true. In all other cases, the signal state at the output of the check validity (OK) instruction is false.

The check validity instruction can be used together with the EN of another instruction. If you connect the OK box to an EN enable input, the enable input is set only when the result of the validity query of the value is true. You can use this function to ensure that an instruction is enabled only when the value of the specified tag is a valid floating-point number.

The example shown in Figure 6-25 uses two OK instructions. The first OK instruction checks to be sure Num_1 is a valid floating-point number. The second OK instruction checks to make sure that Num_2 is a valid floating-point number. If both instructions are true, the EN of the DIV instruction in Figure 6-23 will be true.

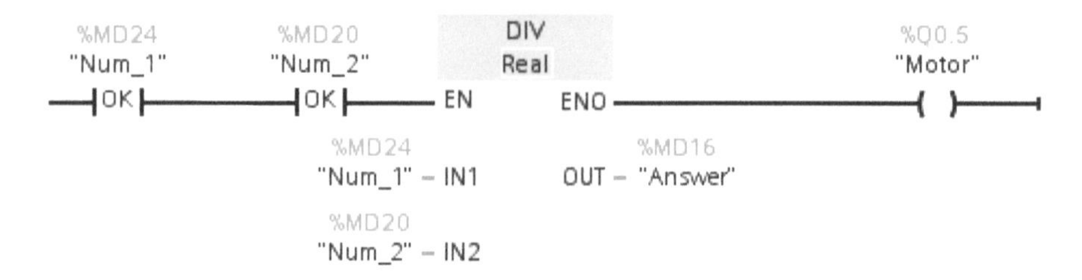

Figure 6-25. Use of two OK instructions.

The check invalidity (NOT_OK) instruction checks if the value of an operand is an invalid floating-point number (see Figure 6-26). The instruction executes in each program cycle when the signal state at the input of the instruction is true.

The output of the instruction is true when the value of the operand is an invalid floating-point number and the input of the instruction is true. In all other cases, the signal state at the output of the NOT_OK instruction is false.

If the value at the operand (Num_1 in this example) is an invalid floating-point number, the output coil will be false.

%I0.2
"Inp_2"
%MD4
"Num_1"
NOT_OK
%M0.0
"Stat"

Figure 6-26. Use of a NOT_OK instruction to check to see if a value is an invalid floating-point number.

Conversion Instructions

Convert Instruction

The convert instruction (CONV) takes the value at the IN parameter and converts it to the data type that is specified.

The convert instruction can only be started when the EN (enable input) is true. If no error occurs during the execution of the instruction, the ENO output is true.

The enable output ENO is false if: the EN input is false or errors occur during the execution of the instruction. Figure 6-27 shows an example of the use of a CONV instruction to convert a Real value to a Dint value. If Num_1 is equal to 78.1256 it will be converted to a Dint type and stored in the tag named Result with a value of 78.

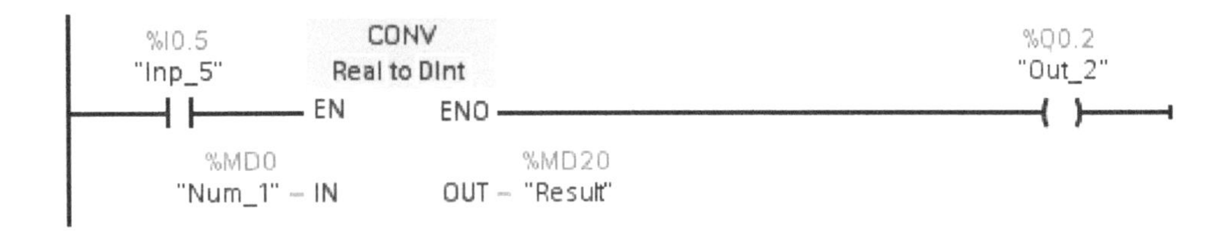

Figure 6-27. Use of a convert (CONV) instruction.

Round Instruction

The ROUND instruction is used to round the number at the IN input and store it in the OUT tag. The instruction interprets the value at input IN as a floating-point number and converts this to the type chosen in the instruction. If the input value is exactly between an even and odd number, the even number is selected. The result of the instruction is stored in the OUT output.

The instruction is executed if the enable input (EN) is true. If there are no errors, the enable (ENO) output will be true.

The ENO enable output will be false if: the enable input EN is false or errors such as an overflow occur during the execution.

Figure 6-28 shows an example of the use of a ROUND instruction to round a Real value to a DINT rounded value. If Num_5 is equal to 5.23 it will be rounded and stored in Num_Out with a value of 5.

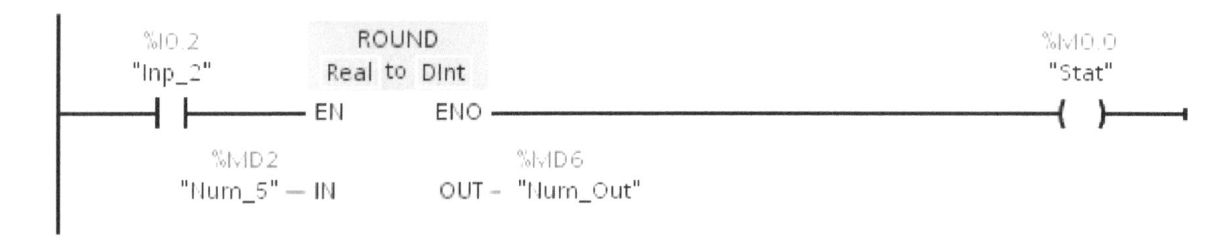

Figure 6-28. Use of a round instruction.

Ceiling (CEIL) Instruction

The ceiling (CEIL) instruction is used to round the value at the IN input to the next higher integer and sore it in an output tag. The Instruction evaluates the value at input IN as a floating-point number and converts it to the next higher integer. The result of the instruction is put in the OUT tag.

The CEIL instruction will execute if the signal state is true at the enable input (EN). If there are no errors during the execution of the instruction, the Enable Output (ENO) output will be true.

Figure 6-29 shows the use of a CEIL instruction to round a Real value to the next highest integer value. Note the output tag is a single integer (Sint) type.

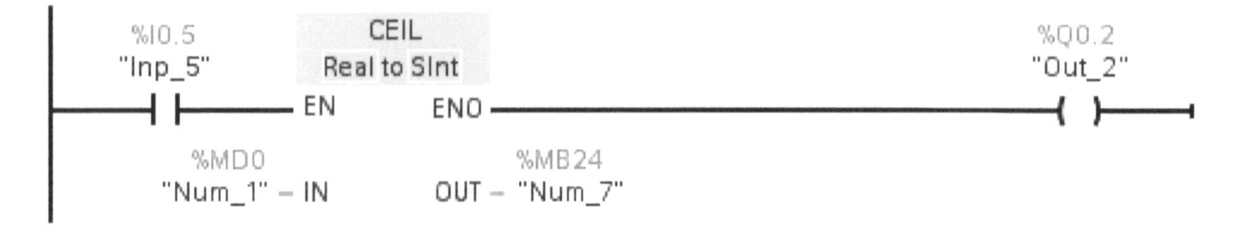

Figure 6-29. Use of a ceiling (CEIL) instruction.

FLOOR Instruction

The Floor instruction can be used to round the value at the IN input to the next lower integer. The Instruction evaluates the value at input IN as a floating-point number and converts it to the next lower integer. The result of the instruction is put in the tag (address) at the OUT.

The instruction is only executed if the enable input (EN) is true. The Enable Output (ENO) output is also true if no errors occur during the execution of the instruction.

The enable output ENO will have a signal state of false if: the EN input is false or errors occur during the execution of the instruction.

Figure 6-30 shows an example of the use of a FLOOR instruction. If Inp_5 is true the instruction will execute. In this example, the value in tag Num_1 (a real number) is 12.78. The FLOOR instruction will round 12.78 to the next lower integer (12) and store it in the OUT tag (Result).

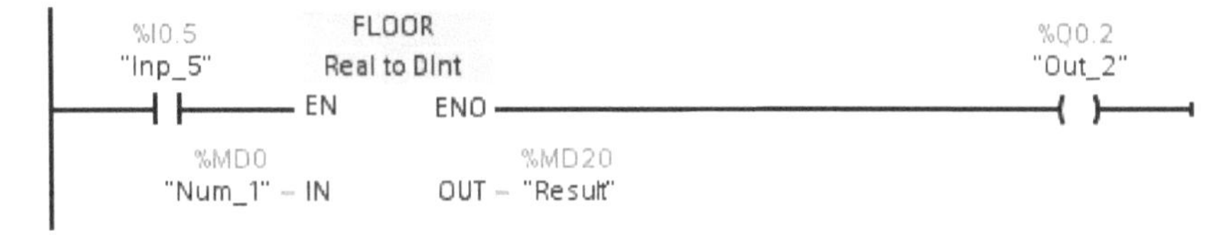

Figure 6-30. Use of a FLOOR instruction.

Truncate (TRUN) Instruction

You can use the truncate numerical value instruction (TRUN) to create an integer without rounding the value at the IN-input tag. The value at input IN is interpreted as a floating-point number. The instruction takes the integer part of the floating-point number and stores it in the OUT tag without the decimal portion of the number. The instruction is only executed if enable input (EN) is true. If no errors occur during execution of the instruction, the ENO output is also true.

The enable output ENO is false if: the EN input is false or errors occur during processing.

In the example in Figure 6-31 if Inp_5 is true the instruction will execute. If Num_1 contains the number 7.456, the instruction will take the integer portion of the number (7) and store it in the OUT tag (Num_6). The ENO will be true when the EN is true unless and error occurs during execution.

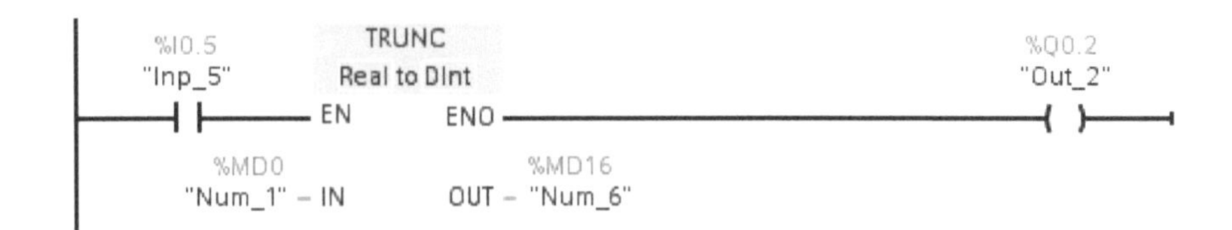

Figure 6-31. Use of a truncate (TRUNC) instruction.

Normalize (NORM_X) Instruction

The normalize instruction (NORM_X) can be used to normalize the value of the tag at the VALUE input by mapping it to a linear scale (a decimal number between 0.0 and 1.0). A graph of a normalized value versus an input values range is shown in Figure 6-32.

The MIN and MAX parameters are used to define the limits of the input range. The result at the OUT output is calculated and stored as a floating-point number between 0.0 and 1.0 depending on the actual input value compared to the specified input range. If the actual input value is equal to the value at the MIN input, the OUT-output value will be set to 0.0. If the input value is equal to the value at the MAX input, the OUT output will be set to 1.0.

For example, imaging an analog input card that is taking an input of 4mA to 20 mA. At 4 mA the bit count would be 6242 in our example (see Figure 6-31). The bit count for 20 mA would be 16383. So, the actual input bit count would be between 6242 and 16,383. A Normalize instruction could be used to convert the actual input value to a number between 0.0 and 1.0. So, if the actual number (bit count) at the value input were 9350, the normalize instruction would output the number .3065 to the OUT tag (Num_2) or address. The difference between 6242 and 16383 (range) would be 10141 bits. 9350 -6242 would be 3108. If we divide 3108 by the range of 10141, we get .3065. This is because 9350 would be about 1/3 (.3065) of the distance between 6242 and 16,383.

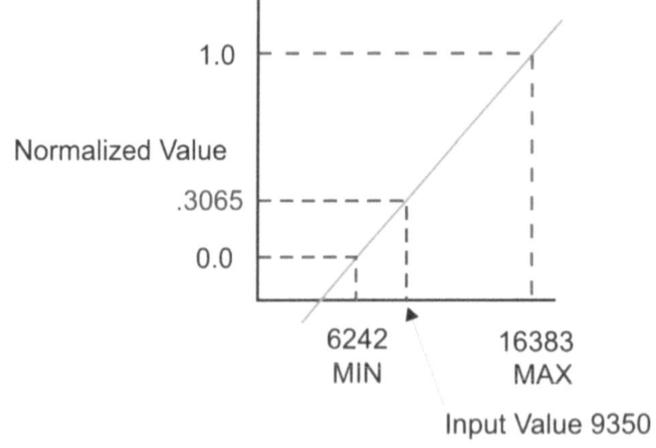

Figure 6-32. Graph of how an input value is normalized. 9350

Figure 6-33 shows an example of the use of a NORM_X instruction. The normalize instruction is executed if the EN input is true. The enable output (ENO) will be true when the instruction is enabled. NORM_X instructions are commonly used with a SCALE_X instruction to scale input or output values.

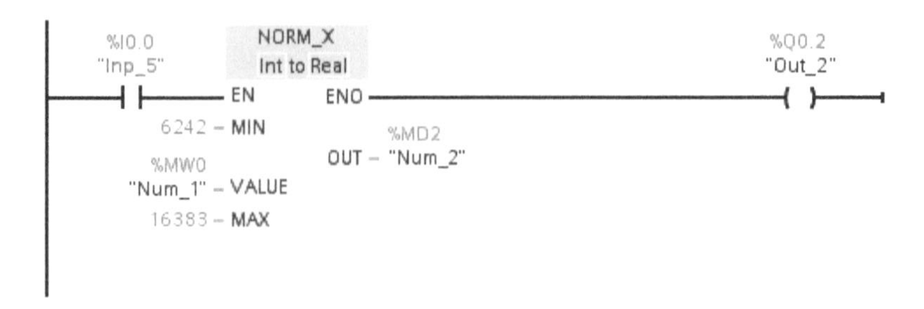

Figure 6-33. Use of a Normalize (NORM_X) instruction.

SCALE_X Instruction

The SCALE_X instruction scales the value at the VALUE input by mapping it to a specified value range. The input VALUE must be between 0.0 and 1.0. When the SCALE_X instruction is executed, the floating-point value at the VALUE input is scaled to the value range that was defined by the MIN and MAX values. The result of the scaling is an integer, which is stored in the OUT output.

Figure 6-34 shows an example of the use of a SCALE_X instruction. If the input VALUE in tag In_Val is .65 the output would be 10649 (.65 * 16384).

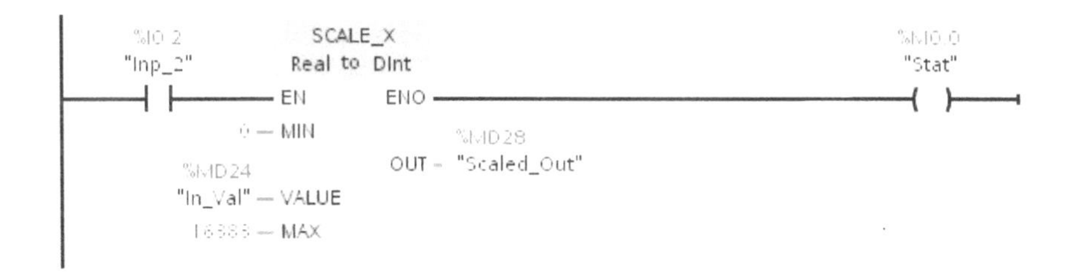

Figure 6-34. Use of a SCALE_X instruction.

Figure 6-35 shows an example of the conversion instructions that are available.

Instruction	Description
CONVERT	The CONVERT instruction reads the content of the IN parameter and converts it according to the data types selected in the instruction box.
ROUND	The ROUND instruction rounds the value at the IN input to the nearest integer.
CEIL	The CEIL instruction rounds the value at the IN input to the next higher integer.
FLOOR	The FLOOR instruction rounds the value at the IN input to the next lower integer.
TRUNC	You can use the TRUNC instruction to form an integer from the value at the IN input. The value at the IN input is interpreted as a floating-point number.
SCALE_X	The SCALE_X instruction scales the value at input VALUE by mapping it to a specified value range.
NORM_X	The NORM_X instruction normalizes the value of the tag at input VALUE by mapping it to a linear scale.

Figure 6-35. Conversion instructions.

Fraction Instructions

You can use the return fraction (FRAC) instruction to find the decimal places of the value at the IN input. The result of the instruction is stored at the OUT-output tag. If the value at input IN is, for example, 123.4567, output OUT tag will be assigned the value 0.4567. The instruction is executed when EN input is true. If the EN is true the enable output (ENO) will also be true. The enable output (ENO) is false if the EN input is false or errors occur during the execution of the instruction.

Figure 6-36 shows an example of the use of a FRAC instruction. If Inp_5 is true the FRAC instruction will execute. The instruction will take the value at the IN input from the tag named Num_1. If the number in Num_1 is 123.4567, the instruction will take the decimal portion (.4567) and store it in the OUT-tag Answer. The ENO will be true because EN is true, unless there is an error during execution.

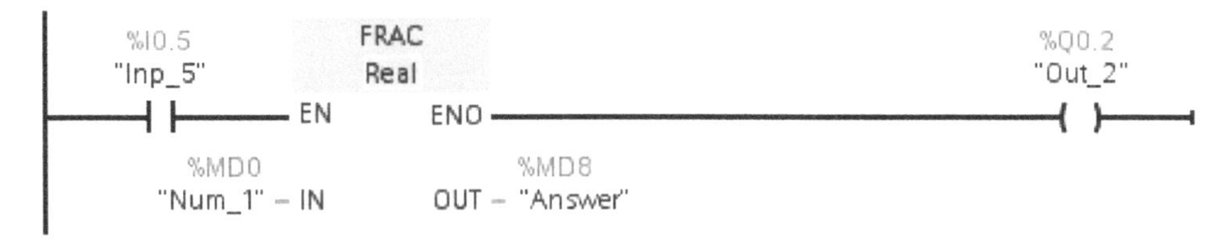

Figure 6-36. Use of a fraction (FRAC) instruction.

MIN

The get minimum instruction (MIN) compares the value at the IN1 input to the value at the IN2 input and writes the lower value to the OUT output. All tags must be the same data type. Figure 6-37 shows the use of a MIN instruction.

The MIN instruction will execute if the enable input (EN) is true. The enable output (ENO) will be true if the EN is true and no errors occur during execution of the instruction.

The enable output (ENO) will be false if: the EN input is false or the tags are not of the same data type or an input tag of the REAL data type has an invalid value.

Study Figure 6-37. If the tag at IN1 (num_5) has a value of 28 and the tag at IN2 (num_6) has a value of 17, 17 (the lower of the two) will be stored in the OUT-tag Result.

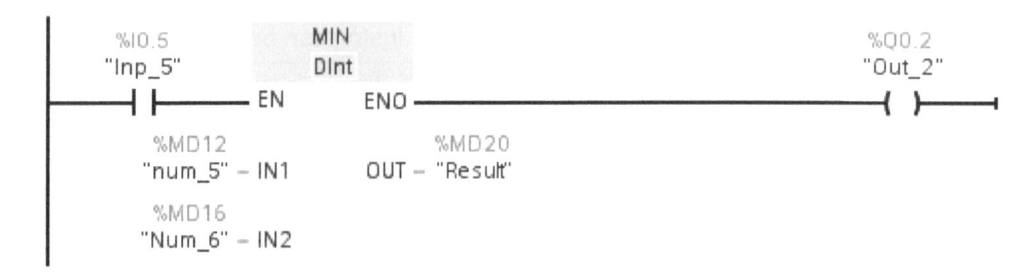

Figure 6-37. Use of a minimum (MIN) instruction.

MAX

The get maximum (MAX) instruction is used to compare the value at the IN1 input to the value at the IN2 input. The higher of the two values is then written to the OUT output. All tags must be the same data type.

The enable output (ENO) will be true if no errors occur during execution of the instruction.

The enable output will be false if:

The EN input is false.

All tags are not the same data type.

An input tag of the REAL data type has an invalid value.

In the example shown in Figure 6-38 if the tag at IN1 (num_1) has a value of 13.567 and the tag at IN2 (num_2) has a value of 12.8, 13.567 (the higher of the two) will be stored in the OUT tag Num_3.

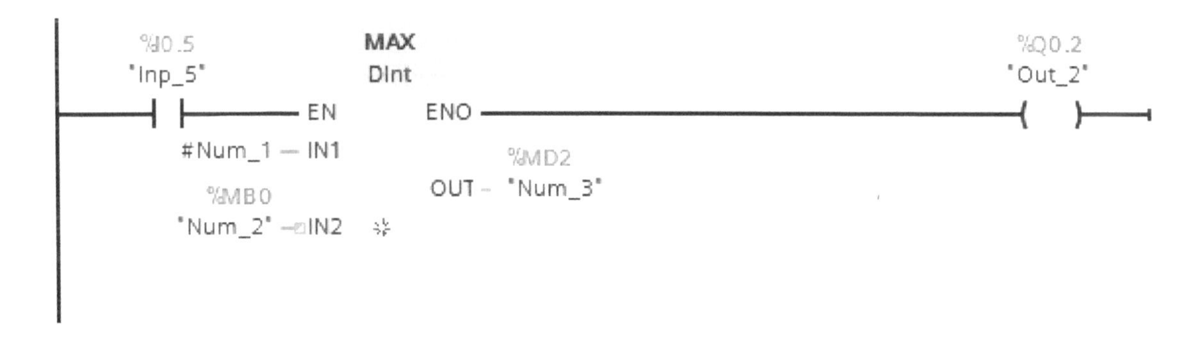

Figure 6-38. Use of a maximum (MAX) instruction.

Exponentiate (EXPT)

The exponentiate (EXPT) instruction can be used to raise the value at the IN1 input by the power specified with the value at the IN2 input. The result of the execution is written to the OUT tag ($OUT = IN1^{IN2}$).

The IN1 input must be a valid floating-point number. Integers can be used for the IN2 input.

If the EN input is true the ENO enable output will also be true.

The enable output ENO will be false if: the EN input is false or errors occur during processing of the instruction.

In the example shown in Figure 6-39, if Inp_5 is true the instruction will take the value from IN1 (Assume it is 2.0) and raise it to the power of the value at IN2 (3). So, in this example it would be 2.0^3 or 2*2*2 = 8. The number 8.0 would be stored in the OUT-tag Answer.

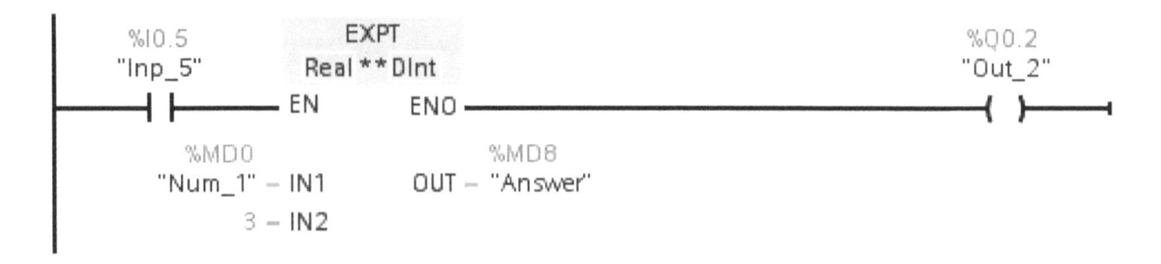

Figure 6-39. Use of an exponent (EXPT) instruction.

Logical Instructions

There are several logical instructions available (see Figure 6-40). They can be very useful to the innovative programmer. They can be used, for example, to check the status of certain inputs while ignoring others.

Instruction	Description
AND	The AND logic instruction can be used to link the value at the IN1 input to the value at the IN2 input bit-by-bit by AND logic and store the result at the OUT output.
OR	The OR logic instruction can be used to link the value at the IN1 input to the value at the IN2 input bit-by-bit by OR logic and store the result at the OUT output.
XOR	The EXCLUSIVE OR logic instruction can be used to link the value at the IN1 input to the value at the IN2 input bit-by-bit by EXCLUSIVE OR logic and store the result at the OUT output.
INVERT	You can use the "Create ones complement" operation to invert the signal status of the bits at the IN input.
DECO	You can use the "Decode" operation to set a bit in the output value specified by the input value.
ENCO	You can use the "Encode" operation to read the bit number of the least significant set bit in the input value and send it to the OUT output.
SEL	The "Select" operation selects one of the inputs IN0 or IN1 depending on a switch (parameter G) and copies its content to the OUT output.
MUX	You can use the "Multiplex" operation to copy the content of a selected input to the OUT output. The number of selectable inputs of the MUX box can be expanded.

Figure 6-40. Some of the available logical instructions.

AND Instruction

There are several logical operator instructions available. The AND instruction is used to perform an AND instruction using each of the bits from two source addresses. The bits are ANDed and a result occurs. An AND instruction needs two sources (numbers) to work with. Each of the bits from these two sources is ANDed and the result is stored in a third address. Figure 6-41 shows how a bit is ANDed. There are four possible combinations of bits. For example, if both bits have a value of 0 the result will be 0. If both bits are a 1, the result will be 1.

IN1	IN2	OUT Value
0	0	0
1	0	0
0	1	0
1	1	1

Figure 6-41. Results of an AND instruction on the four possible bit combinations.

Figure 6-42 shows an example of an AND instruction. IN1 (Val_1) and IN2 (Val_2) are ANDed. The result is placed in the OUT-tag Output_Val.

Examine the bits in the Val_1 and Val_2 so that you can understand how the AND produced the result in the tag named Output_Val.

The AND logic instruction can be used to perform an AND with the value at the IN1 input and the value at the IN2. The result of the AND is stored at the OUT output.

When the instruction is executed, bit 0 of IN1 input is ANDed with bit 0 of the value at the IN2 input.

The result is stored in bit 0 of the OUT output. The same logic operation is executed for the rest of the bits of IN1 and IN2 see Figure 6-43).

The result bit will only be 1 when both of the bits in the logic operation are also 1. If one of the two bits of the logic operation has signal state "0", the corresponding result bit is reset.

The number of inputs can be increased by right-clicking on the instruction and choosing- add input. When the instruction is executed, the values of all available input parameters are ANDed. The result is stored in the OUT output.

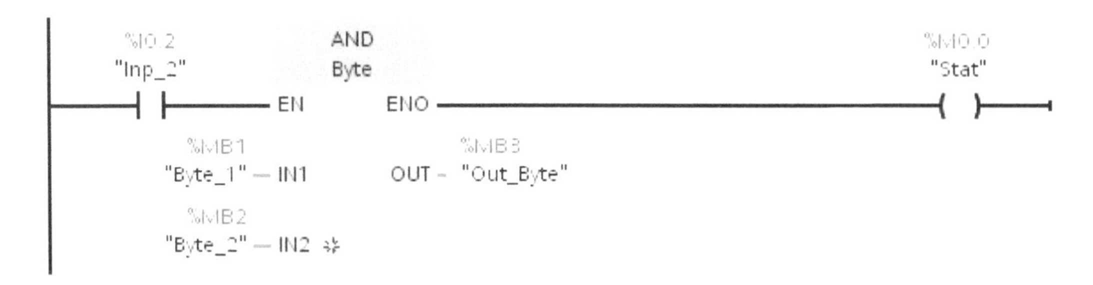

Figure 6-42. Use of an AND instruction.

Val_1	1	0	0	0	0	1	1	0
Val_2	0	0	0	0	0	0	0	0
Output_Val	0	0	0	0	0	0	0	0

Figure 6-43. Table showing the result of an AND operation.

OR Instruction

Bitwise OR instructions are used to compare the bits of two numbers. Figure 6-44 shows how each Source A bit is ORed with each Source B bit.

IN1	IN2	OUT Value
0	0	0
1	0	1
0	1	1
1	1	1

Figure 6-44. Result of an OR instruction on bit states.

The OR instruction can be used to perform an OR on the values at the IN1 input and the value at the IN2 input. The result of the bi-by-bit OR is written to the OUT output.

When the instruction is executed, bit 0 of the value at the IN1 input is ORed with bit 0 of the value at the IN2 input. The result is written in bit 0 of the OUT output. The same logic operation is executed for all bits of the specified tags.

The result bit is true when at least one of the two bits in the logic operation is true. If both of the bits of the logic operation are false, the corresponding result bit is reset.

In this example, the ENO output will be true. If the signal state at the enable input EN is false, the enable output ENO is reset to false.

See Figure 6-45, for an example of how an OR instruction functions. Byte was chosen for the number type in this example. In this example IN1 (Byte_1) and IN2 (Byte_2) are ORed and the result is put in Out_Byte. The result of the instruction execution is shown in Figure 6-46 for this example. Note that if the instruction is enabled at the EN input the ENO output will be true unless an error occurs.

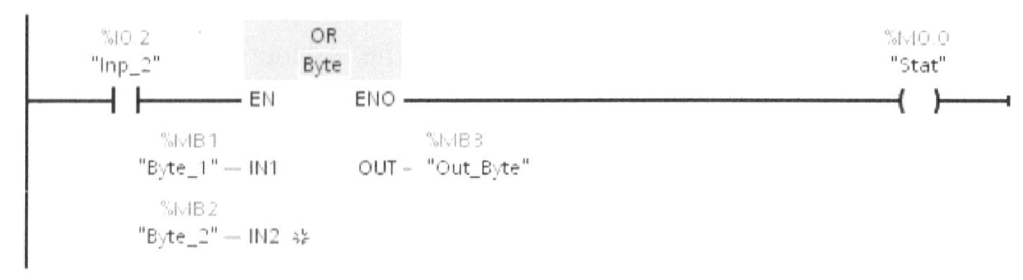

Figure 6-45. OR instruction example.

Byte_1	1	0	0	0	0	1	1	0
Byte_2	0	0	0	0	0	0	1	0
Out_Byte	1	0	0	0	0	1	1	0

Figure 6-46. Truth table for the OR example in Figure 6-36.

The number of inputs can be expanded in the instruction by right-clicking it. The added inputs are numbered in ascending order in the box. When the instruction is executed, the values of all available input parameters are ORed. The result is stored in the OUT output.

Exclusive OR (XOR)

The exclusive OR (XOR) instruction can be used to perform an exclusive OR with the value at the IN1 input and the value at the IN2 input. The result of the bit-by-bit EXCLUSIVE OR operation will be stored in the OUT output. A truth table for the XOR instruction is shown in Figure 6-47.

IN1	IN2	OUT Value
0	0	0
1	0	1
0	1	1
1	1	0

Figure 6-47. Result of an XOR instruction on bit states.

The result bit has the signal state 1 when one of the two bits in the logic operation is true. If both of the bits of the logic operation have signal state 1 or 0, the corresponding result bit is reset.

When the instruction executes, bit 0 of the value at the IN1 input and bit 0 of the value at the IN2 input are exclusively ORed. The result is stored in bit 0 of the OUT output (see Figure 6-48). The same logic operation is executed for the rest of the bits. The number of inputs can be expanded in the instruction box. When the instruction is executed, the values of all available input parameters are combined with exclusively ORed. The result is stored in the OUT output. If the enable input (ENO) is false, the enable output (ENO) is also false.

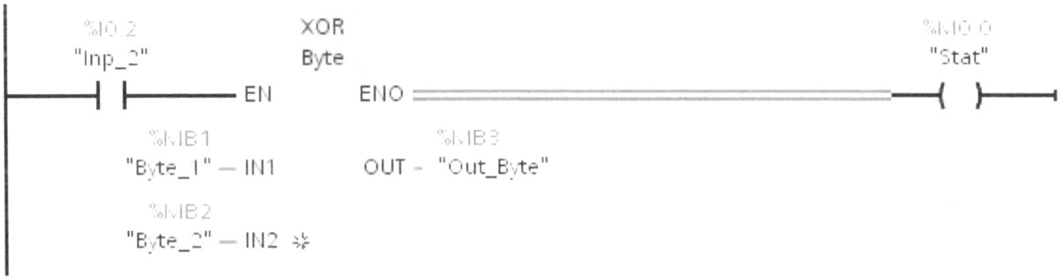

%I0.2 "Inp_2" — EN — XOR Byte — ENO — %M0.0 "Stat"

%MB1 "Byte_1" — IN1 — OUT — %MB3 "Out_Byte"

%MB2 "Byte_2" — IN2

Figure 6-48. Use of an XOR instruction.

Figure 7-49 shows the result of the execution of the XOR instruction in this example.

Byte_1	1	0	0	0	0	1	1	0
Byte_2	0	0	0	0	0	0	1	0
Out_Byte	1	0	0	0	0	1	0	0

Figure 6- 49. Truth table for the XOR example in Figure 6-47.

Trigonometric Instructions

There are a variety of trigonometric instructions available. Trigonometric instructions are used to calculate angles or lengths of side for right triangles. These can be very helpful in applications that require motion. Trigonometry can be used to calculate positions and distances. Figure 6-50 shows some of the trigonometric instructions.

Instruction	Use
SIN	The SIN instruction calculates the tangent of an angle.
COS	The COS instruction calculates the tangent of an angle.
TAN	The TAN instruction calculates the tangent of an angle.
ASIN	The ASIN (Arc sine) instruction calculates the angle that corresponds to the sine value specified at the IN input.
ACOS	The ACOS (Arc cosine) instruction calculates the angle that corresponds to the cosine value specified at the IN input.
ATAN	The ATAN (Arc tangent) instruction calculates the angle that corresponds to the tangent value specified at the IN input.

Figure 6-50. Trigonometric conversion instructions.

SIN Instruction

The sine instruction is used to calculate the sine of an angle. Figure 6-51 shows how the SINE function is applied to a right triangle. The size of the angle is specified in radians at the input IN. The result of the instruction is sent to the OUT output and can be queried there.

Figure 6-51. Sin function.

The instruction is executed if the enable input is true. If the instruction is executed without errors, the ENO output is also true.

The enable output (ENO) is false if: the enable input (EN) is false or the value at the IN input is not a valid floating-point number.

In the example shown in Figure 6-52 if Inp_5 is true, the SIN instruction is executed. The instruction calculates the sine of the angle specified at the Num_1 input and stores the result in the OUT tag named Answer. If the instruction is executed without errors, the ENO output is set.

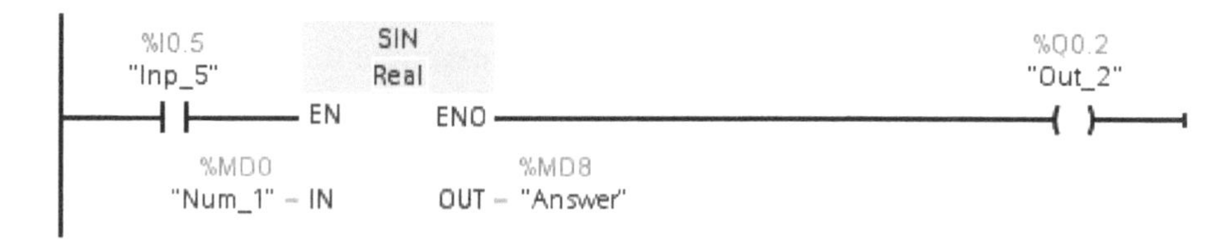

Figure 6-52. Example of a sine instruction.

TAN Instruction

The TAN instruction calculates the tangent of an angle. Figure 6-53 shows how the TAN function is applied to a right triangle. The size of the angle is specified in radians at the IN input. The result of the instruction is stored in the OUT output.

The instruction is only executed if the enable input (EN) is true. If the instruction executes without errors, the enable output (ENO) is also true.

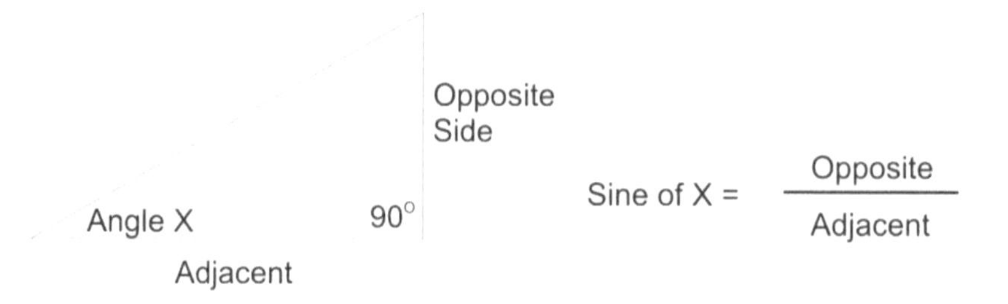

Figure 6-53. Tangent function.

Figure 6-54 shows an example of the use of a TAN instruction. If the enable input (EN) is true, the tangent instruction is executed. The instruction calculates the tangent of the angle specified at the IN input and stores the result in the OUT tag. If the instruction is executed without errors, the ENO output is true.

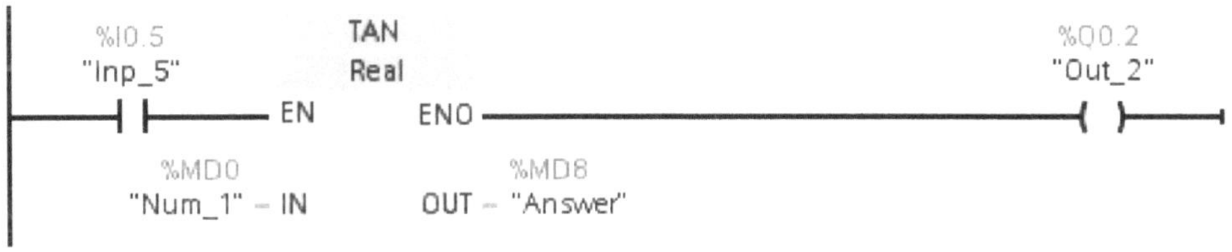

%I0.5
"Inp_5"
TAN
Real
EN ENO
%Q0.2
"Out_2"
%MD0
"Num_1" - IN OUT - "Answer"
%MD8

Figure 6-54. Use of a tangent instruction.

Chapter Questions

1. Explain the terms normalize and scale

2. Why might a programmer use an instruction that would change a number to a different number system?

3. Explain the following logic.

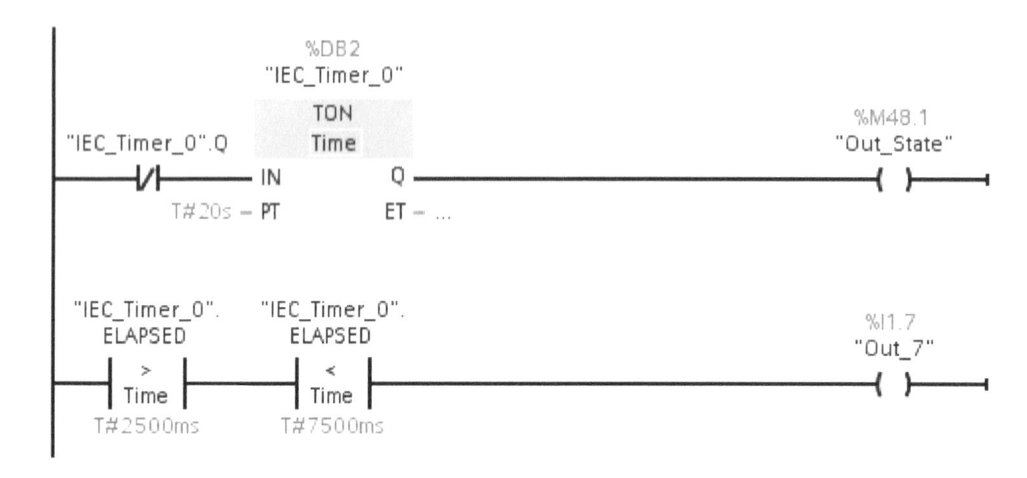

4. Explain the following logic.

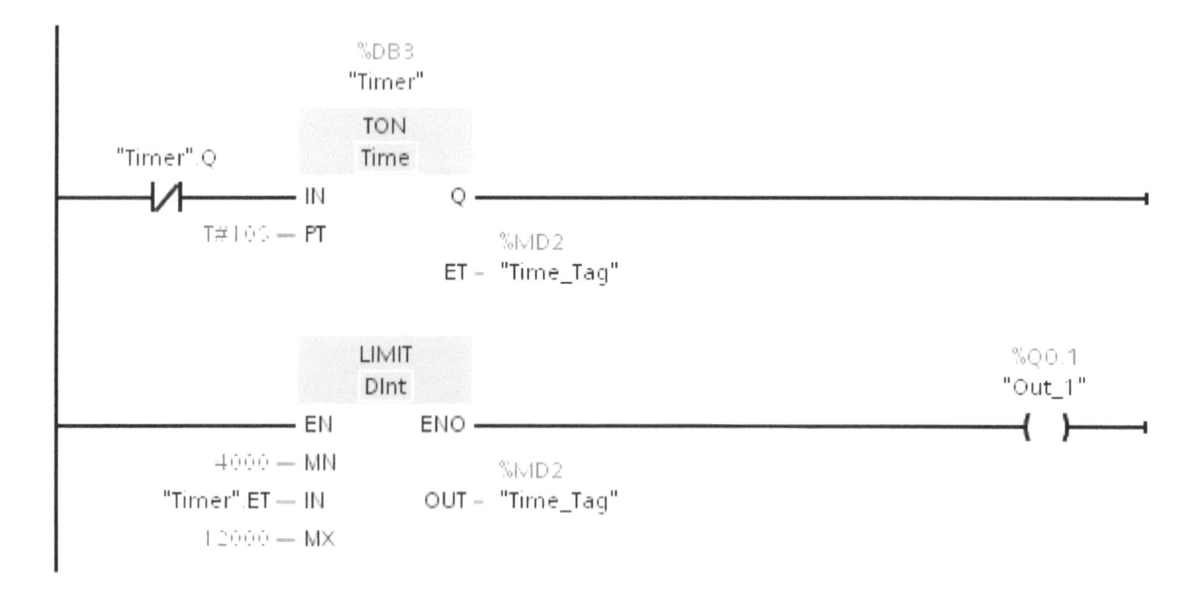

5. Write a rung of ladder logic that would compare two values to see if the first is less than the second. Turn an output on if the statement is true.

6. Write a rung of logic that checks to see if one value is equal to or greater than a second value. Turn on an output if true.

7. Write a rung of logic that checks to see if a value is less than 212 or greater than 200. Turn on the output if the statement is true.

8. Write a rung of logic to check if a value is less than 75 or greater than 100 or equal to 85. Turn on an output if the statement is true.

9. Utilize math instructions and any other instructions that may be helpful to program the following application. A machine makes coffee packs and puts 8 in a package. There is a sensor that senses each pack of coffee as it is produced. It would be desirable to show the number of packs that have been produced and the total number of packages of 8 that have been produced.

10. Write a ladder diagram program to accomplish the following. A tank level must be maintained between two levels. An ultrasonic sensor is used to measure the height of the fluid in the tank. The output from the ultrasonic sensor is 0 to 10 volts. This directly relates to a tank level of 0 to 5 feet. It is desired that the level be maintained between 4.0 and 4.2 feet. Output 1 is the inflow valve. The sensor output is an analog input to an analog input module. Utilize math comparison instructions to write the logic. Calculate the correct analog counts for the instruction. Use the tagnames shown in the table.

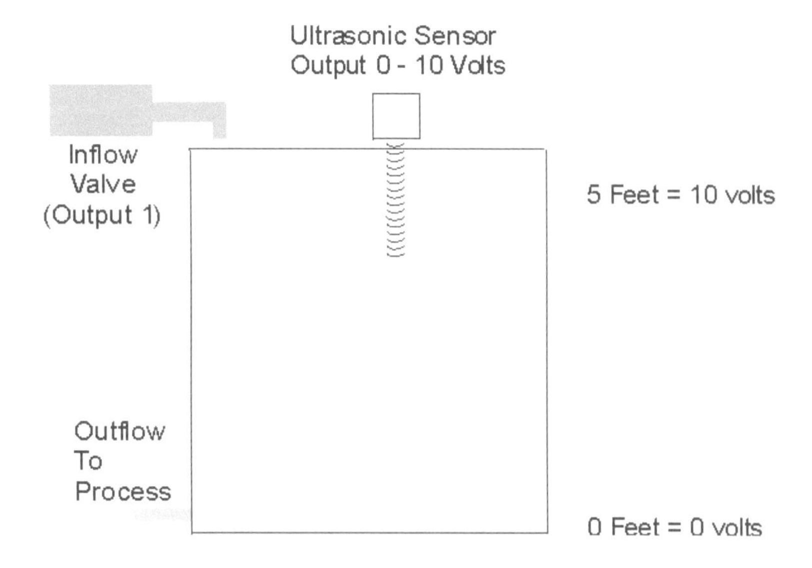

I/O	Tagname	Description	Analog
Ultrasonic Sensor	Level_Sensor	Analog Output (0-10 Volts)	0-32767
Inflow Valve (Output1)	Input_Valve	On or Off	
Start	Start	Momentary Normally-Open Switch	
Stop	Stop	Momentary Normally-Closed Switch	
Run	Run	BOOL	

11. Utilize math statements to convert Fahrenheit temperature to Celsius temperature.

Tc = (5/9)*(Tf-32); Tc = temperature in degrees Celsius, Tf = temperature in degrees Fahrenheit

Utilize math instruction to convert radians to degrees.

12. Write a ladder logic routine for the following application. Utilize comparison instructions.

This is a simple heat treat machine application. The operator places a part in a fixture then pushes the start switch. An inductive heating coil heats the part rapidly to 1500 degrees Fahrenheit. When the temperature reaches 1500 the coil turns off and a valve is opened which sprays water on the part to complete the heat treatment (quench). The operator then removes the part and the sequence can begin again. Note there must be a part present or the sequence should not start.

I/O	Type	Description
Part_Present_Sensor	Discrete	Sensor used to sense a part in the fixture.
Temp_Sensor	Analog	Assume this sensor outputs 0-2000 degrees.
Start_Switch	Discrete	Momentary normally-open switch.
Heating_Coil	Discrete	Discrete output that turns coil on.
Quench_Valve	Discrete	Discrete output that turns quench valve on.

Chapter 7

Special Instructions

Objectives

Upon completion of this chapter, the reader will be able to:

Explain some of the special instructions that are available and their use.

Utilize file instructions.

Utilize sequencer instructions.

Understand the use of special instructions so that new instructions can be quickly learned.

Introduction

Siemens has a wide variety of instructions available to perform special functions. This chapter will sample a few of the many instructions that are available. Once you understand how to use a few of them it will be easy to learn and utilize new ones.

Move and Fill Instructions

There are many reasons why we need to move data in a PLC. Imagine a system where we have different recipes for a particular product. Let's imagine a company that makes several varieties of bread. The company might have 10 different recipes for the bread, there might be different ingredients, times, and temperatures involved for different types of loaves and many other parameters that might vary between different products. One can imagine a manufacturing line that can produce several of the types depending on what is needed at the time. All that would need to change would be the specific parameters for that product. The company could setup the parameters for all of the products in memory. When the operator receives his orders for the day the operator could choose the product to be made on a touchscreen. The PLC program could then perform a move instruction to move the parameters for that product to the PLC program to make the desired product. There are various types of Move instructions available. You can move blocks of values from one memory location to another or fill a block of memory with one value.

Move Value (MOVE) Instruction

A MOVE value instruction can be used to transfer the value of the tag or constant at the IN input to the tag or memory address at the OUT1 output. The transfer is always made in ascending address order.

The tags at IN input and at the OUT1 output must be the same data type. Arrays can be copied as long as the data type of the IN and the OUT1 are the same. Remember that an array is a set of values that must all be the same number type. For example, we could have an array of 10 temperatures that are all real numbers.

The ENO output is true when the EN Input is true. The ENO output is false when the EN Input is false. In the example in Figure 7-1, 210 will be moved to the OUT1 tag named Temp_5.

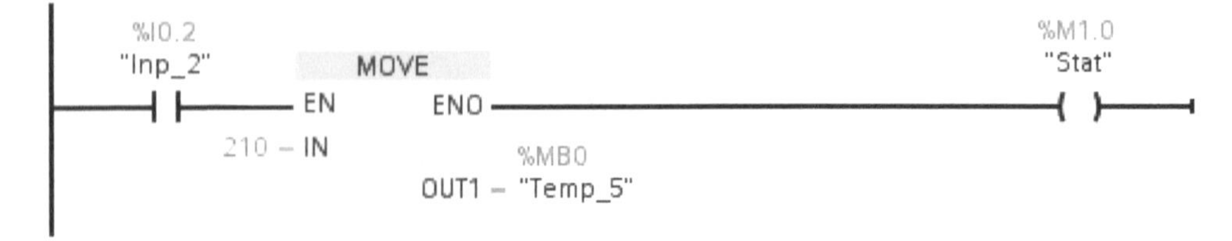

Figure 7-1. A MOVE value instruction copies data from one place in memory to another place in memory.

Additional outputs can be added to this instruction. If additional outputs are added, the constant or the content of the tag or address at the IN input is transferred to all of the outputs.

Move Block (Move_BLK) Instruction

A move block (MOVE_BLK) instruction can be used to copy a block of memory to another block of memory specified by the OUT tag (see Figure 7-2). The COUNT parameter is used to determine how many memory locations should be copied. The width of the elements to be copied is defined by the type of the tag at input IN. In this example 10 values beginning at Recipe_A.Specifications[0] through Recipe_A.Specifications[9] will be copied to 10 OUT tag array elements beginning at Run_Parameters.Specification[0] through Run_Parameters.Specification[9].

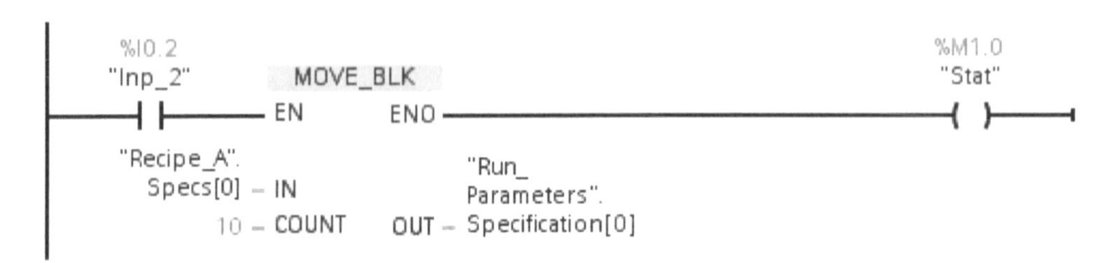

Figure 7-2. A move block (MOVE_BLK) example.

The instruction is executed when the enable input (EN) is true. The ENO output is true when the EN input is true as long as there are no errors during the instruction execution.

If the Move Block instruction tries to copy more data than is available in the memory area at the OUT-output memory the enable output (ENO) will be false.

Move Block Uninterruptible (UMOVE_BLK) Instruction

The move block uninterruptible instruction is used to copy a block of memory to another block of memory specified by the OUT tag (see Figure 7-3). The instruction cannot be interrupted by other operating system activities. The COUNT parameter is used to determine how many memory locations should be copied. The width of the elements to be copied is defined by the type of the tag at input IN. The size of the memory type at the IN input determines the size that must be used for the output (OUT) tag.

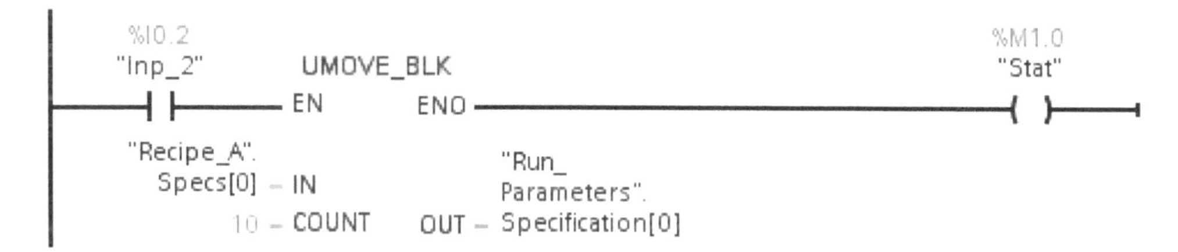

Figure 7-3. Example of a move block uninterruptible (UMOVE_BLK) instruction.

This instruction copies the content of the source area to the destination area in ascending order. For this example, the instruction will copy 10 addresses beginning at Recipe_A_Specs[0] up to Recipe_A_Specs[9]. The alarm reaction times of the CPU actually increase during the execution of the Move block uninterruptible instruction because this instruction is uninterruptible.

The instruction is only executed if the enable input (EN) is true. If no errors occur during execution of the instruction, the ENO output is also true.

The enable output ENO will be false if: the EN input is false or more data is copied than is made available at output OUT.

Fill Block (FILL_BLK) Instruction

The Fill block (FILL_BLK) instruction is used to fill a memory area with the value found at the IN input. The out tags are filled, beginning with the address specified at the OUT output. The number of repeated copy operations is specified with the COUNT parameter. When the instruction is executed, the value at input IN is selected and copied to the OUT tags as many times as specified by the value of the COUNT parameter.

In the example shown in Figure 7-4 count is equal to 5. The IN in this example is a constant value of -99. If Inp_2 is true, -99 will be copied to 5 tags, in order in the OUT-tag array named “Data_block_1”.Product_Spec[5] starting at element 5 (5, 6, 7, 8, and 9 will contain -99 after execution).

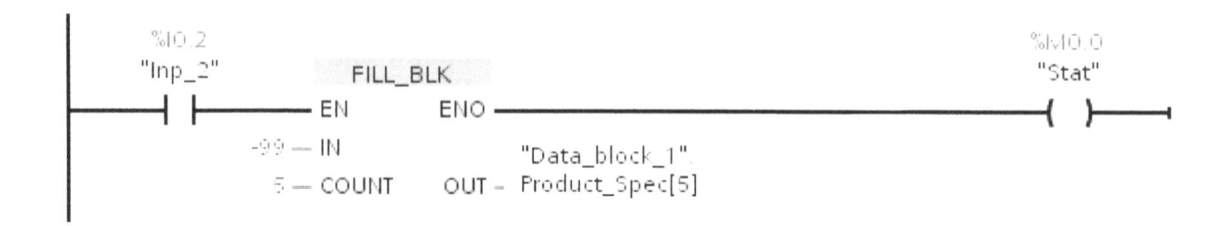

Figure 7-4. Example of the use of a fill block (FILL_BLK) instruction.

The instruction is only executed if enable input (EN) is true. The ENO output will be true if the EN is true, unless an error occurs.

The enable output ENO will be true if: The EN input is false or more data is copied than is available in the memory area at output OUT.

Fill Block Uninterruptible (UFILL_BLK) Instruction

The fill block uninterruptible (UFILL_BLK) instruction can be used to fill a memory area with the value of the IN input, without interruption. The destination area is filled beginning with the address specified at the OUT output. The number of repeated copy operations is specified with the COUNT parameter. When the instruction is executed, the value at input IN is selected and copied to the Out tag[s] as many times as are specified by the value of the COUNT value.

In the example shown in Figure 7-5 count is equal to 5. The IN in this example is a constant value of -99. If Inp_2 is true, -99 will be copied to 5 tags without interruption, in order to the OUT-tag array named "Data_block_1".Product_Spec[5] starting at element 5 (5, 6, 7, 8, and 9 will contain -99 after execution).

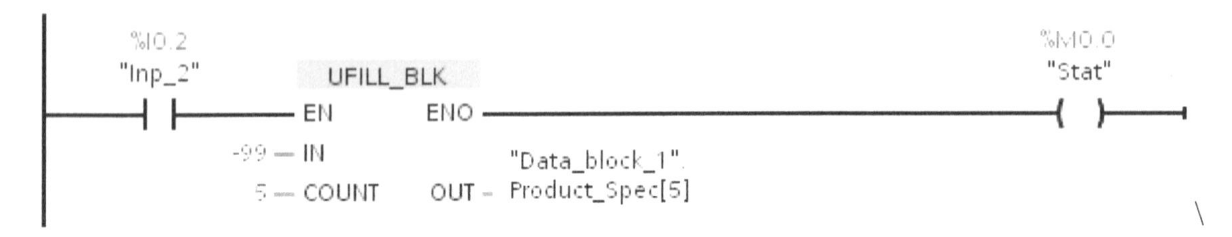

Figure 7-5. Example of a UFILL_BLK uninterruptible instruction.

This instruction cannot be interrupted by other operating system activities. This is why the alarm reaction times of the CPU increase during the execution of the Fill block uninterruptible instruction.

This instruction executes if the enable input (EN) is true. The enable output (ENO) is true if the EN is true unless an error occurs.

The enable output (ENO) is false if: the EN input is false or more data is copied than is available in the memory area at output OUT.

Memory Manipulation Instructions

It is often desirable when programming an application to be able to manipulate memory locations. In other words, be able to move bits and bytes around in a memory location. For example, if we have a very sequential type of application with several processing stations, we might want to track product as they move through the sequence. Moving bits in a memory location would make this possible. There are several instructions available for this.

Swap (SWAP) Instruction

The SWAP instruction can be used to change the order of the bytes within the tag at the IN input and store them in the tag at the OUT output. An example of a SWAP instruction is shown in Figure 7-6. If Inp_2 is true, the instruction will change the order of bytes from the IN input named Inp_Word and store them in the OUT tag named Out_Word.

The SWAP instruction is only executed if the EN (enable) input is true. If the EN input is true, the ENO enable output will also be true.

The ENO enable output is reset when the EN enable input has signal state 0 or errors occur during execution of the instruction.

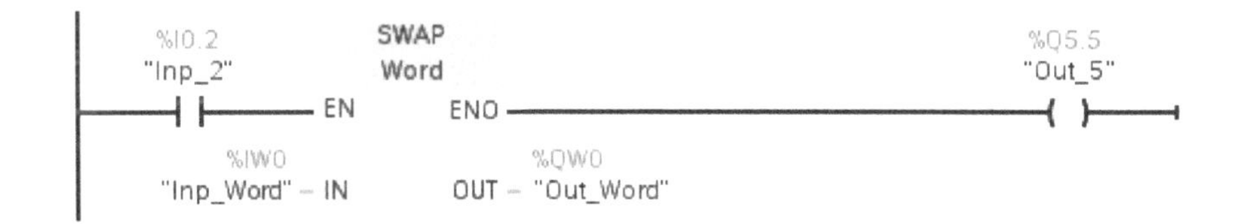

Figure 7-6. Example of a swap (SWAP) instruction

Figure 7-7 shows how the bytes of a DWORD data type tag are swapped using the SWAP instruction.

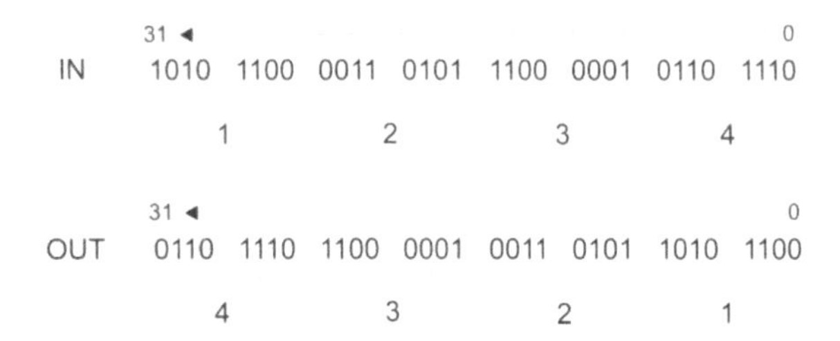

Figure 7-7. Effect of a SWAP instruction on bytes in memory.

Shift and Rotate Instructions

A shift register is a storage location in memory. Assume a tag with 32 bits of data. In other words, 32 bits that are either a 1 or a 0. Each 1 or 0 could be used to represent good or bad parts, presence or absence of parts, or the status of outputs. Bits can be used to represent many industrial processes. Many manufacturing processes are very linear in nature.

Shift registers essentially shift bits through a register to control I/O. Think of a bottling line. There are many processing stations. Each station could be represented by a bit in the shift register. Each station should operate only if there is a part present that requires this station. As the bottles enter the line a one is entered into the first bit. Processing takes place. The stations then release their product and each product moves to the next station. The shift register also increments. Each bit is shifted one position. Processing takes place again. Each time a product enters the system a 1 is placed in the first bit. The 1 follows the part all the way through production to make sure that each station processes it as it moves through the line. There is also a bit shift right (SHR) instruction.

The table in Figure 7-8 shows some shift instructions. There are several shift instructions available.

Instruction	Description
Shift Right (SHR)	The shift right instruction is used to shift the content of the tag at the IN-input bit-by-bit to the right and store the result at the OUT-output.
Shift Left (SHL)	The shift left instruction to shift the content of the tag at the IN-input bit-by-bit to the left and store the result at the OUT-output.
Rotate Right (ROR)	The rotate right instruction to rotate the content of the tag at the IN-input bit-by-bit to the right and store the result at the OUT-output.
Rotate Left (ROL)	The rotate left instruction to rotate the content of the tag at the IN-input bit-by-bit to the left and store the result at the OUT-output.

Figure 7-8. Shift and rotate instructions.

Figure 7-9 shows an example of the use of a bit Shift Right (SHR) instruction in ladder logic.

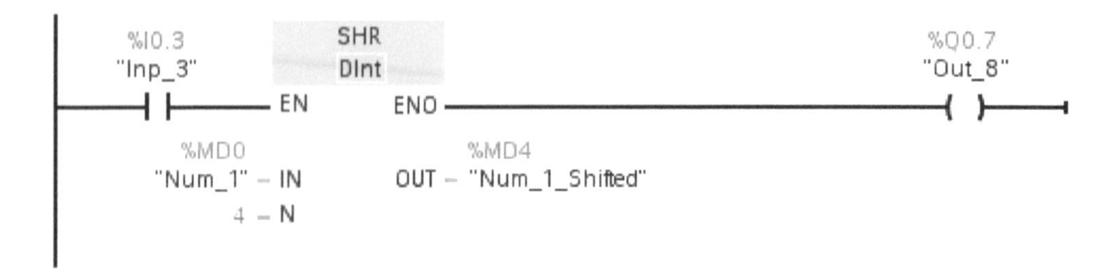

Figure 7-9. A shift right instruction.

The shift right instruction can be used to shift the bits of the tag at the IN-input bit-by-bit to the right and store the result at the OUT tag. You can use the N parameter to specify the number of bit positions by which the specified value should be shifted.

When the N parameter has the value 0, the value at the IN input is copied without change to the tag at the OUT output.

If the value at the N parameter is greater than the number of bit positions, the value at IN input tag is shifted by the available number of bit positions to the right.

The bits in the left area that were shifted are now free and are filled by zeroes when values without signs are shifted. If the specified value has a sign, the free bit positions are filled with the signal state of the sign bit.

The shift right instruction is only executed if the signal state is 1 at the enable input EN. If the EN bit is true, the enable output (ENO) is also true.

If the enable input (EN) is false, the enable output (ENO) is also false.

Figure 7-10 shows how the content of an integer data type tag is shifted four-bit positions to the right. N was 4 in this example.

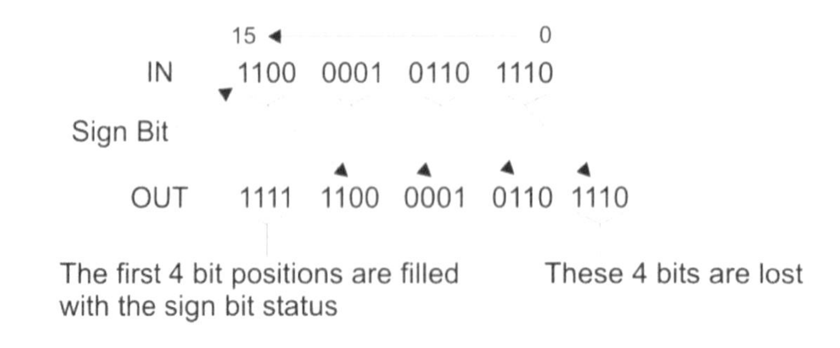

Figure 7-10. Example of a SHR instruction on memory.

Shift Left (SHL) Instruction

There is also an instruction available to shift bits to the left. Figure 7-11 shows a SHL instruction. The Shift Left (SHL) bit instruction can be used to shift the content of the tag at the IN-input bit-by-bit to the left and store the result at the OUT output. You can use the N parameter to specify the number of bit positions by which the specified value should be shifted.

If the value of N is 0, the value at the IN input is copied to the tag at the OUT output.

If the value of the number at input N is greater than the number of bit positions, the tag value at IN input is moved by the available number of bit positions to the left.

The bit positions in the right part of the tag freed by the shift are filled with zeros.

In the example in Figure 7-11 the bits in the value at the IN (Tag_1 would be shifted 3 positions to the left (N = 3) and stored in the OUT (Tag_3).

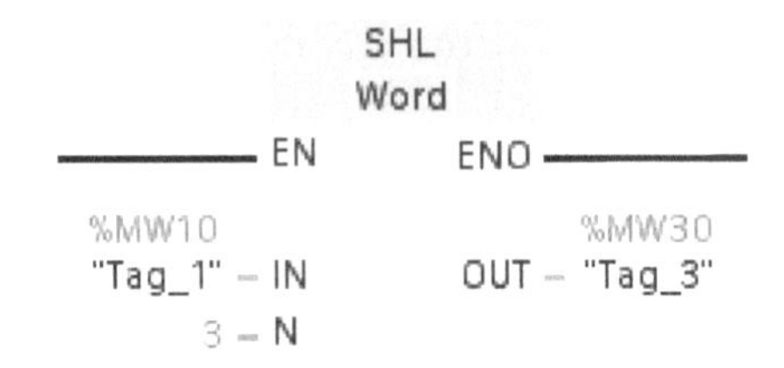

Figure 7-11. Shift left (SHL) instruction.

Rotate Right Instruction (ROR)

The Rotate Right (ROR) instruction rotates the content of the value at the IN-input bit-by-bit to the right and stores the result at the OUT output. A ROR instruction is shown in Figure 7-12. The N parameter is used to specify the number of bit positions by which the specified value is to be rotated. The bit positions freed by rotating are filled with the bit positions that are pushed out.

When the value at the N parameter is 0, the value at the IN input is copied to the operand at the OUT output.

When the value at the N parameter is greater than the number of available bit positions, the operand value at the IN input is nevertheless rotated by the specified number of bit positions.

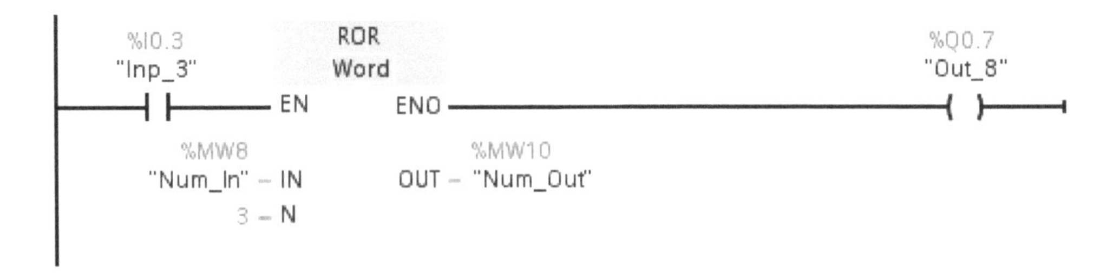

Figure 7-12. An example of a ROR instruction.

Figure 7-13 shows how the content of an operand of DWORD data type is rotated three positions to the right. The bits in the IN double word (32 bits) are shifted 3 bits to the right (N=3 in this example). The three least significant bits that are shifted out of the double word are then moved to the 3 most significant bits to replace the bits that were shifted to the right. Note that the IN word was not changed.

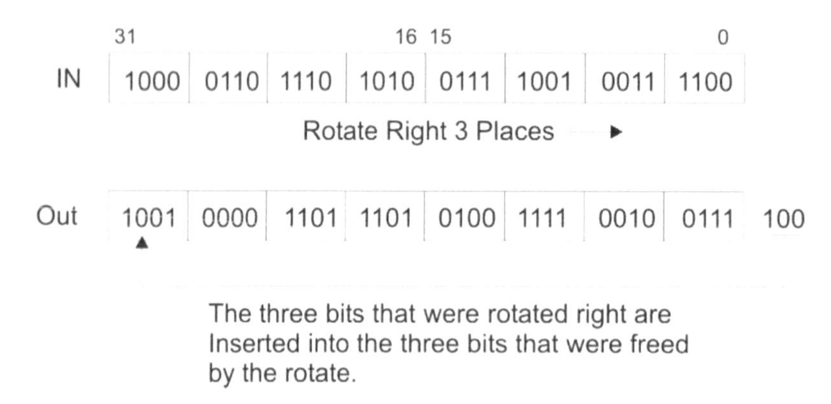

Figure 7-13. Illustration of how a ROR instruction works in memory.

The Rotate Right instruction is only executed if the EN enable input is true. If the EN input is true, the EN output is also true.

If the signal state at the enable input (EN) is false, the enable output (ENO) is also false.

Bit Logic Instructions

Set/Reset Flip/Flop (SR) Instruction

The set reset set flip-flop (SR) instruction can be used to set or reset the bit of a specified operand based on the signal state of inputs S and R1 (see Figure 7-14). If the signal state at input S is 1 and is 0 at input R1, the specified operand is set to 1. If the signal state at input S is 0 and is 1 at input R1, the specified operand is reset to 0.

Input R1 takes priority over input S. If the signal state is 1 at the two inputs S and R1, the signal state of the specified operand is reset to 0.

The instruction is not executed if the signal state at the two inputs R1 and S is 0. The signal state of the operand then remains unchanged.

The operands Output_Bit and Out_5 are set when the following conditions are fulfilled:

- Sensor_5 (Input S) is true.
- Sensor_6 (Input R1) is false

The operands Output_Bit and Out_5 (Q bit) are reset when one of the following conditions is fulfilled:

- Sensor_5 (Input S) is false and the operand Sensor_6 (Input R1) is true.
- Sensor_5 (Input S) and Sensor_6 (Input R1) are both true.

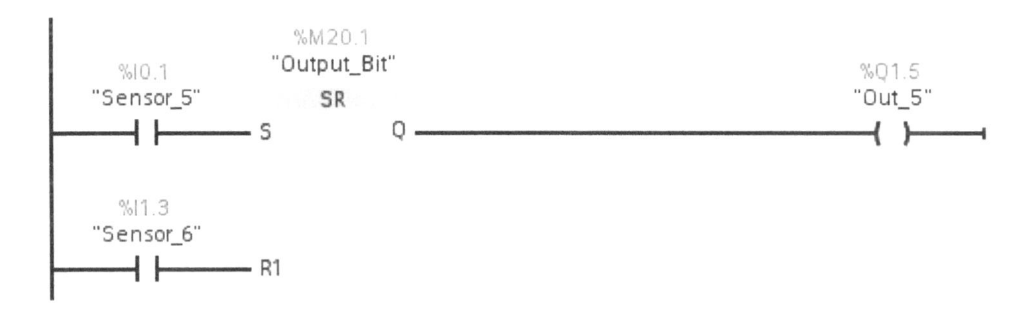

Figure 7-14. Example of a set/reset flip/flop (SR) instruction.

Reset/Set (RS) Flip-Flop Instruction

The reset/set flip-flop instruction can be used to reset or set based on the signal state of the inputs at R and S1 (see Figure 7-15). If the signal state at input R is 1 and is 0 at input S1, the specified operand is reset to 0. If the signal state at input R is 0 and is 1 at input S1, the specified operand is set to 1.

Input S1 takes priority over input R. If the signal state is 1 at inputs R and S1, the signal state of the specified operand is set to 1.

If the signal state at the two inputs R and S1 is 0, the instruction is not executed. The state of the output (Q) then remains unchanged.

Bit memory M 0.0 (Output_Bit) and output Q 0.0 (Output_5) are reset when the following conditions are fulfilled:

- Inp_1 is true.
- Inp_2 is false.

Tags Output_Bit and Output_5 are set when the following conditions are fulfilled:

- Input Inp_1 is false and Inp_2 is true.
- Both inputs Inp_1 and Inp_2 are true.

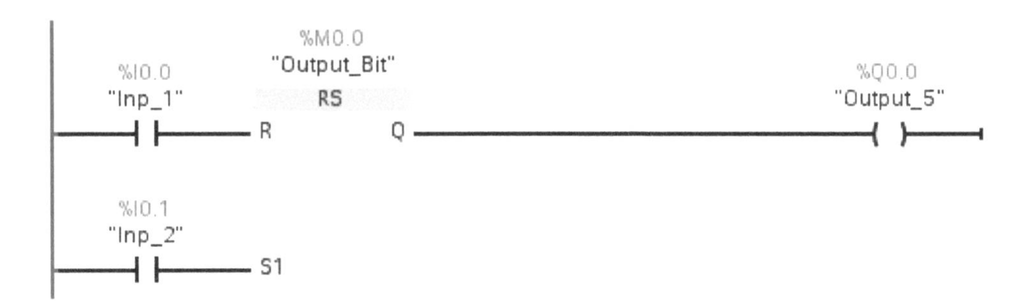

Figure 7-15. Example of a set/reset (RS) instruction.

Set Bit Field (SET_BF)

Figure 7-16 shows an example of the use of set and reset bit field instructions. You can use the Set bit field instruction (SET_BF) to set several bits beginning with a specified address (Tag_1 in this example). The number of bits to be set determines the value of Operand 2: 4 in this example. The address of the first bit to be set is defined by Operand1 (Tag_1 in this example). If the value of operand 2 is greater than the number of bits in a selected byte, the bits of the next byte are set. The bits remain set until they are explicitly reset, for example, by another instruction.

The instruction is executed if enable input is true. If the input is false, the instruction is not executed.

Reset Bit Field (RESET_BF)

Figure 7-16 shows an example of the use of set and reset bit field instructions. The reset bit field (RESET_BF) instruction to reset several bits beginning with a specified address (Tag_1 in this example). The number of bits to be reset are determined by the value of Operand 2 (5 in this example). The address of the first bit to be reset is defined by Operand 1 (Tag_1 in this example). If the value of Operand 2 is greater than the number of bits in a selected byte, the bits of the next byte are reset. The bits remain reset until they are explicitly set, for example, by another operation.

The instruction is only executed if the enable input is true. If the enable input is false, the instruction is not affected.

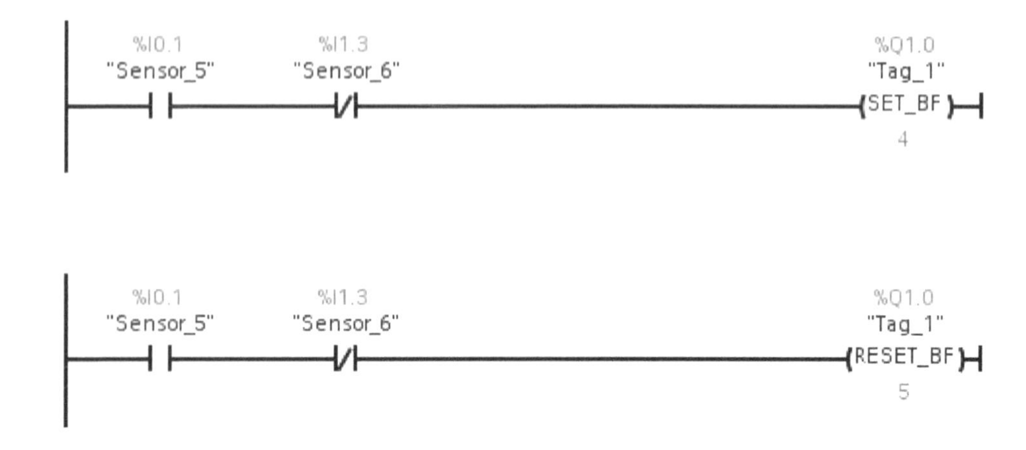

Figure 7-16. Use of SET_BF (set bit field) and RESET_BF (reset bit field) instruction.

Logical Instructions

Select (SEL) Instruction

The select instruction selects one of the inputs IN0 or IN1 depending on a switch (parameter G) and copies its content to the OUT output (see Figure 7-17). If parameter G has signal state 0, the value at input IN0 is copied. When the G parameter has the signal status 1, the value at the IN1 input is copied to the OUT output.

This can be a very useful instruction. An operator could choose which of 2 values should be used, for example, on a touch screen that would change control the value of input G. Or depending on what is happening in an application the desired value could be used based on the value at input G.

The instruction is only executed if the signal state is 1 at the enable input EN. If no error occurs during execution, the ENO output also has signal state 1.

The ENO enable output is false when the EN enable input is false or errors occur during execution of the instruction.

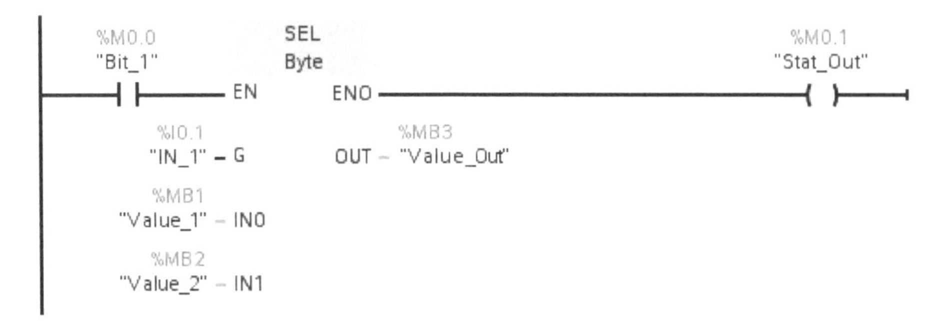

Figure 7-17. Example of a select (SEL) instruction.

Multiplex (MUL)

The multiplex instruction can be used to copy the content of a selected input to the OUT output (see Figure 7-18). The number of selectable inputs of the MUX box can be expanded. Inputs are automatically numbered in the box. Numbering starts at IN0 and is incremented continuously with each new input. The K parameter can be used to determine the input whose content should be copied to the OUT output. If the value of the K parameter is greater than the number of available inputs, the content of the ELSE parameter is copied to the OUT output and the enable output ENO is assigned signal state 0.

In this example if the value in in the K input (NUM) is 2 the value of Val_3 at IN2 would be copied into the tag at the OUT output (Out_Val in this example). If the value of K is larger than the number of inputs the value at the ELSE input would be copied into the OUT tag.

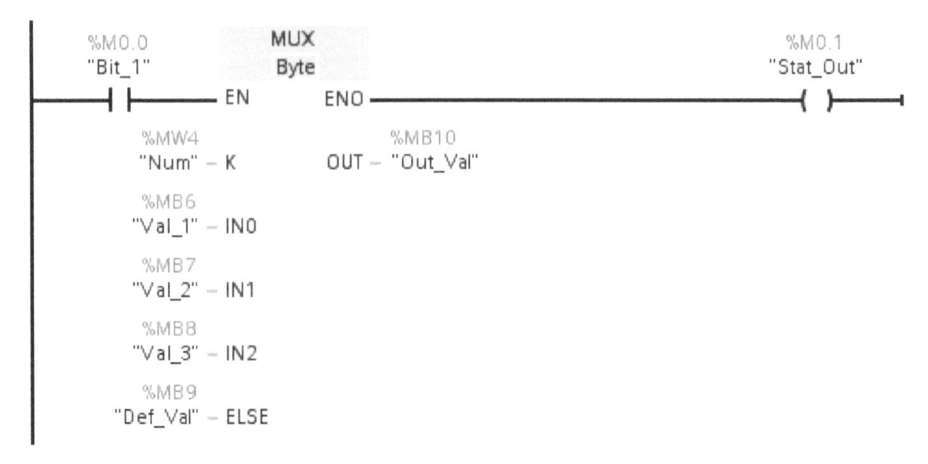

Figure 7-18. Example of a multiplex (MUL) instruction.

In a multiplex instruction the tags at all inputs (other than the K input) and at the OUT output must be the same data type. The K parameter is an exception, since only integers can be specified for it.

The instruction is only executed if the enable input (EN) is true. If no error occurs during execution, the ENO output is also true.

The enable output ENO is reset if: the enable input EN is false; the value of K is greater than the number of available inputs; or errors occurred during the execution of the instruction.

Encode (ENCO) Instruction

The encode instruction can be used to determine the bit number of the least significant bit that is a 1 in the input value and to send it to the OUT output (see Figure 7-19).

The encode instruction is executed when the enable input (EN) is true. If there are no errors during the execution of the instruction, the enable output (ENO) output will be true.

If the enable input (EN) is false, the enable output (ENO) will also be false.

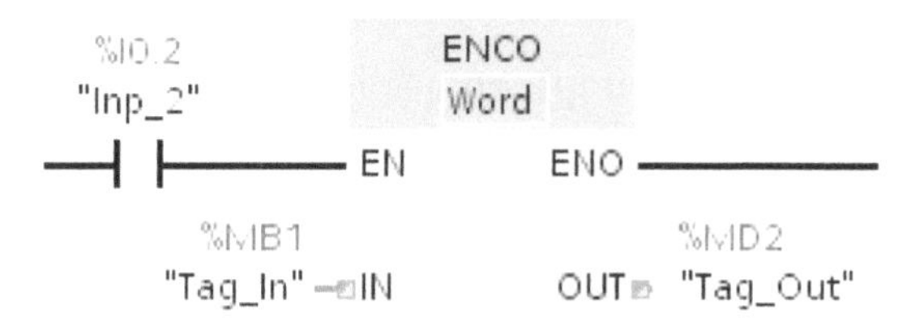

Figure 7-19. An ENCO instruction.

If the IN input (Tag_In) is true, the encode instruction will be executed. Study the example in Figure 7-20. The instruction determines the least significant bit of the number in the IN input (Tag_In in this example) input and writes bit position "4" to the tag in the Tag_Out output tag.

If the instruction executes without errors, the enable output (ENO) will be true.

Figure 7-20. How the ENCO instruction operates in the example.

Decode (DECO) Instruction

The decode instruction can be used to set a bit in the output tag as determined by the input tag value.

Study the example in Figure 7-21. The decode instruction can be used to read the value at the IN input (Tag_In in this example) and set the bit in the OUT-output tag (Tag_Out in this example) whose bit position corresponds to the read value from the IN input. The other bits in the output tag will be overwritten with zeroes. When the value at the IN input is greater than 31, a modulo 32 instruction is executed on the IN value.

The decode instruction is only started when the signal state at the EN enable input is true. If no errors occur during execution, the ENO output also has the signal state true. If the enable input (EN) is false, the enable output (ENO) will also be false.

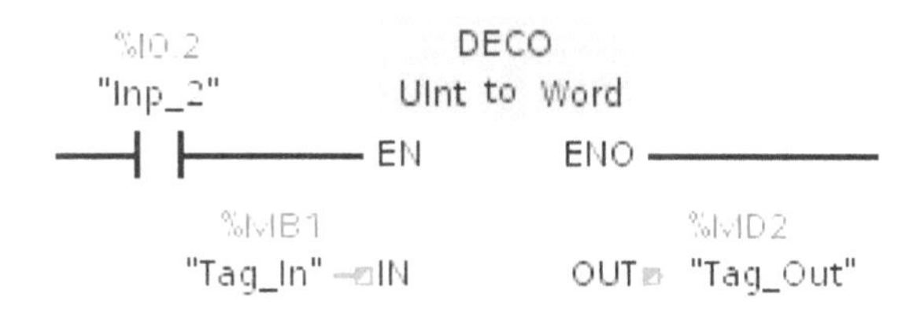

Figure 7-21. An example of a DECO instruction.

If the IN input (Tag_In in this example) is true, the decode instruction will be executed. Study the example in Figure 7-22. The instruction reads bit number 4 from the value at the input named Tag_In and sets the fourth bit in the value at the output named Tag_Out.

If the instruction executes without an error, the enable output (ENO) output will be true.

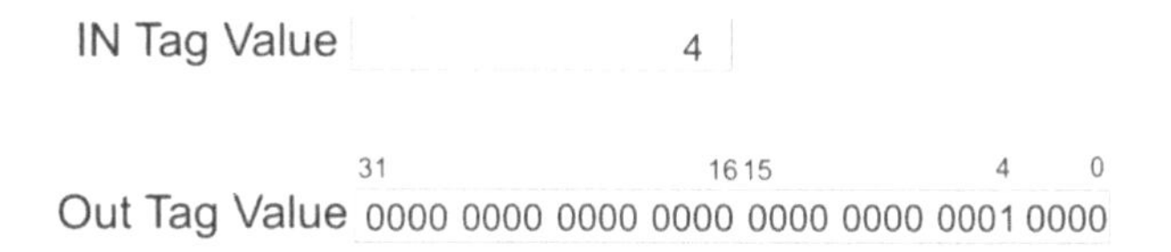

Figure 7-22. How the DECO instruction operates in this example.

Demultiplex (DEMUX) Instruction

The demultiplex instruction can be used to copy the content of the IN input to a selected output (see Figure 7-23). The number of selectable outputs can be expanded in the instruction box. The outputs are automatically numbered in the box. Numbering starts at OUT0 and continues consecutively with each new input. You use the K parameter to define the output to which the content of the IN input is to be copied. The other outputs are not changed. If the value of the K parameter is greater than the number of available outputs, then the content of the IN input will be copied to the ELSE parameter and the signal state false is stored to the enable output.

The tag types for a demultiplex instruction in the IN input and in all outputs must have the same data type except for the K input. The K input is an exception, since only integers can be specified for it. If the value at K is greater than the number of outputs the value from the IN input will be stored in the ELSE output. The instruction is executed if the enable input (EN) is true. If no errors occur during execution, the enable output (ENO) is also true.

The ENO enable output is reset if one of the following conditions is fulfilled: the EN enable input is false; the value of the K parameter is greater than the number of outputs; errors occur during execution of the instruction.

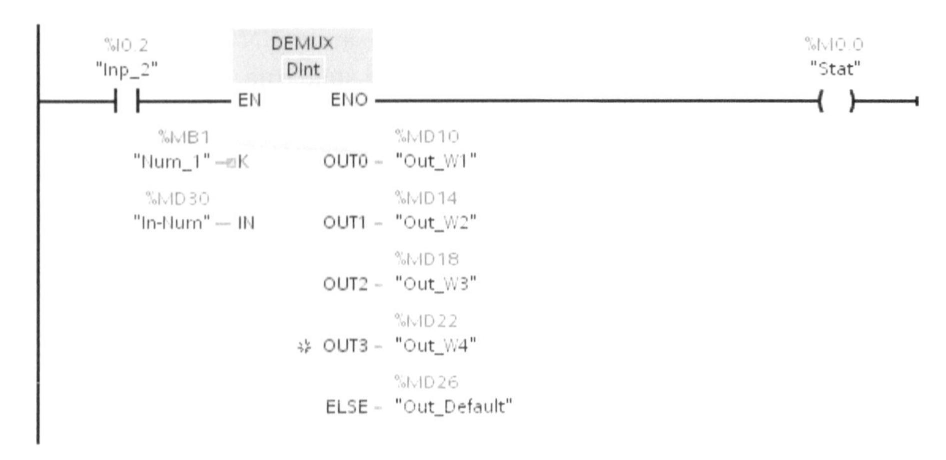

Figure 7-23. Example of the use of a demultiplex instruction.

Extended Instructions

System Functions (SFCs) and System Function Blocks (SFBs)

Some functions that are commonly used are integrated into the operating system of the S7-CPUs and can be called from there. Some of these functions are, for example, clock and calendar, string and character, program control, communications, interrupts, PID, and motion control SFBs and SFCs (see Figure 7-24). They are also called extended instructions.

In Figure 7-24 the Date and time-of-day extended instructions tab was expanded to show the available extended instructions under clock and calendar.

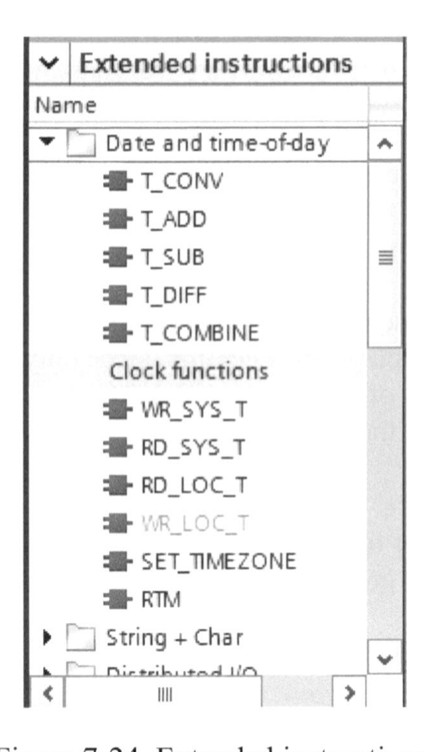

Figure 7-24. Extended instructions.

String and character instructions are shown in Figure 7-25. These instructions are used to work with character and string information. For example, if we wanted a user to enter their user name and password, we could check them against a table of authorized users with string and character instructions.

String and Character Instructions

Instruction	Description
S_MOVE	This instruction is used to write the content of a character string (W)STRING from parameter IN to the data area that you specify at parameter OUT.
S_CONV	Using STRG_VAL, you convert a character string to a numeric value.
STRG_VAL	Using VAL_STRG, you convert a numeric value to a character string.
VAL_STRG	You can use the "LEN" instruction to query the current length of the string specified at the IN input and output it as a numerical value at the OUT output.
Strg_TO_Chars	This instruction converts an ASCII string into an array of CHAR/BYTE starting at a given position.
Chars_TO_Strg	This instruction converts an array of CHAR/BYTE into a ASCII string starting at a given array element.
MAX_LEN	This instruction is used to determine the length of a character string
ATH	The "ATH" instruction is used to convert the ASCII character string specified at the IN-input parameter into a hexadecimal number. The result of the conversion is output to the OUT-output parameter.
HTA	The instruction "HTA" instruction is used to convert the hexadecimal number specified at the IN input into an ASCII character string. The result of the
LEN	CONCAT joins the string parameters IN1 and IN2 to form one string provided at OUT.
CONCAT	CONCAT joins the string parameters IN1 and IN2 to form one string provided at OUT.
LEFT	You can use LEFT to extract a partial string beginning with the first character of the string at the IN input.
RIGHT	You can use RIGHT to extract a partial string beginning with the last character of the string at the IN input.
MID	You can use MID to extract a portion of the string at the IN input.
DELETE	You can use DELETE to delete a portion of the string at the IN input.
INSERT	Inserts string 2 into string 1 beginning at character position and stores the result in the string destination.
REPLACE	You can use REPLACE to replace a string at the IN1 input with the string at the IN2 input.
FIND	You can use FIND to search through the string at the IN1 input to locate a specific character or a specific string of characters.

Figure 7-25. String and character functions.

There are also clock and calendar instructions available. Figure 7-26 shows clock and calendar instructions. These could be used to track times when events occurred, and so on.

Clock and Calendar Instructions

Figure 7-27 shows clock and calendar instructions and a description of each.

Instruction	Description
T_CONV	Using T_CONV, you convert the value at the IN input to the data format specified at the OUT output.
T_ADD	Using T_ADD, you add the time at the IN1 input to the time at the IN2 input.
T_SUB	Using T_SUB, you subtract the time at the IN2 input from the time at the IN1 input.
T_DIFF	Using T_DIFF, you subtract the time at the IN2 input from the time at the IN1 input.
T_COMBINE	This instruction combines the value of a date with the value of a time and converts this into a combined date and time value.
WR_SYS_T	You can use WR_SYS_T to set the date and time of the CPU clock.
RD_SYS_T	You can use RD_SYS_T to read the current date and current time of the CPU clock.
RD_LOC_T	You can use RD_LOC_T to read the current local time from the CPU clock and output this in DTL format at the OUT output.
WR_LOC_T	The instruction "WR_LOC_T" is used to set the date and time of the CPU clock.
SET_TIMEZONE	The instruction "SET_TIMEZONE" is used to set the parameter for the local time zone and the daylight saving / standard time changeover.
RTM	The RTM instruction can be used to set, start, stop, and read out a 32-bit operating hours counter of your CPU.

Figure 7-26. Clock and calendar instructions.

There is a variable called Time_Of_Day (TOD) that holds the number of milliseconds since the beginning of the day (0:00 o'clock) in the form of an unsigned integer.

Program Control/Runtime Control

Figure 7-27 shows program control instructions. These can be used to start and stop the CPU execution, and to query errors in a block.

Instruction	Description
ENDIS_PW	The "Limit and enable password legitimation" instruction is used to specify whether configured passwords may be legitimated or not for the CPU.
RE_TRIGR	With RE_TRIGR, you restart the CPU cycle monitoring.
STP	With STP, you change the CPU to STOP mode and therefore terminate program execution.
GetError	With the "GetError" instruction, you can query the occurrence of errors within a block.
GetErrorID	With the "GetErrorID" instruction, you can query the occurrence of errors within a block.
INIT_RD	The "Initialize all retain data" instruction can be used to reset the retain data of all data blocks, bit memories and SIMATIC timers and counters at the same time.
WAIT	The "Configure time delay" instruction pauses the program execution for a specific period of time.
RUNTIME	The "Measure program runtime" instruction is used to measure the runtime of the entire program, individual blocks or command sequences.

Figure 7-27. Program control instructions.

Some of the available communications instructions are shown in Figure 7-28. These can be used to send and receive data.

Instruction	Description
TSEND_C	TSEND_C is asynchronous and has the following functions: Setting up and establishing a communications connection: Sending data via an existing communications connection: Terminating the communications connection:
TRCV_C	TRCV_C is asynchronous and has the following functions: Setting up and establishing a communications connection: Receiving data via an existing communications connection: Terminating the communications connection:
TCON	With TCON, you set up an Ethernet communications connection and establish the connection. After the connection is set up and established, it is automatically maintained and monitored by the CPU. TCON is asynchronous.
TDISCON	Using TDISCON, you terminate an Ethernet communications connection.
TSEND	With TSEND, you send data over an existing communications connection.
TRCV	With TRCV, you receive data over an existing communications connection.
PORT_CFG	PORT_CFG allows dynamic configuration of communications parameters for a point-to-point communications port.
SEND_CFG	SEND_CFG allows dynamic configuration of serial transmission parameters for a point-to-point communications port.
RCV_CFG	RCV_CFG allows dynamic configuration of serial receive parameters for a point-to-point communications port.
SEND_PTP	With SEND_PTP, you start the transmission of data. The "SEND_PTP" instruction does not execute the actual transmission of the data. The data of the transmit buffer is transferred to the relevant communications partner. The communications partner handles the actual transmission.
RCV_PTP	With RCV_PTP, you enable receipt of a sent message. Each message must be enabled individually. The sent data is only available in the receive area when the message has been acknowledged by the relevant communications partner.
RCV_RST	With RCV_RST, you delete the receive buffer of a communications partner.
SGN_GET	With SGN_GET, you query the current state of several signals of an RS-232 communications module.
SGN_SET	With SGN_SET (set RS-232 signals), you set the status of the output signals of an RS-232 communications module.

Figure 7-28. Communication instructions.

Saving Process Values (Recipe and Data Logging)

Recipe functions can be used to import recipes or export recipes. The instructions are RecipeImport and RecipeExport.

Data logging instructions can be used to save process values to data logs. Data logs can be saved in the internal load memory or on a SD (Secure Digital) memory card (see Figure 7-29). The data logs are saved in Comma Separated Value (CSV) format.

There are data logging instructions to create or open a data log, to write an entry and to close the data log file.

The programmer chooses which values are logged when the data buffer is created. The data buffer is used as a memory location for new data log entries. New values must be written to the buffer before the DataLogWrite instruction is executed. During the execution of the DataLogWrite instruction, data is written from the buffer into a data log record.

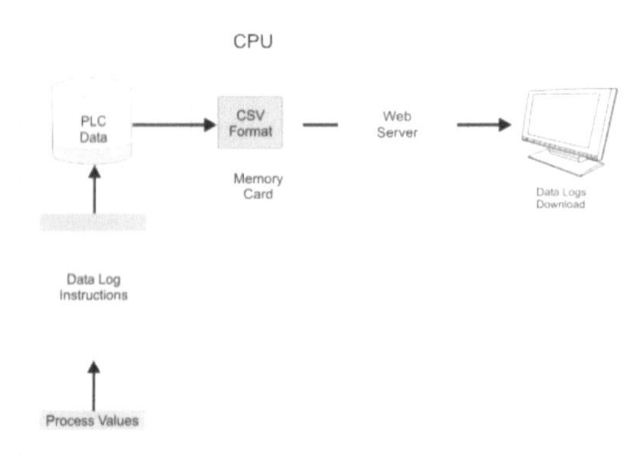

Figure 7-29. Saving process values.

Data log files can be copied to the PC as follows:

A web browser can be used to access the data logs via the Web server if the PLC PROFINET interface is connected to a PC. The CPU can be in RUN or STOP mode for accessing the data logs. If the CPU is in RUN mode, the program will continue to run while the web server is transferring the data.

If there is a memory card in the S7-1200 CPU, you can remove this card and insert it into a standard slot for Secure Digital (SD) card or MulitMediaCard (MMC) on a PC. Use File Manager to transfer the data log files from the memory card to the PC. The CPU switches to STOP mode when the memory card is removed.

New data records are added until the maximum number of data records is reached (RECORD parameter). The next data record then overwrites the oldest data record in the data log. If you want to protect data records from being overwritten, the DataLogNewFile instruction can be used to create a new data log file based on the current data log. New data records are then written into the new data log.

Copying Data Log Files

The memory card in the S7-1200 CPU can be used to transfer the log files to a computer. When the memory card is removed from the PLC the CPU will go into STOP mode.

If the PROFINET interface is connected to the PC, a web browser can be used to access the data logs. The CPU can be in RUN or STOP mode for this. If the CPU is in RUN mode, the program continues running while the Web server is transferring data.

Data Log

New data records are added until the maximum number of data records is reached (RECORD parameter). The next data record then overwrites the oldest data record of the data log.

The DataLogNewFile instruction can be used to prevent data records from being overwritten, by creating a new data log file based on the current data log. New data records are then written into the new data log.

Creating a Data Log

The DataLogCreate instruction is used to create a new data log file in the \DataLogs directory of the CPU load memory. The name assigned at the NAME parameter is the designation for the data log and is also used the file name for the CSV file. The file is stored in the directory DataLogs.

The block parameter DATA determines the data buffer for the new data log object, the columns, and data types in the data log. The columns and data types of a data record in the data log are generated from the elements of the structure declaration or the array declaration of this data buffer. Each element of a structure or of an array corresponds to a column in a line in the data log.

The HEADER block parameter can be used to assign a header text in the header to each column. The DataLogCreate instruction returns an ID that references the created data log and is used by the other data logging instructions to reference the created data log.

Writing to Data Logs

As a prerequisite to writing a data record to a data log, a data log must be open The DataLogOpen instruction is used to open a data log. The DataLogWrite instruction writes a data record to the data log.

DataLogClose is used to close the data file. DataLogNewFile is used to log data in a new file. The "DataLogNewFile" instruction is used to create a new data log with the same properties as an existing data log. This enables the contents of an existing data log to be retained.

This chapter has covered a small sample of the many special instructions that are available. If you understand and use the instructions in this chapter, you will be able to learn new instructions very easily. If you have a special need in an application, the odds are good that an instruction is available to meet the need. The instruction help file in TIA Portal is a great source of available instructions and their use.

Chapter Questions

1. What instruction could be used to fill a range of memory with the same number?

2. What instruction could be used to move an integer in memory to an output module?

3. What are rotate and shift instructions used for and how do they function?

4. Explain an RS instruction.

5. What does a SEL instruction do?

6. What does a MUL instruction do?

7. What is a DEMUX instruction?

8. How can process values be saved?

Chapter 8

Function Block Diagram Programming (FBD)

Objectives

After completing this chapter, the reader will be able to:

Explain the types of applications that FBD is well suited to.

Develop FBD logic.

Introduction

Function block diagram (FBD) programming is one of the IEC-61131 languages. IEC-61131 is an international standard that specifies PLC-type languages. FBD is a powerful and friendly language once you have learned the basics. An instruction in an FBD can take one or more inputs, make decisions or calculations and then generate one or more outputs. An instruction within an FBD can also provide outputs to one or more instructions' inputs. There are many types of instructions available to perform different tasks. Function block programming can simplify programs and make them more understandable.

FBD programming is very useful for applications where a lot of information/data flow occurs. Process control is a good example of an application where there are a lot of data calculations. Function block diagrams can be broken into multiple networks. Networks are like separate pages of a program (see Figure 8-1). This helps you organize your program and make it easier to understand.

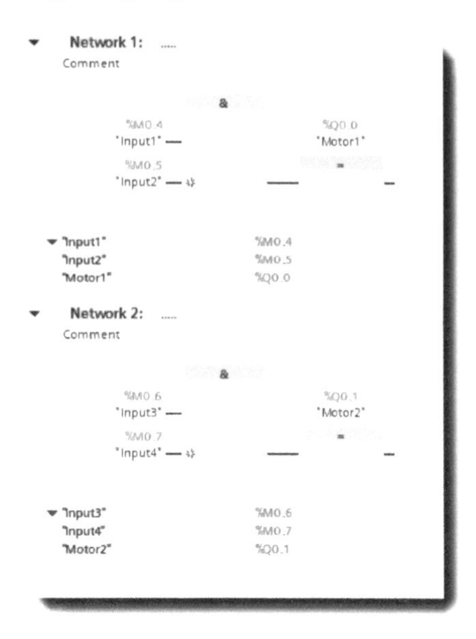

Figure 8-1. Multiple networks in a function block diagram.

When a function block diagram executes, all networks execute. It is a good idea to use one network for each device that is to be programmed. In Figure 8-1 there are two networks of FBD programming. In this example each network controls one device. Note that this is about as simple as it gets. Normally there might be multiple blocks within a network to perform different, but related tasks.

Language Selection

In order to program in Function Block Diagram (FBD) you must first switch the language setting of an OB. In Figure 8-2 under the language selection pull-down, we can choose between ladder logic (LAD) and FBD. By changing the language style from LAD to FBD, any logic already within the OB will be converted to FBD style. Figure 8-3 gives an example of LAD converted to FBD.

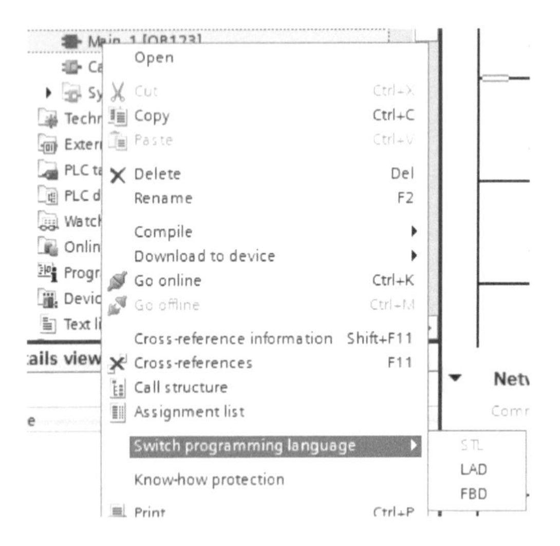

Figure 8-2. FBD language selection (general tab under OB properties).

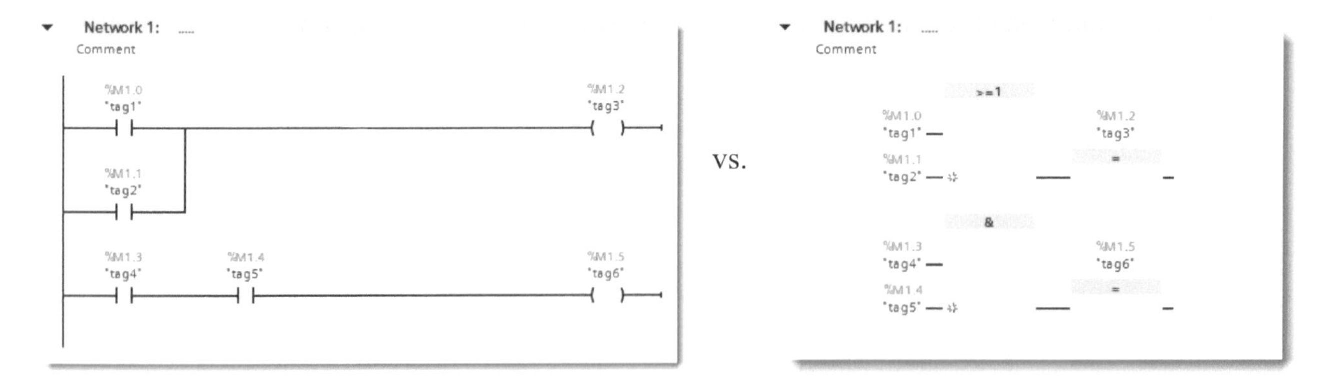

Figure 8-3. Ladder vs. function block diagram.

In Figure 8-3 we can see the exact same logic represented in both Ladder Logic and Function Block Diagram. In this example, the existing logic was written in the language style LAD. By changing the language style to FBD, logic remains the same, however, how it is represented is altered. Yet we still have the same logical result of tag1 or tag2 being true makes tag3 true. In addition, tag4 and tag5 both need to be true for tag6 to be true. Notice the difference in the output assignment instructions in each example.

In LAD programming the typical coil symbol is used for outputs. This symbol is known as an Assignment instruction in FBD. In both LAD and FBD the Assignment instruction works the same. In FBD the assignment instruction is represented with a box containing an equal symbol. Tags assigned to the Assignment instruction evaluate true as long as the logic leading up to the assignment instruction is true. With the differences in logic representation, one language may be more desirable to program with than the other. FBD is very useful for process control where the flow of data is easily represented. LAD is useful for complex bit relational decisions such as logical and/or conditions.

Working with Instructions

Once you've changed the style of an OB from LAD to FBD you can start inserting instructions into the program. When you insert an instruction into an FBD, depending on the type of instruction, a set of tags may need to be created. Remember Functions (FCs) do not have memory assigned to them so no tags are needed. Function Blocks (FBs); however, have a Data Block (DB) associated with them. Tags from Data Blocks are available to the programmer to use anywhere in logic. Figure 8-4 shows the comparison of an ADD instruction and timer TON instruction. Note the timers default name for its data block is IEC_Timer_0_DB.

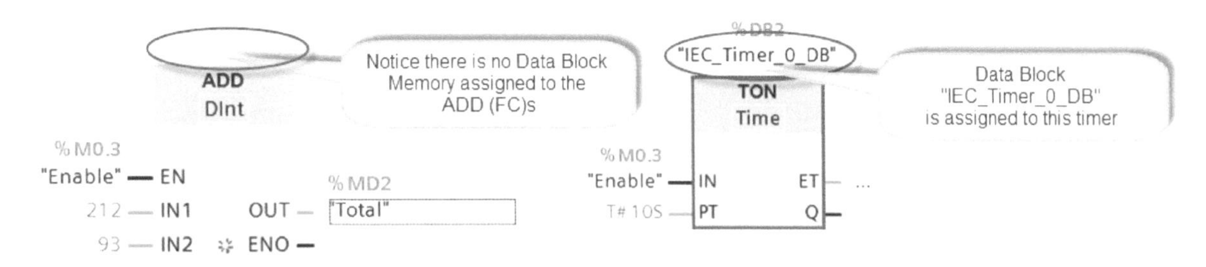

Figure 8-4. Example of an ADD function (FC) and a timer (TON) function block (FB).

Figure 8-5 shows the Data Block and its members that are generated when you create a timer single instance DB. In this example, six tag members were created all IEC_Timer_0_DB with an appended mnemonic. These tags can be used anywhere in logic.

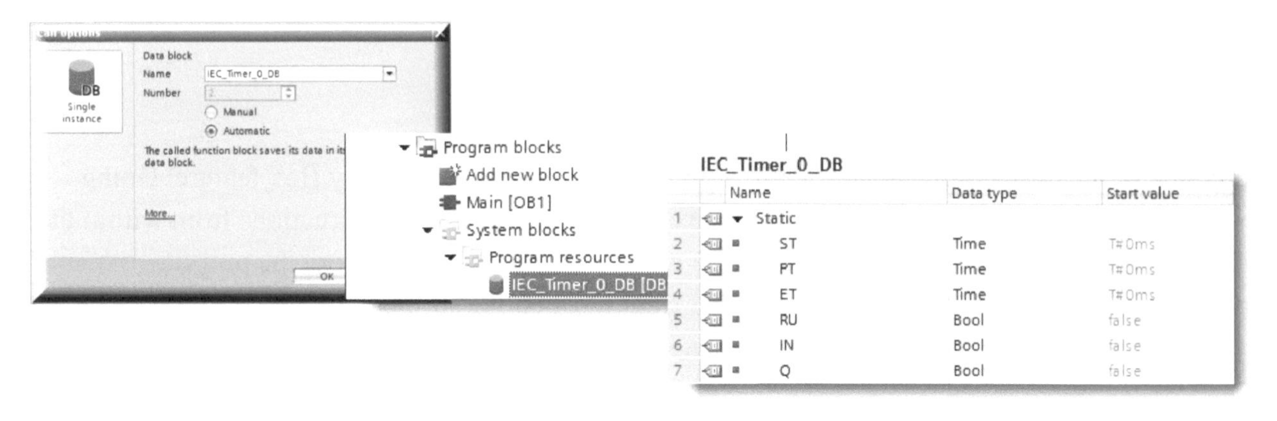

Figure 8-5. Creating a timer single-instance data block (DB).

Inserting an Instruction

There are a couple of different ways an instruction can be inserted into a network. The first way is to use the Instructions window. Within the Instructions window you can choose from the three main categories of instructions (see Figure 8-6). Under Favorites, the most used common bit logic instructions and programming structure elements are available. Beneath Basic Instructions, a variety of instructions are available to satisfy most application solutions. Beneath Extended instructions, specialty instructions are available to provide solutions that go beyond simple application programming. Within the folders of the basic and extended instructions sets, specific instructions for each group are available. By clicking the arrow adjacent to the folder, a more detailed list of instructions is revealed. In our example in Figure 8-6, under Timer operations, we can see the available instructions for timers. As you can see timers are just one of the many instructions sets available to write applications. Once within a network an instruction can be inserted by either right-clicking, then selecting Insert Instruction or drag-and-drop from the instruction window to the open network.

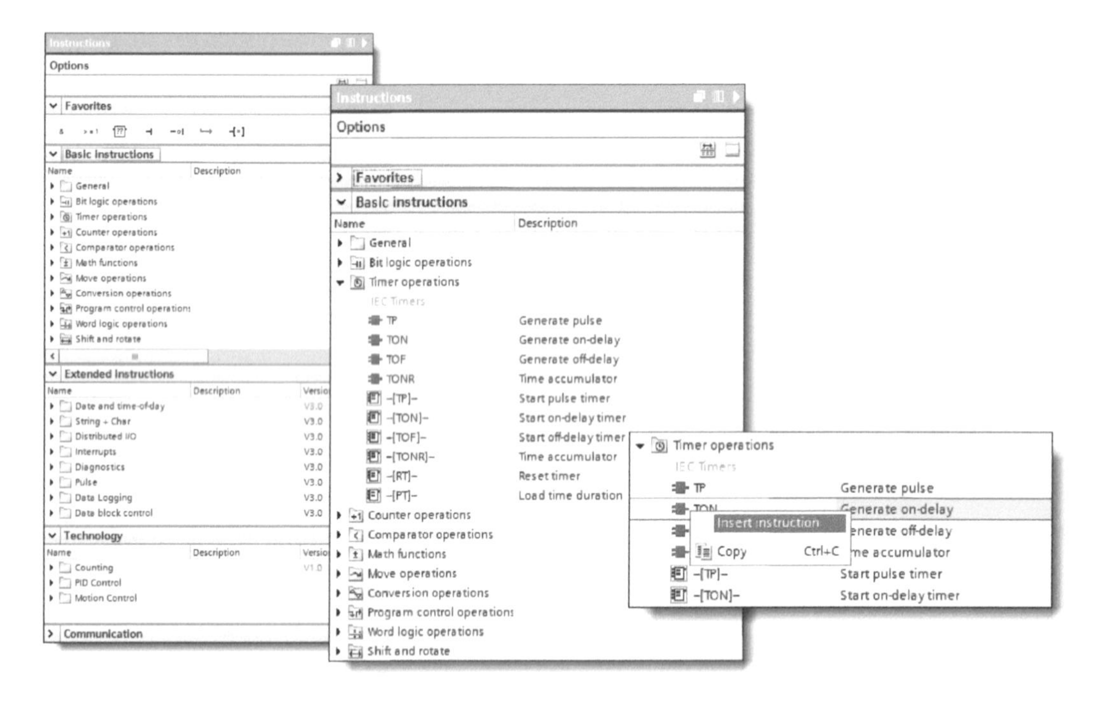

Figure 8-6. Instructions window.

The other way we can insert an instruction into a network is by using the Empty Box feature. Going back to our Favorites category we see the empty box icon. A way to assign instructions from within the network is by first placing an empty box. Then you can select the instruction from the pull-down menu. The pull-down menu is found by clicking the upper right most corner of the empty box. In Figure 8-7 we demonstrate the insertion of a AND instruction using the empty box feature.

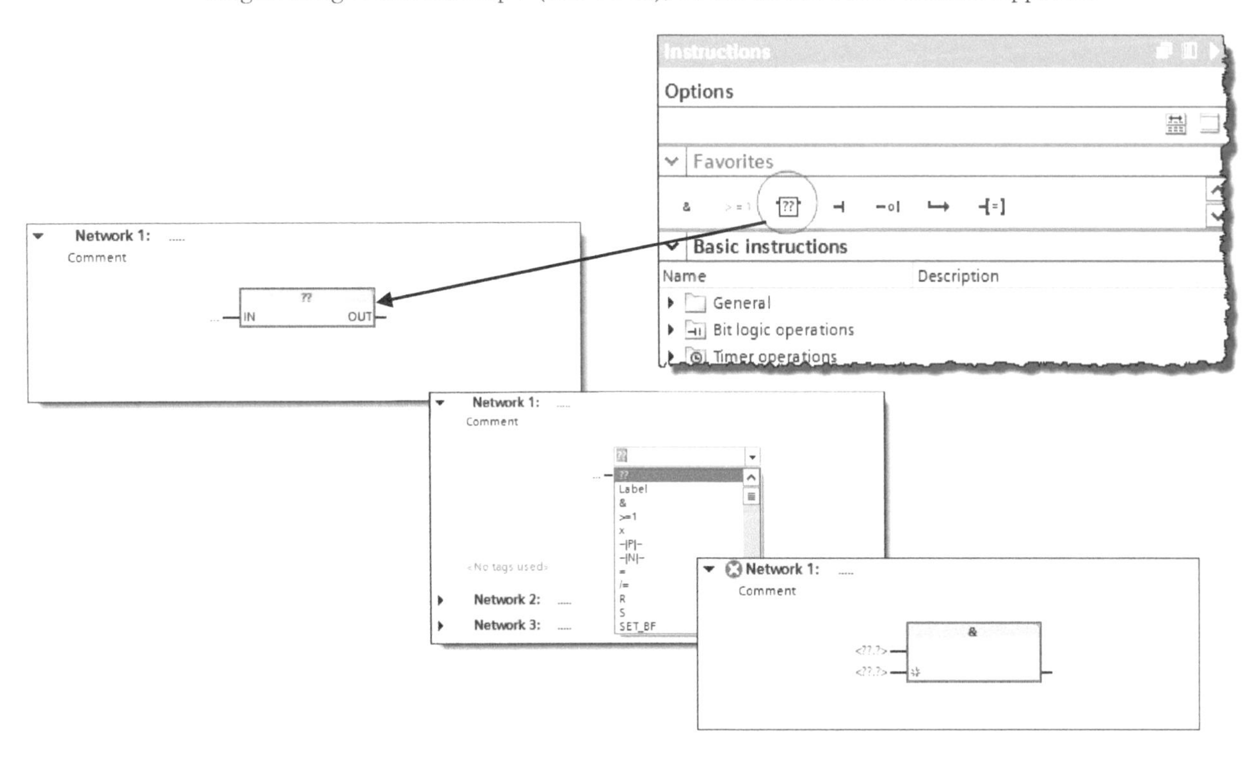

Figure 8-7. Using empty box feature to assign instruction.

Wiring Instructions

Wiring instructions together can be a helpful tool when programming process control applications. By wiring instructions together a user can see how values flow in and out of instructions. Figure 8-8 shows how to wire instructions together. Start by clicking on an instruction's output, then drag it toward the input of the instruction you want to pass a value to. Once the green square appears, release the mouse click and the instructions will be wired and organized together. This process can be repeated as many times as the output values are needed to control an application. In addition, a single output can be wired to multiple inputs the same way branch circuits are configured in LAD. Wiring instructions allows the user to show the relationship between instructions by having a graphical connection between instructions. Using wires instead of tags as placeholders between instructions minimizes tag creation.

Figure 8-8. Wiring instructions together.

Referencing Data Block Values

As mentioned earlier, DB objects can be referenced and assigned beyond their main instruction. In the next example, Figure 8-9, we show how the elapsed time of a Timer DB can be used in a compare instruction. This would be a good way to use a timer beyond its own capacity to make additional decisions.

In this example, we are comparing the elapsed time of our timer to a fixed time of 1 minute 30 seconds. Once the elapsed time has become greater than 1 minute 30 seconds the output of our comparison instruction will evaluate true. This could be used to control some action in our application prior to the timer finishing.

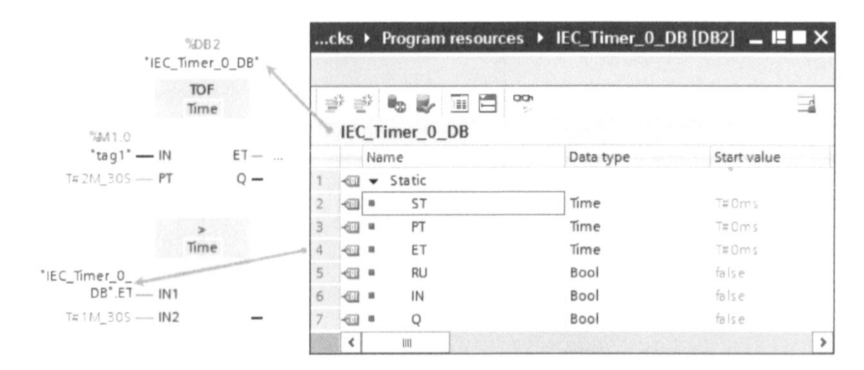

Figure 8-9. DB referencing to multiple instructions.

Bit Programming

In this next example, in Figure 8-10, we are going to show how AND instructions can be used to control the diverting of boxes on a conveyor. We are also going to demonstrate the use of the Invert RLO function. We will also use the Set Output and Reset Output instructions to turn on and off our diverters. In this example, we are diverting boxes of three different heights. We will control two outputs that turn on or off two separate diverters.

Looking at our first AND instruction we can see that if the TopSensor, MiddleSensor, and BottomSensor are all true, both Diverter_Left and Diverter_Right are turned off by using the "Reset Output" instructions. In this example, the tallest box continues straight on the conveyor. We chose to use Set Output and Reset Output instructions instead of Assignment instructions. The reason for this is simple; Assignment instructions only allow one line of returning logic to control a single output. This limits the flexibility to write separate conditions to affect the same output differently. In addition, assignment instructions revert outputs back to their low state as soon as the input conditions are no longer met. This means in our conveyor example if a box is detected and the appropriate diverters are turned on, they would only remain on while the box is at the sensors. This poses a problem once the box has passed the sensors and has not yet made it to the diverters. The Set Output and Reset Output instructions leave outputs in the last state that they were set to; even though, the input conditions that set them are no longer true. This is of course, as long as no other set or reset conditions for that output is met. Set Output and Reset Output instructions are usually used in pairs but can have as many sets and resets as needed for an application. In our example, we are using four Reset Output instructions and two Set Output instructions. The reason for this imbalance in sets and resets is the result of the application. Set Output instructions turn on outputs and Reset Output instructions turn off outputs. When deciding between different output instructions, it is up to the user to select the best method that provides both flexibility and ease of understanding.

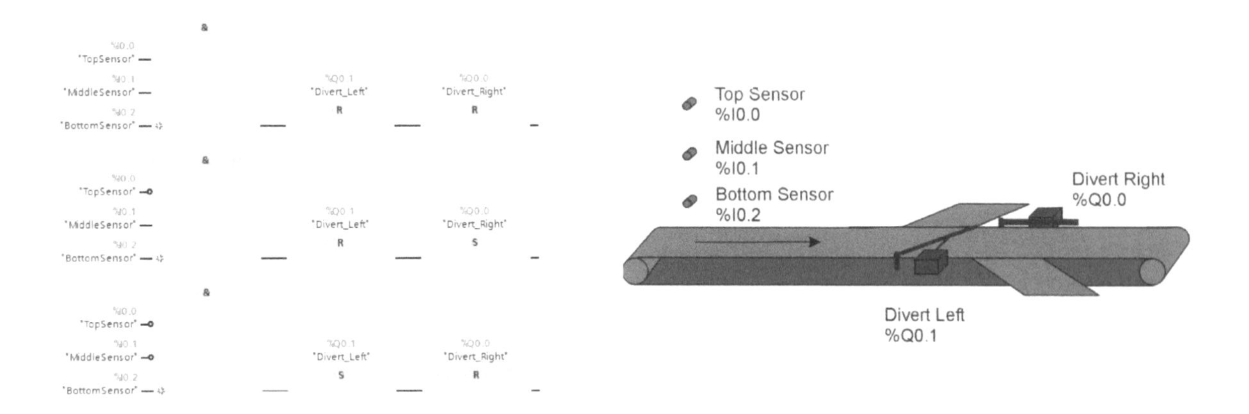

Figure 8-10. Conveyor diverter logic.

To understand the result of the second AND instruction it is important to go over the Invert RLO function. The invert result of logic operation or Invert RLO function is used to take the result of an input and invert the signal to the opposite state. In our example, our second and last AND instructions show inputs with small circles on them. These designate the Invert RLO function. As we can see in our second AND instruction the TopSensor Input has an Invert RLO after it.

The logical result of TopSensor being true, or 1, would result in a false, or 0, into the AND instruction. So, in order to get a logical true output from our second “AND” instruction we would need TopSensor False, MiddleSensor True, and BottomSensor True. A true output from our second AND instruction turns on the Diverter_Right and turns off the Diverter_Left. In this example, a mid-size box is diverted to the right of the conveyor. Finally, our last AND instruction has two Invert RLOs on it.

The logical result of TopSensor and MiddleSensor both being false and BottomSensor being true would result in a true output from our AND instruction. A true output from the last AND instruction turns off the Diverter_Right and turns on the Diverter_Left. In this example, a small-size box is diverted to the left of the conveyor.

To insert an Invert RLO simply click on the input you wish to invert then double-click on the Invert RLO from the Favorites area (see Figure 8-11).

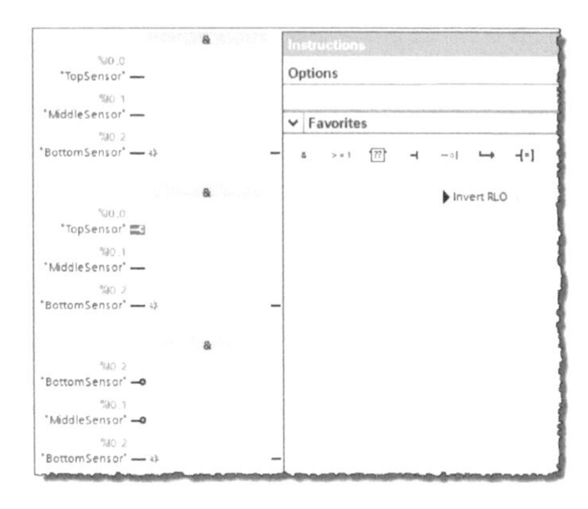

Figure 8-11. Assigning an invert RLO.

Compare Instructions

In Figure 8-12, a simple example of tank level control shows how a less-than comparison instruction opens the fill valve. In the same respect, we show how the Empty_Valve is turned on at greater than 51% of tank level. The resulting process could be described as: fill the tank as long as Tank_Level is less than 95%.

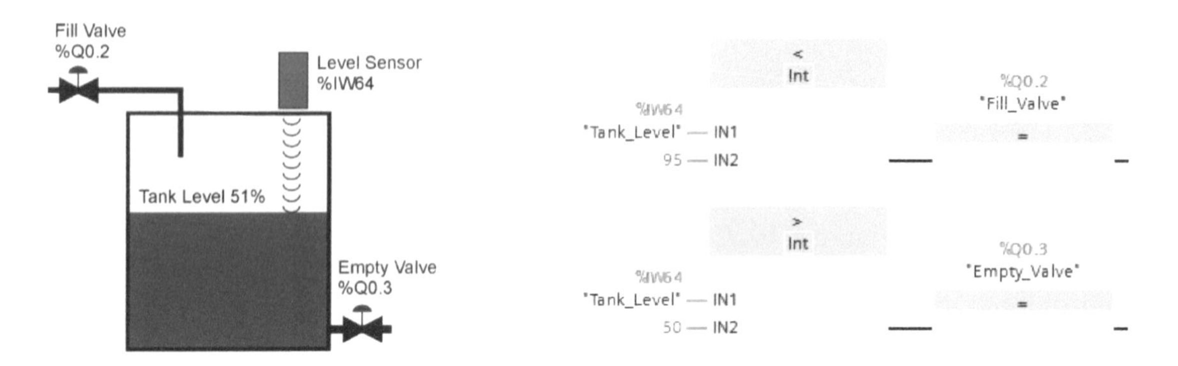

Figure 8-12. Tank level control.

In addition, only open the Empty_Valve if Tank_Level is greater than 50%. The result is a maintained Tank_Level of 50% that is to never exceed 95%. Compare instructions allow users to monitor changing values and make appropriate decisions to control the process.

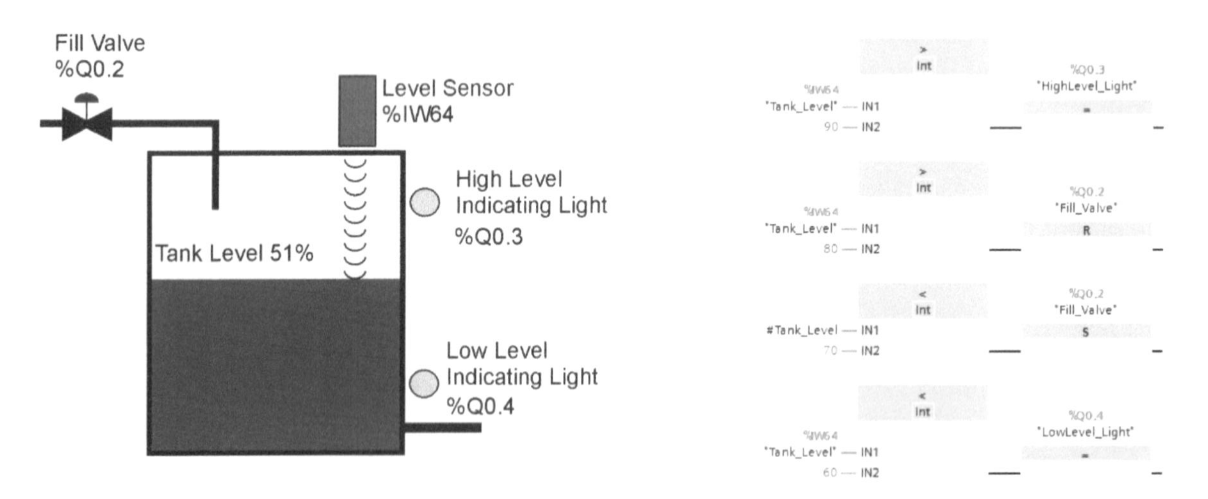

Figure 8-13. Tank level control with indicators.

In Figure 8-13, we show a slightly different tank filling application. In this application, we are working only with a fill valve. By monitoring the tank level, we can use logic to turn on and off our fill valve. By having different monitoring conditions, we can make multiple decisions on both controlling tank level as well as notification to an operator. By having different conditions for set (S) and reset (R) output conditions, we can provide some dead-band between valve switching. Dead-band is the area of control where there is no change in the output state. In this example between 70% and 80% of tank fill there is no effect on the valve output. When the tank goes below 70%, the fill valve is set and the tank begins to fill again. Once tank level rises above 80%, the fill valve is reset and the tank stops filling. If we did not have a dead-band built into our tank application, we could experience valve chatter from changes as small as 1% around our filling and stop filling levels.

The indicators become a deciding factor for the experienced operator to make a decision. Indicators and alarms are often used to notify operators when the conditions of a process are beyond their desired operating characteristics. If the tank level goes below 60% the low-level indicator is turned on. In addition, if the tank level goes above 90% the high-level indicator will be turned on. In either low or high state, it would be the operator's responsibility to take appropriate action.

Stop Light Application

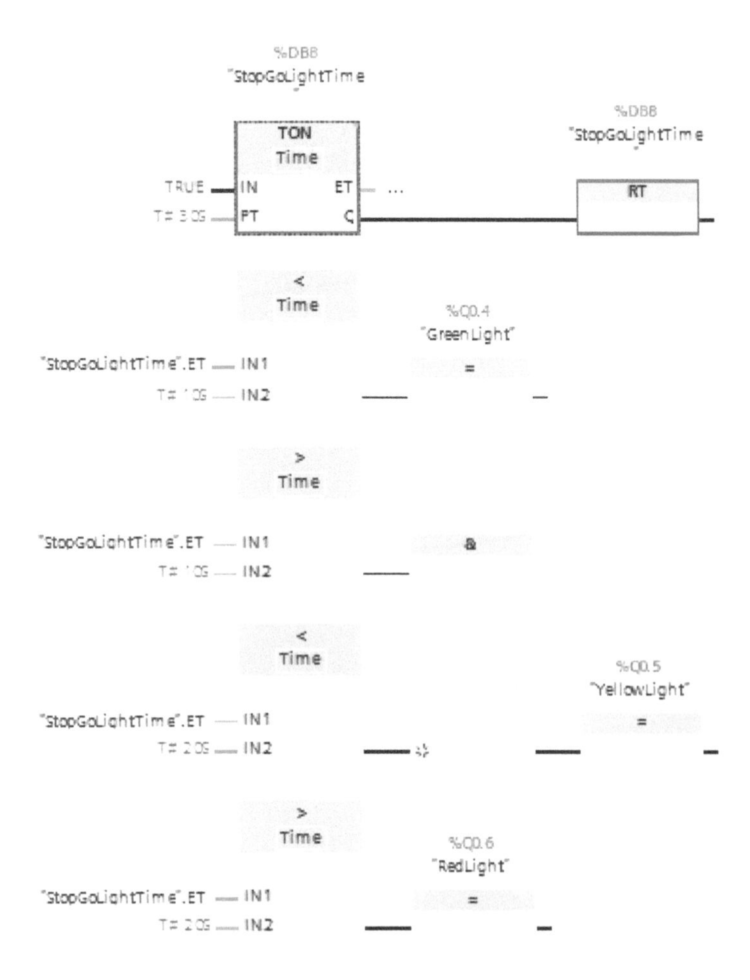

Figure 8-14. Simple stop light logic.

In Figure 8-14 we use comparison instructions along with a timer to assign outputs for a stop light application. Here we use a single timer that is reset (RT) every 30 seconds. While the timer is timing, the comparison instructions check the elapsed time on the timer to decide when to illuminate each light. In this example, when the elapsed time is less than 10 seconds the green light will turn on. After the timer has passed 10 seconds, the yellow light will turn on and stay on until the timer has passed 20 seconds. Finally, the red light will turn on for the remaining 10 seconds until the timer resets at 30 seconds, at which time the green light will turn back on.

Summary

FBD programming can provide many solutions for today's process control applications. Remember it is up to the user to choose a style that best suits the application for ease of programming as well as troubleshooting. Instructions in LAD and FBD operate the same. This chapter demonstrated the differences between LAB and FBD programming. Understanding the FBD programming environment along with the function specific differences, allows users to choose which style is most appropriate for an application.

Chapter Questions

1. IEC-61131 is an international standard for:
   a. Robotics
   b. PLCs
   c. PLC programming languages
   d. PLC communications
2. How can LAD style programming be converted to FBD Style?
3. What is the difference in the exact same logic written in either LAD or FBD?
4. What types of applications are best suited to FBD Style programming?
5. Is it best to put all logic in one Network location?
6. How can you invert the output from a FBD instruction?

Chapter 9

Technology Objects – PID

Objectives

Upon completion of this chapter, the reader will be able to:

Describe what PID is.

Describe how a PID technology object is configured.

Describe how a PID instruction can be used to control a system.

Introduction

Siemens has Technology Objects that make it easy to configure and program some special technologies. This chapter will cover one of the available technology objects: proportional, integral, and derivative (PID) control.

PID is used for process control. It is used to control properties such as flow rate, level, pressure, temperature, density, and so on. The PID instruction takes input from the process, compares it to the setpoint, and controls the output to try and make the actual process value equal to the setpoint. The input is normally from an analog input module. The output is usually an analog output. PID instructions are used to keep a process variable at a commanded setpoint.

Figure 9-1 shows a tank level example. In this example there is a level sensor that outputs an analog signal between 4 and 20 milliamps. It outputs 4 milliamps if the tank is 0% full and 20 milliamps if the tank is 100% full. The output from this sensor becomes an input to an analog input module in the PLC. The analog output from the PLC is a 4-20 mA signal to a valve. The variable valve on the tank is used to control the inflow to the tank. The valve is 0% open if it receives a 4-mA signal and 100% open if it receives 20 milliamps. The PLC takes the input from the level sensor and uses the PID equation to calculate the proper output to control the valve.

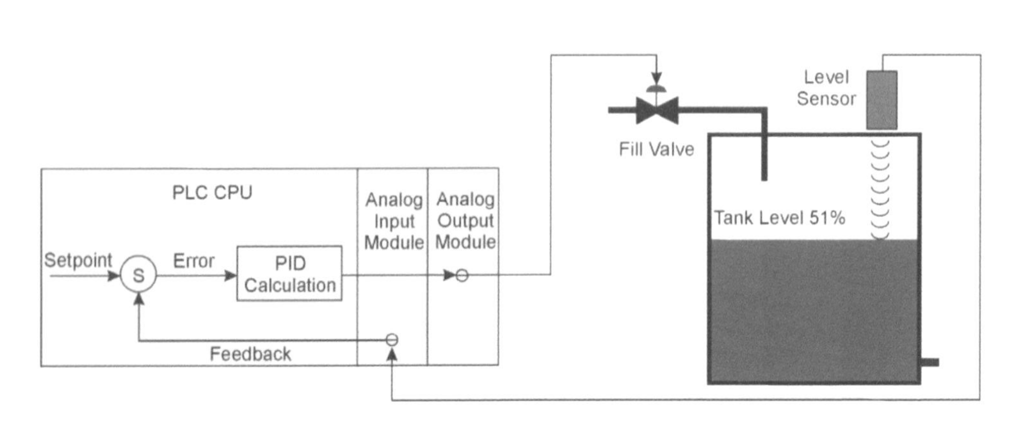

Figure 9-1. A level control system.

Figure 9-2 shows how a PID system functions. The setpoint is set by the operator and is an input to a summing junction. The summing junction is the circle labeled S in the figure. The output from the level sensor (process feedback) becomes the feedback to the summing junction. The summing junction sums the setpoint and the feedback and generates an error. The PLC then uses the error as an input to the PID equation. There are three gains in the PID equation. The P gain is the proportional function. It is the largest gain. It generates an output that is proportional to the error signal. If there is a large error the proportional gain generates a large output. If the error is small the proportional output is small. The proportional gain is based on the magnitude of the error. The proportional gain cannot completely correct an error. There is always a small error if only proportional gain is used.

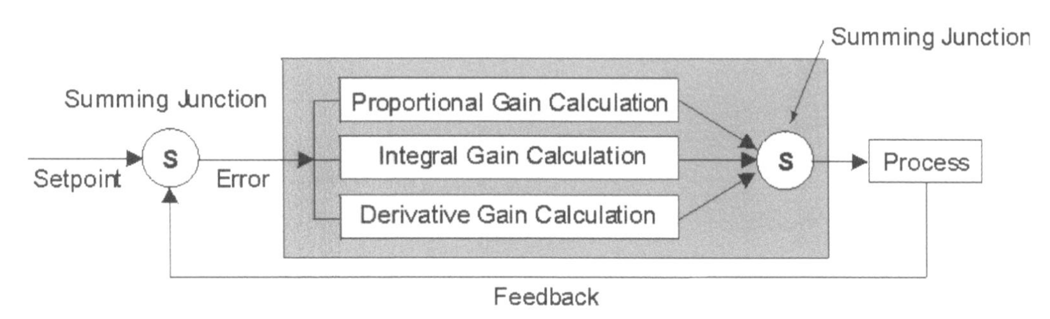

Figure 9-2. A block diagram of the use of a PID instruction.

The I gain is the integral gain. The integral gain is used to correct for small errors that persist over time. The proportional gain cannot correct for very small errors. The integral gain is used to correct for these small errors over time.

The D gain is the derivative gain. It is used to help correct for rapidly changing errors. The derivative looks at the rate of change of the error. When an error occurs, the proportional gain attempts to correct for it. If the error is changing rapidly (for example, maybe someone opened a furnace door). The proportional gain is insufficient to correct the error and the error continues to increase. The derivative would see the increase in the rate of change in the error and add a gain factor. If the error is decreasing rapidly the derivative gain will damp the output. The derivative's damping effect enables the proportional gain to be set higher for quicker response and correction.

As you can see from Figure 9-2 the output from each of the P, I, and D equations are summed and a control variable (output) is generated. The output is used to bring the process variable back to the setpoint.

Refer back to Figure 9-1. As you can see in this figure the feedback from the process (tank level) is an input to an analog input module in the PLC. This input is used by the CPU in the PID equation and an output is generated from the analog output module. This output is used to control a variable valve that controls the flow of liquid into the tank. Note that there are always disturbances that affect the tank level.

The temperature of the liquid affects the inflow and outflow. The outflow varies also due to density, atmospheric pressure, and many other factors. The inflow varies because of pressure of the fluid at the valve, density of the fluid and many other factors. The PID instruction is able to account for disturbances and setpoint changes and control processes very accurately.

Figure 9-3 shows how a PID technology object can be added to a program from the Portal view. Note on the left of the figure there is a choice to add a technology object. In the center of the figure a Motion control Technology object or a PID Technology object can be added. The PID technology object was chosen in this example. Note also that there are two choices a PID_Contact instruction or a PID_3Step instruction. The example in this chapter will use a PID_Compact instruction.

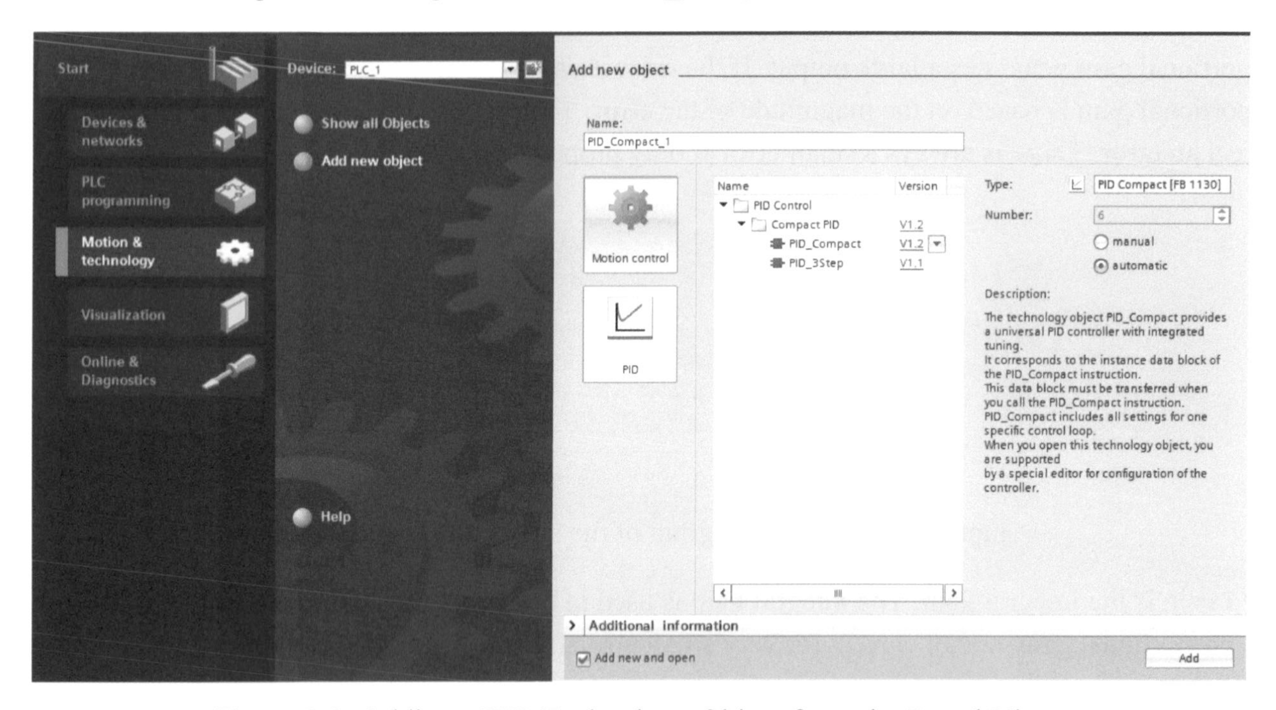

Figure 9-3. Adding a PID Technology Object from the Portal View.

Figure 9-4 shows how a Technology object can be added from the Project view.

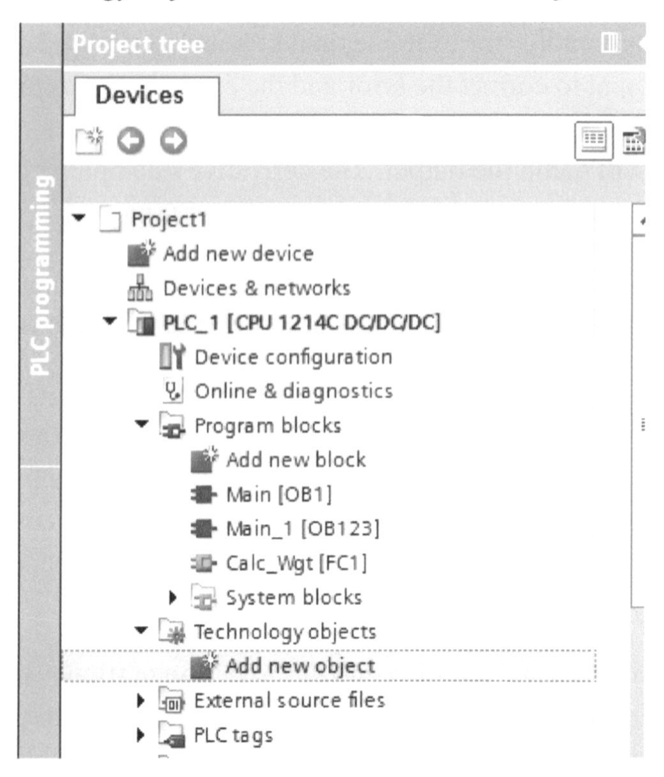

Figure 9-4. Adding a PID Technology Object from the Project view.

Note that when a PID_Compact instruction is added to network logic, it will appear in the organizational tree on the left of the screen under the Technology Object icon (see Figure 9-5). The PID_Compact instruction (technology object) acts as a PID controller with optimizing self-tuning for automatic and manual mode.

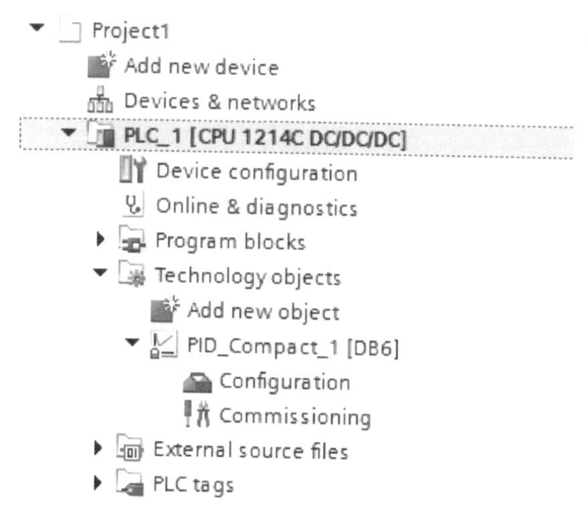

Figure 9-5. Note that a PID_Compact technology object has been added under the Technology Objects icon.

If you double-click on the Compact PID Instruction in the Portal view the screen in 9-6 will appear.

Figure 9-6 shows the basic settings screen for a PID Technology object. The controller type is chosen in this screen. There are many choices for the type of system. Temperature was chosen in this example, as well as degrees F. Note that there is also a checkbox to invert the control logic.

Select the "Inversion of the control direction" check box if an increase in the manipulated value causes a decrease of the actual value. We might do this if the system were a chilling process instead of a heating process. An example of this might be increasing the opening of a chill valve actually resulting in a decreasing temperature.

Note that Basic Settings was chosen which makes the display show both the Controller type and Input/output parameters screens for configuration. Note they could have been selected individually.

Input/Output Parameters Configuration

Figure 9-5 also shows the Input/output parameter setting screen. The output of a PID technology object uses a variable of the user program as the variable output. There are two choices of output type: Output_PER and Output_PWM.

An Output_PER output uses an analog output as the variable output.

An Output_PWM output uses a digital switching output and controls it by means of a pulse width modulation. The result at the PWM output is variable turn-on and turn-off times.

The output type chosen in this example was an analog output.

Setpoint

The user must select whether the value at the function block or the value of the instance DB is to be used. The user can enter a constant value or the tag to be used for the setpoint

Actual value

The user must select whether the input parameter "Input" or "Input_PER" is to be used.

Use the Input choice if you want to use an actual value from the user program. You must also select whether the value at the function block or the value of the instance DB is to be used.

Use the Input_PER choice if you want to use the actual value of an analog input. In the field below this specify the tag or address of the actual input value. The input would generally be the address of the analog input that represents the actual temperature in this example.

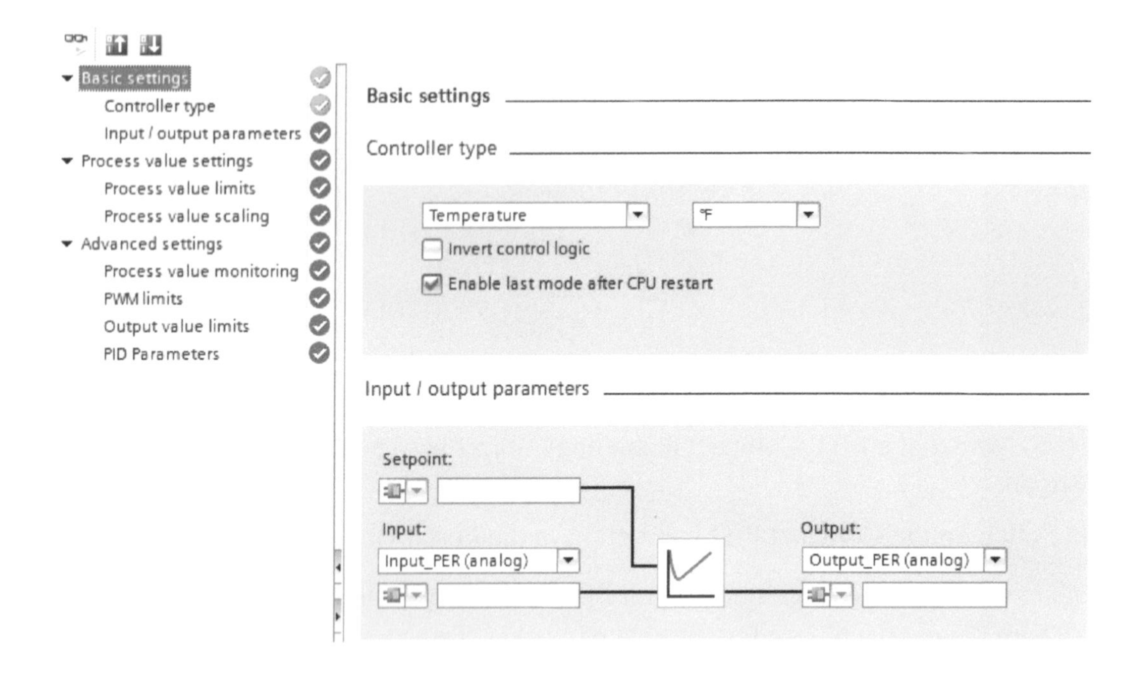

Figure 9-6. Configuring the Basic settings for the PID_Compact instruction.

The next configuration screen (Figure 9-7) is used to configure the process value limits and the process value scaling.

Upper and lower limit

Specify the absolute upper and lower limit of the actual desired process value. As soon as these limits are exceeded or undershot during operation, the regulatory control switches off and the manipulated variable is set to 0.0 %. In this example the high limit of the process was set to 120 %. The process low limit was set to 0.0 %.

Scaling

Specify the scaling of your actual value by means of a lower and an upper pair of values. In this example the scaled high value was set to 100.0 % and the low was set to 0.0 %.

Each pair of values consists of the analog value of the analog input and the physical value of the respective scaling point. Depending on the configuration of the default setting, a process value of the user program can also be used instead of the analog value of the analog input.

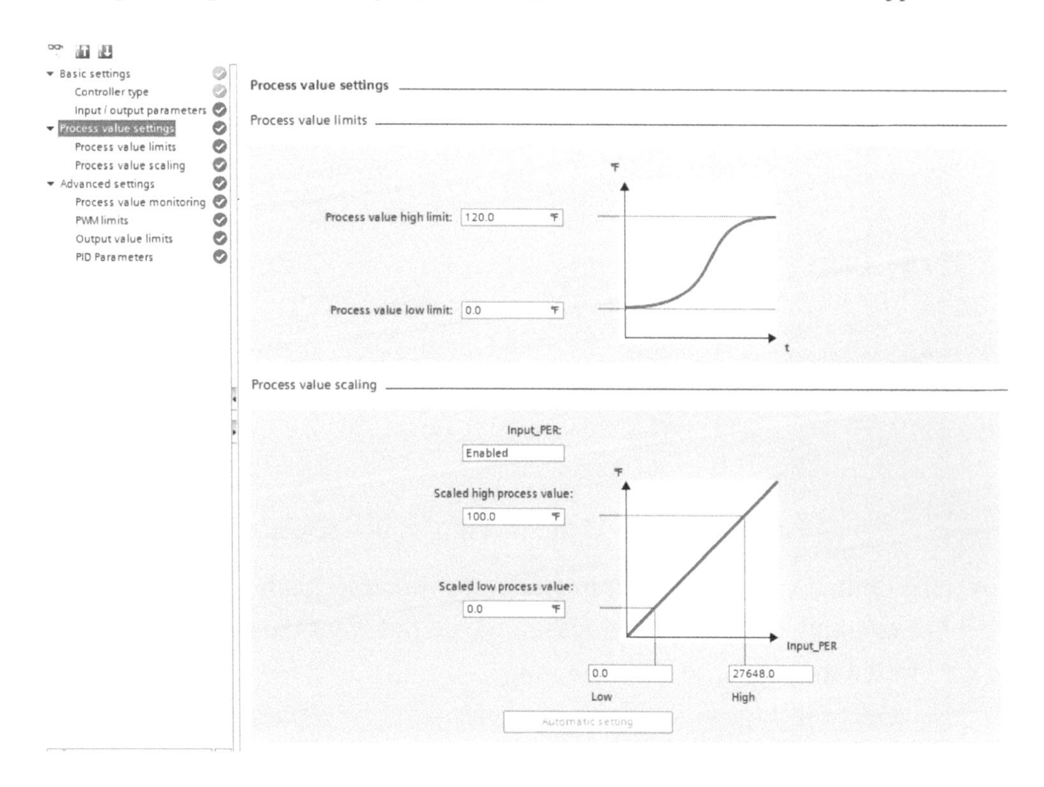

Figure 9-7. Process value settings screen.

Advanced Settings

In the advanced Settings screen (see Figure 9-8) we first set the process value monitoring limits. In this example we would like a warning if the process hits the high limit warning of 98.0 %. We would like a low alarm if the process reaches 65.0 %.

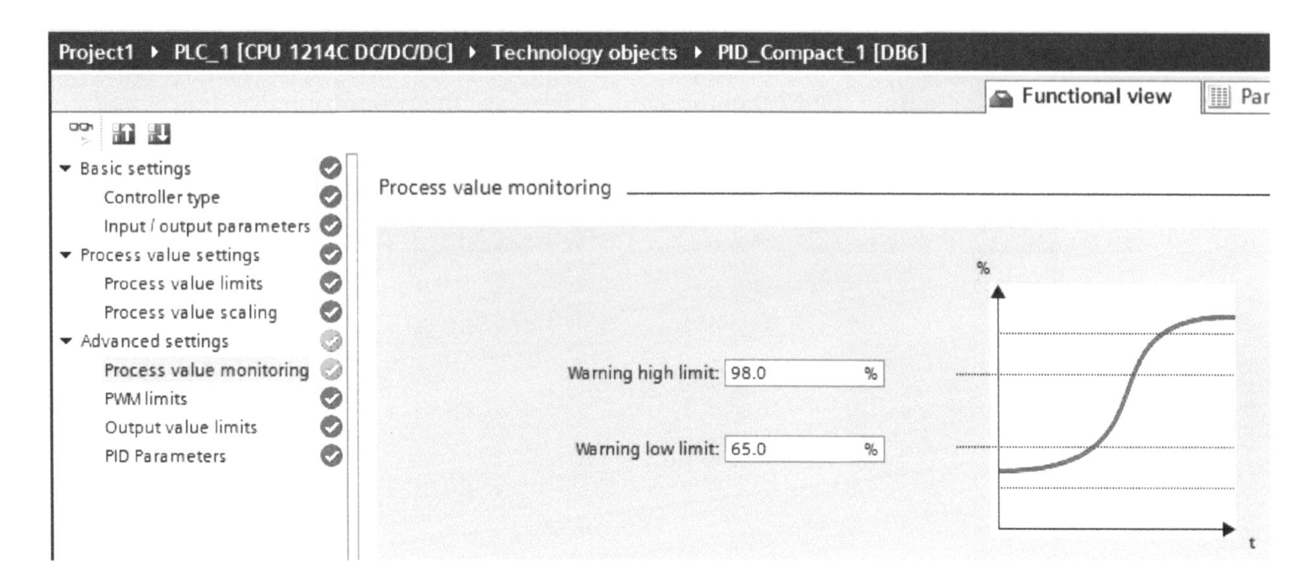

Figure 9-8. Setting the Process value monitoring high and low limit.

Figure 9-9 shows the PWM (pulse width modulation) limits parameters screen. This screen is used to set the minimum on and off time for the output.

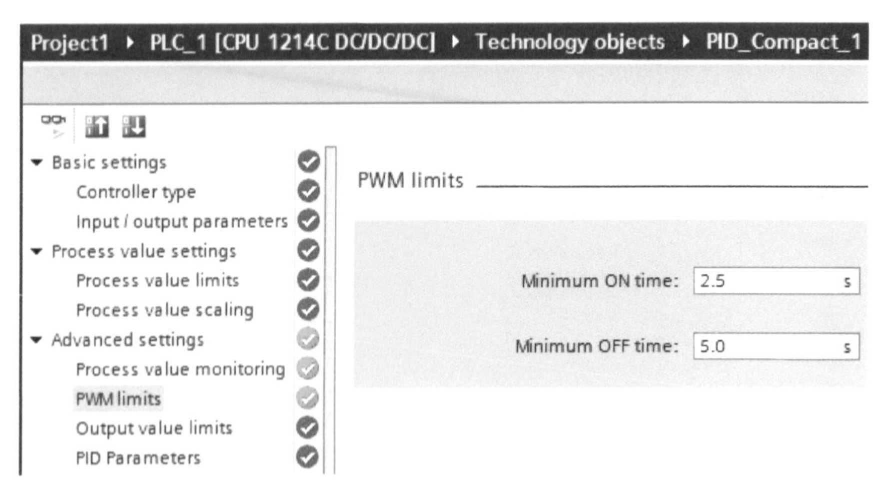

Figure 9-9. PWM limits configuration screen.

Figure 9-10 shows the Output value limits parameter entry screen. This is used to limit how high or low the output can be. For example there may be a reason we do not want the output to be as high as 100%. In this example we limited the output to 80 %.

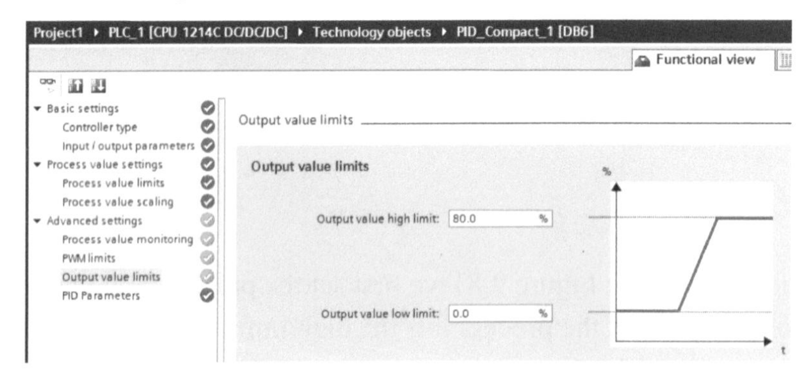

Figure 9-10. Setting Output value limits.

PID Parameters

Figure 9-11 shows the screen for setting the PID parameters. These are the proportional, integral and derivative gains that will be used by the PID Technology object. This is where the gains in the PID process loop can be tuned so that it responds quickly, controls closely, and is stable.

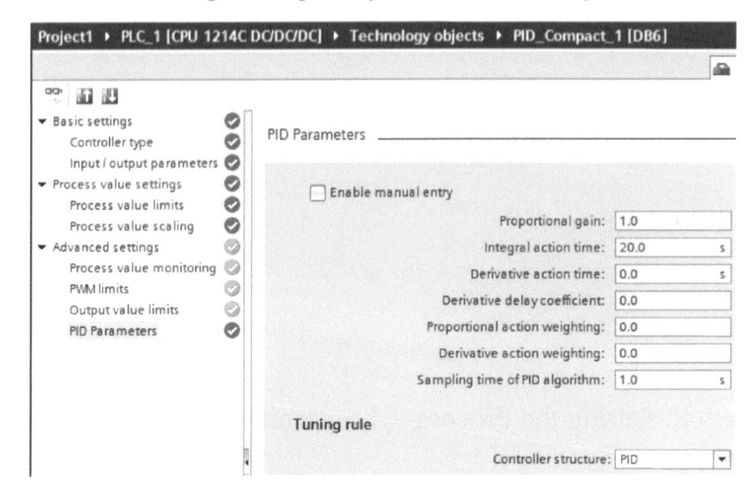

Figure 9-11. PID Parameter entry screen.

Now that the PID technology object has been configured we can add a PID_Compact instruction to the program. Note the down arrow at the bottom of the PID_Compact instruction in Figure 9-12. If you click on the down arrow the instruction will be expanded and more parameters will be available (see Figure 9-13).

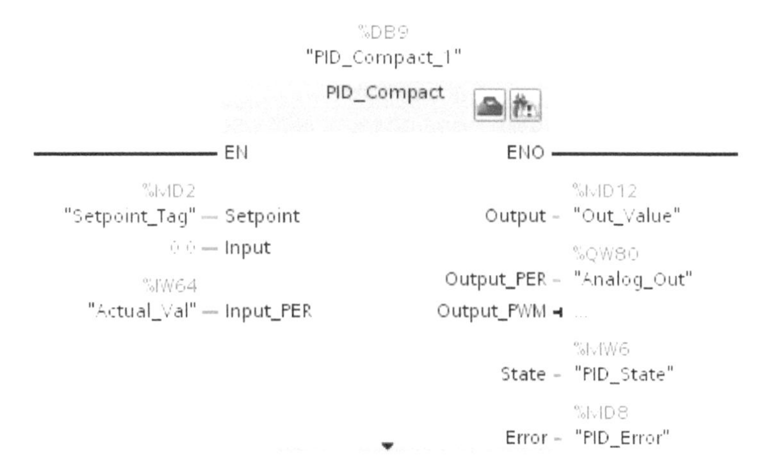

Figure 9-12. A PID_Compact technology object.

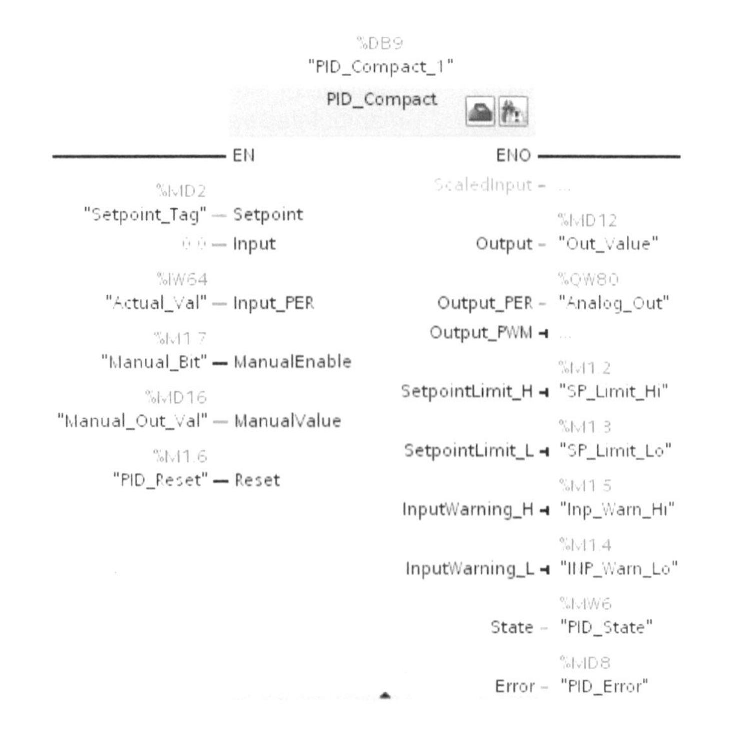

Figure 9-13. A PID_Compact technology object- expanded.

PID Configuration

Figure 9-14 shows the inputs, their type and description of their purpose for the PID instruction.

Inputs to a PID Instruction			
Parameter	Type	Initial Value	Description
Setpoint	REAL	0.0	Setpoint of the PID in auto mode.
Input	REAL	0.0	Variable of the user program that is the source of the actual value.
Input_PER	WORD	W#16#0	Analog input is the source of the actual value.
ManualEnable	BOOL	FALSE	False to True transition selects the manual mode. A true to false transition selects the most recently active operating mode.
ManualValue	REAL	0.0	Manipulated value for manual mode.
Reset	BOOL	FALSE	Restarts the controller. The following rules apply to Reset = True: Inactive operating mode; Manipulated variable = 0; Interim values in the controller are reset; The PID parameters are retained.

Figure 9-14. Inputs to a PID instruction.

Figure 9-15 shows the outputs for a PID instruction, their type, and description of their purpose.

Outputs from a PID Instruction			
Parameter	Type	Initial Value	Description
ScaledInput	REAL	0.0	Output of the scaled actual value,
Output	REAL	0.0	Variable of the user program for output of the manipulated variable,
Output_PER	WORD	W#16#0	Analog output for outputting the manipulated variable,
Output_PWM	BOOL	FALSE	Switching output for outputting the manipulated variable using pulse width modulation,
SetpointLimit_H	BOOL	FALSE	If true, the absolute setpoint high limit has been reached or exceeded. In the CPU the setpoint is limited to the configured absolute high limit of the actual value.
SetpointLimit_L	BOOL	FALSE	If true, the absolute setpoint low limit has been reached or underpassed. In the CPU the setpoint is limited to the configured absolute low limit of the actual value.
InputWarning_H	BOOL	FALSE	If true, the actual value has reached or exceeded the high warning limit,
InputWarning_L	BOOL	FALSE	If true, the actual value has reached or fallen below the low warning limit,
State	INT	0	Operating Mode of the Controller,
			If 0, Inactive- manipulated variable is set to 0,
			If 1, Self tuning during initial start,
			If 2, Self tuning in operating point,
			If 3, Auto mode,
			If 4, Manual mode,
Error	DWORD	W#32#0	Error Message,
			0000 0000 no error,
			> 0000 0000 one or more errors are pending. The PID controller enters the inactive mode. Refer to the error messages to analyze the active error.

Figure 9-15. Outputs from a PID instruction.

Within a control loop the PID controller continuously acquires the measured actual value of the controlled variable and compares it with the desired setpoint. From the resulting error the PID controller calculates a controller output that adapts the controlled variable to the setpoint as rapidly and stably as possible. At the PID controller the calculated value of the controller output consists of three components:

- **Proportional gain**

  The value of the controller output calculated by the proportional component is proportional to the system deviation.

- **Integral gain**

  The value of the controller output calculated by the integral component increases with the duration of the controller output and finally results in the controller output being compensated.

- **Derivative gain**

  The derivative component of the PID controller increases as the rate of change of the system deviation increases. The controlled variable is adapted as fast as possible to the setpoint. When the rate of change of the system deviation decreases, the derivative component decreases as well.

The PID technology object can calculate the proportional, integral, and derivative gains by self-tuning during initial start. The parameters can be optimized further by means of self-tuning or manual-tuning. Note the self-tuning mode can be chosen in the State parameter (Figure 9-15).

Configuration in the Configuration Window

The PID Technology object configuration can be modified and/or commissioned (tuned) easily by selecting the Technology object in the Project tree (see Figures 9-16 and 9-17). PID parameters can be entered manually or set automatically though Auto-Tuning.

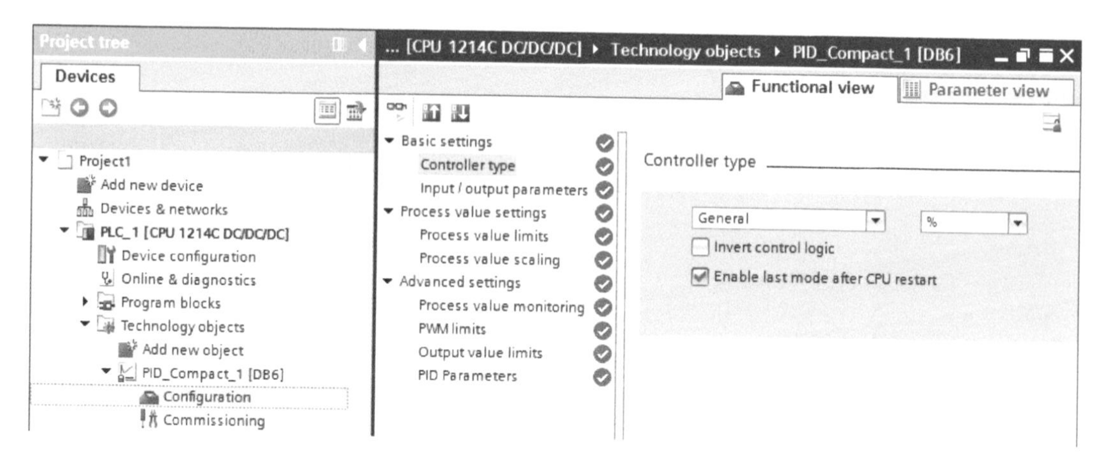

Figure 9-16. Configuration screen.

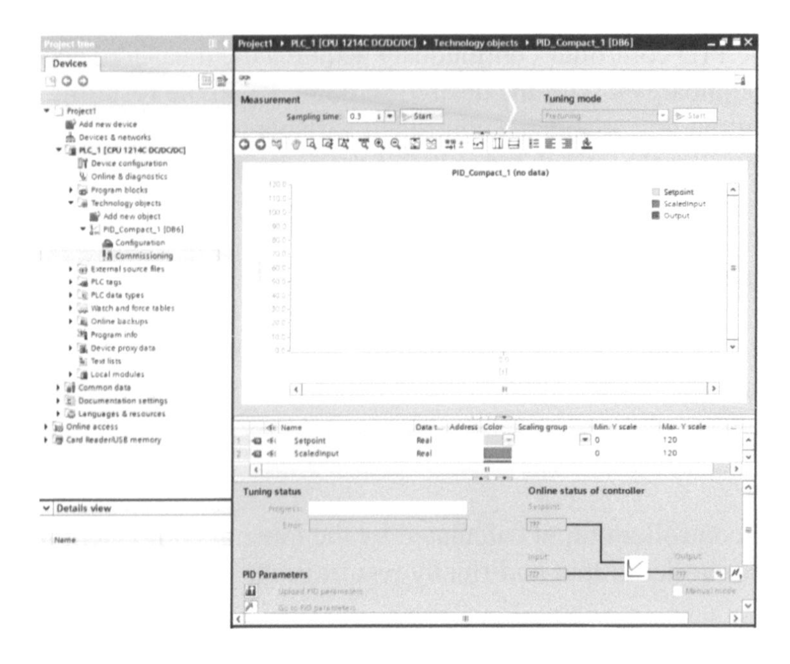

Figure 9-17. Commissioning screen. This screen is used to tune the PID process.

Chapter Questions

1. What is a technology object?
2. What does PID stand for?
3. What are PID instructions typically used for?
4. What are the two types of outputs available in a PID compact instruction?
5. What two types of input values can be used and what is the difference between the two?
6. Describe what the proportional portion of PID control corrects.
7. Describe what the integral portion of PID control corrects.
8. Describe what the derivative portion of PID control corrects.
9. How can the initial gains settings for PID be established?

Chapter 10

Motion Control Using an Axis Technology Object

Objectives

Upon completion of this chapter, the reader will be able to:

Describe the difference between an axis and a drive.

Describe how an axis is configured.

Explain the sequence of motion instructions to enable and home an axis.

Describe the use of other motion instructions.

Overview

Siemens has Technology Objects that make it easy to configure and program some special technologies. This chapter will cover the Axis Technology Object.

The Axis Technology Object creates an image of an axis in the controller and can be used to control stepper motors and servo drives with pulse interface. An axis must first be created and configured before it can be programmed and operated.

Axis Versus Drive

The term "axis" denotes the technical image of the drive by the Axis technology object. The Axis technology object interfaces the user program with the drive. The technology object receives the motion control commands from the user program, executes them and monitors them during runtime. The motion control commands are initiated in the user program by means of motion control instructions.

The term "drive" represents the electromechanical unit of a stepper motor plus power section or a servo drive plus converter with pulse interface. The drive is controlled by the Axis Technology Object via a PLC pulse generator.

Adding and Configuring an Axis

Figure 10-1 shows the Add New Technology Object screen in the Portal view. Figure 10-2 shows the other way to Add New Technology Object screen in the Project tree view. Note that there are two choices: Motion and PID.

The axis technology object TO_Axis_PTO is used to map a physical drive in the control (see Figure 10-1 or Figure 10-2). It makes functions for controlling stepper motors and servo motors available with pulse interface. After an axis is configured in the technology object, the actual drive motion can be programmed with motion control function blocks.

The technology object Command TO_CommandTable_PTO choice can be used to create motion control commands and motion profiles in a table using PLCopen. The created profiles are applied to a physical drive with the axis technology object. In this example TO_Axis_PTO will be used.

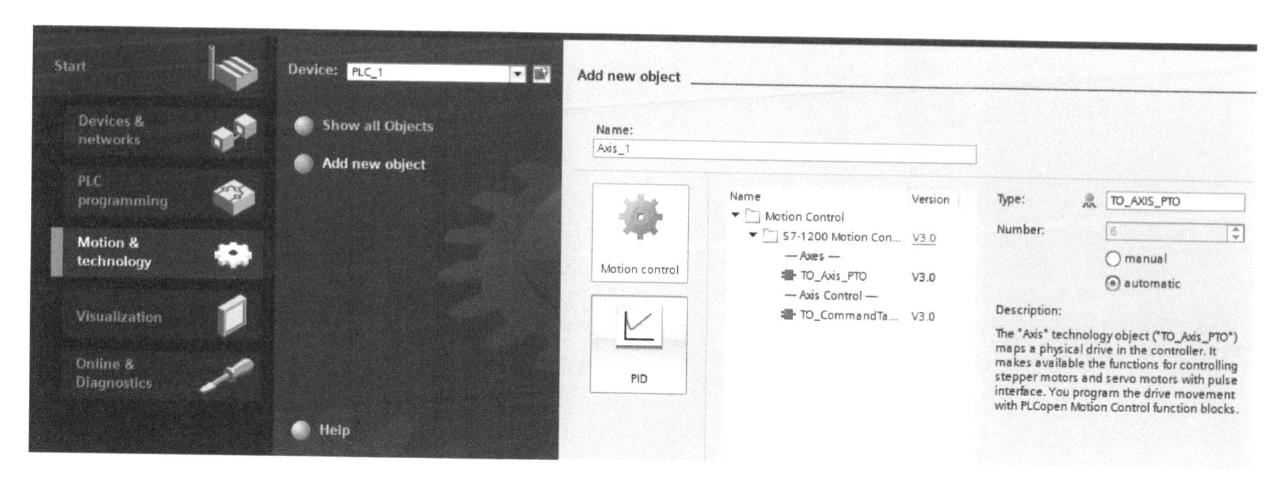

Figure 10-1. Adding a motion Technology object through the Portal view.

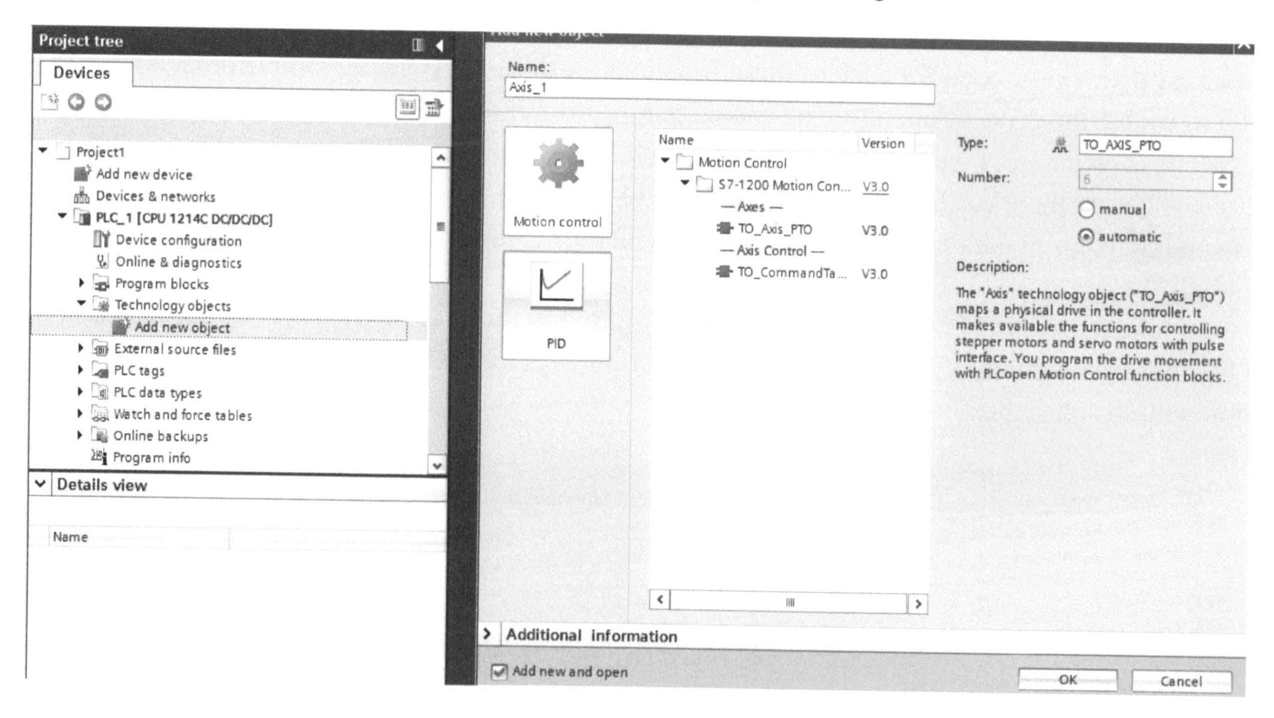

Figure 10-2. Adding a motion Technology object through the Project tree view.

The screen in Figure 10-3 is used to specify the outputs from the PLC to the drive and where the position input information (encoder counts in this example) will come from. The PLC output to be used in this example is Q0.0 and direction output is Q0.1. The position information will come from fast counter HSC_1. The unit of measurement is also set in this screen. Millimeters was chosen for the measurement unit in this example.

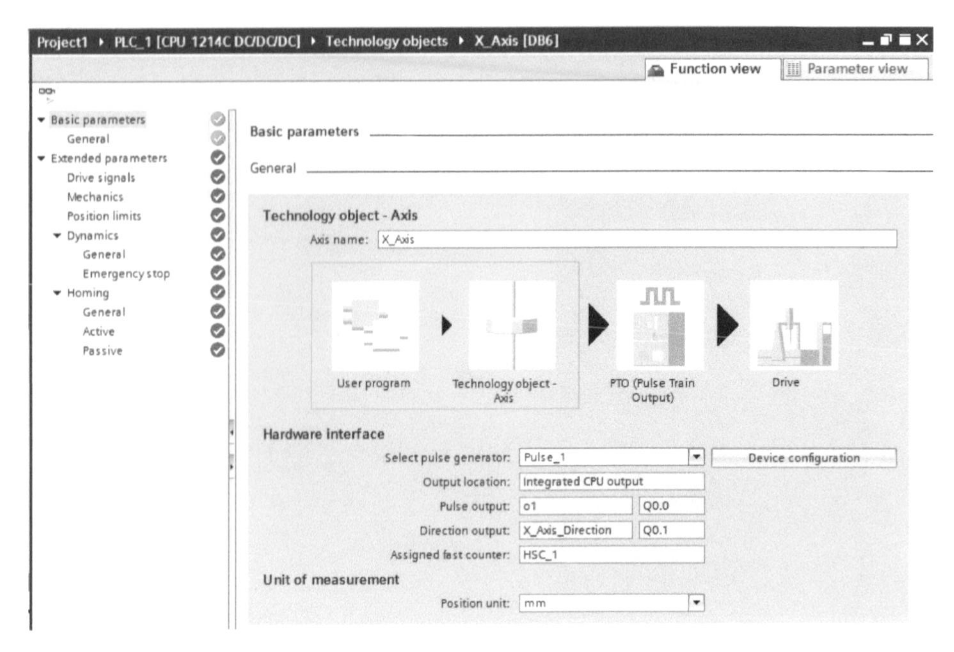

Figure 10-3. Basic drive configuration screen.

Figure 10-4 shows the configuration screen for the enable output and also the ready input. You select the tag address that will be used for each of these functions. A drive must get an input from a controller (PLC) to enable the drive to operate. The enable output from the PLC will be used for this. This is where you specify which output will be used. After the drive receives the enable input it will turn on an output that is used to tell the PLC that the drive is ready to be operated. This is the ready output from the drive and the ready input to the PLC. You must specify which PLC input will be used for this.

The mechanical properties of the drive are also configured in this screen. This sets up the conversion mm based on the number of pulses per motor revolution. It is also possible to invert the direction of motion with the checkbox on this screen.

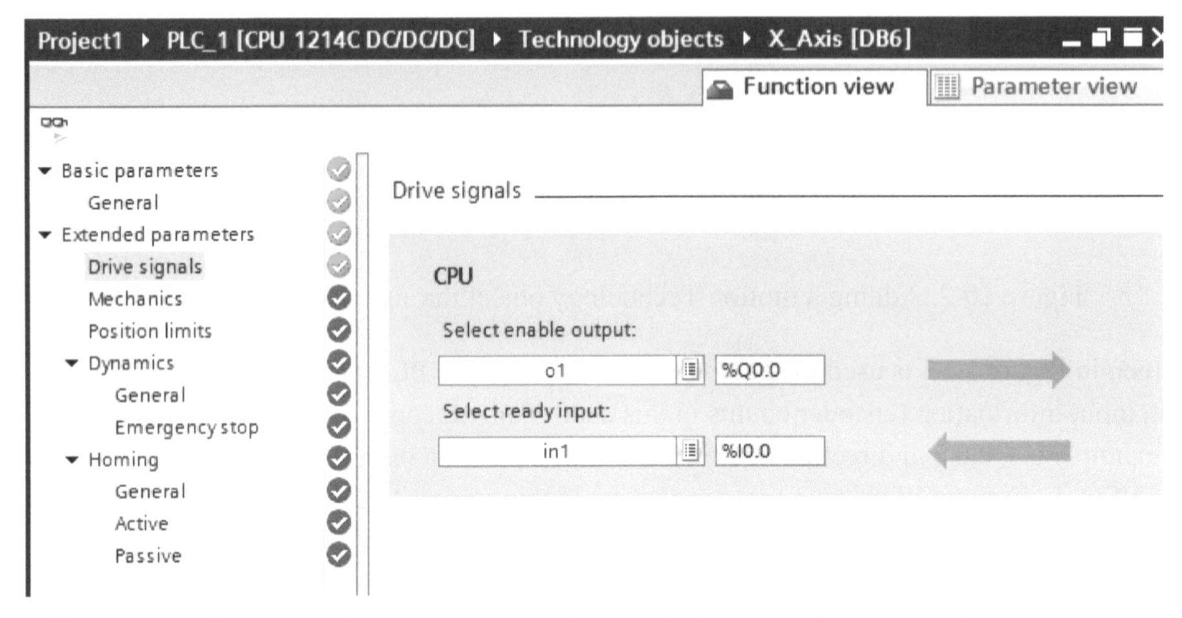

Figure 10-4. Configuration screen for enable, ready input, and motion units conversion.

Figure 10-5 shows how the Mechanics parameters are set in the Motion technology object. In this example there are 1000 pulses per revolution and the nut would advance 10 mm per revolution. This information is used to set the drive resolution so it can calculate moves for this particular axis.

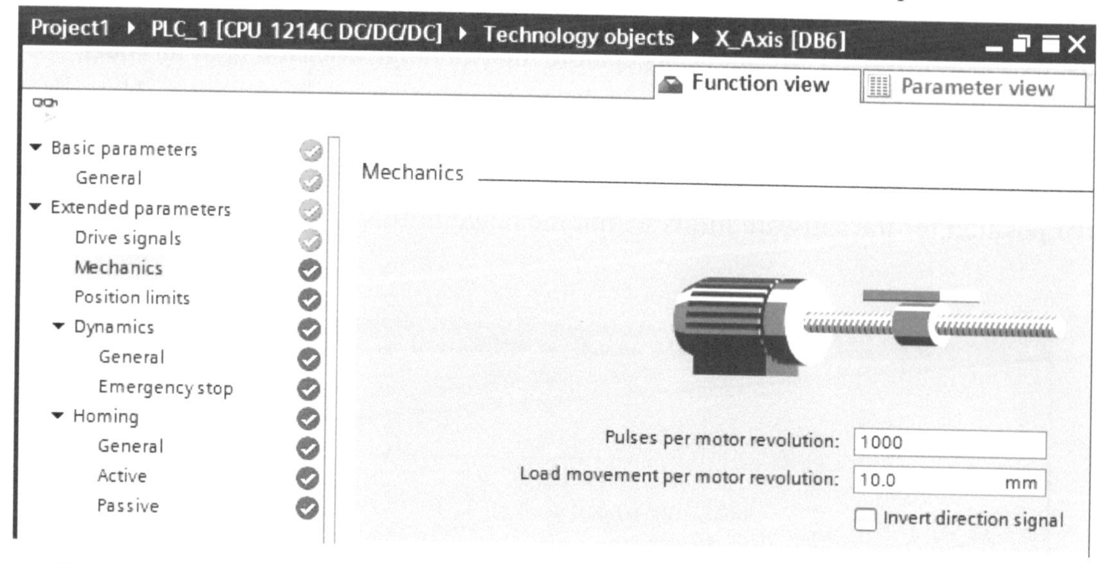

Figure 10-5. Setting the pulses per revolution and the movement per revolution.

Hardware and Software Position Limits

Figure 10-6 shows the screen that is used to configure the hardware and software limits. Hardware limits utilize actual inputs. There is one located where the drive should stop motion in the counterclockwise direction and one in the clockwise direction. If the drive moves far enough in the CCW direction the CCW limit switch would change state. If the software limits are to be used the user sets a low and a high number (position) that will act as software limts. Software limits are set so that the drive will run into the software limits before it would hit the hardware limits. If we would command the drive to move beyond of the software limits it will not make the move.

The hardware overtravel switches are used to stop the axis if it reaches the clockwise or counterclockwise overtravel limit switch. After the axis has energized the hardware limit switch, the axis is ramped down to a standstill by means of emergency stop deceleration. For a homing operation, it can be set up to reverse direction to find the home switch in the opposite direction of travel.

The hardware overtravel inputs must be selected from the drop-down list box (see Figure 10-4). The inputs must support alarm functionality. The integrated CPU inputs and the inputs of a signal board can be selected as inputs for the HW limit switches.

Signal Level Selection

The hardware limit switches can be set for normally-open or normally-closed operation. Select the triggering signal level (Lower level (Normally-closed)/ Upper level (normally-open)) of the hardware limit switch. If Lower level is chosen, the input signal is FALSE after the axis has reached or passed the hardware limit switch. If Upper level is chosen, the input signal is TRUE after the axis has reached or passed the hardware limit switch.

Low and High Software Limits

Software limits are used to stop the axis motion before the axis would reach a hardware limit switch. Motion is stopped when the axis reaches the software limit. The technology object reports a fault if this occurs. The axis can be moved within the acceptable motion range after the fault has been acknowledged. Note that software limit only works on an axis that has been homed. Note also that you do not have to have software limits and just use the hardware limits to protect the axis.

The desired position for the software limits within the range of travel are set in this screen (see Figure 10-6).

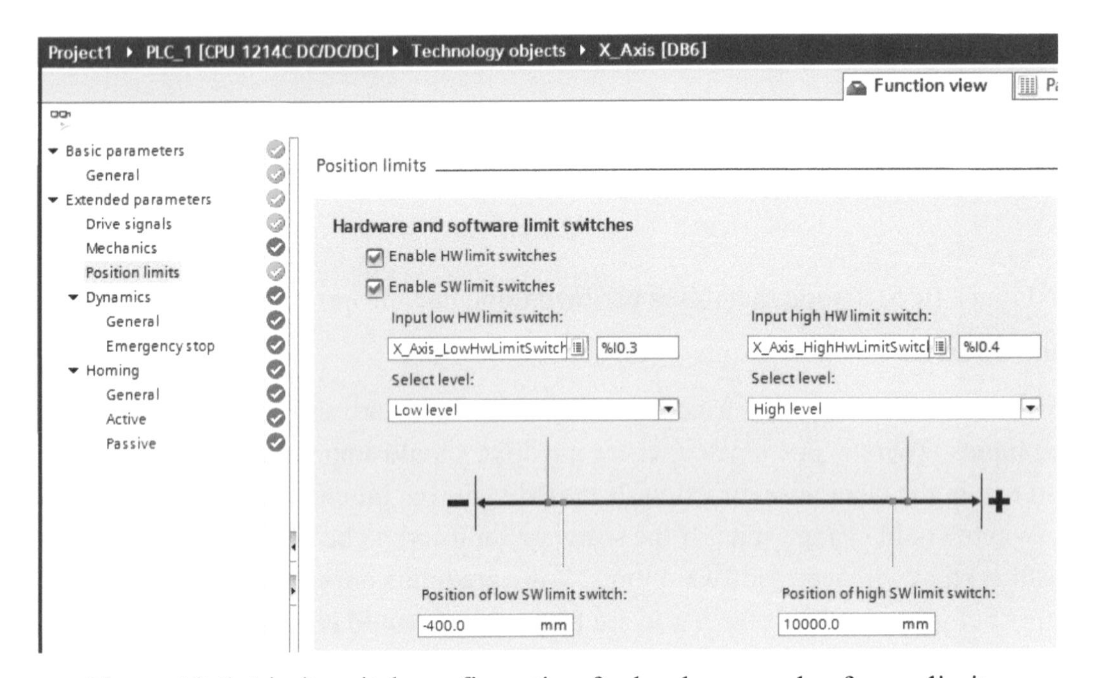

Figure 10-6. Limit switch configuration for hardware and software limits.

Setting the Axis Dynamics

Figure 10-7 shows the configuration screen for setting up the velocity limits and accelleration/decelleration values. This screen is used configure the maximum velocity, the start/stop velocity, and the maximum acceleration/deceleration of the axis. The physical unit for the velocity limits must also be selected from the drop-down list box.

Velocity

The maximum and start/stop velocity of the axis is also configured in this screen.

Acceleration/Deceleration

The Ramp-up time or Acceleration fields must also be configured. Set the deceleration value in the Ramp-down time or Deceleration fields.

Note that all changes to velocity limits have an effect on the acceleration/deceleration values of the axis. The ramp-up and ramp-down times are retained.

You may also activate the jerk limiter in this screen and enter a Smoothing time factor and a Jerk factor in mm/sec^2. Imagine being in a roller coaster ride at startup/stop or a rapid acceleration or deceleration. In the older rides you may have been "jerked" forward or backward in your seat by start/stop or acceleration/deceleration. Jerk factors in motion control are used to reduce or eliminate jerking.

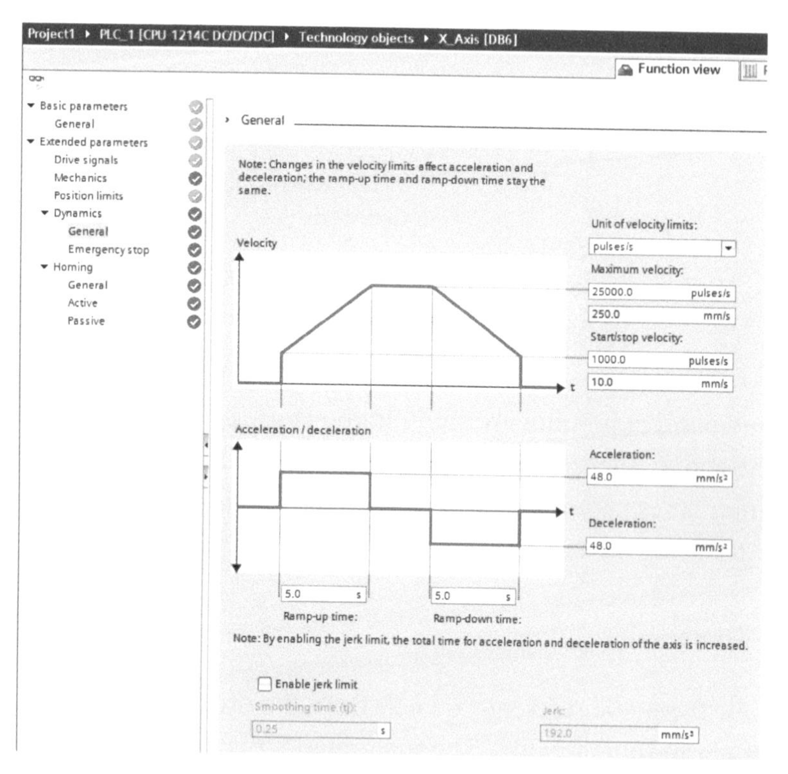

Figure 10-7. Velocity, accelleration/decelleration, and jerk screen.

Figure 10-8 shows the screen for configuring the emergency stop parameters for maximum velocity, start/stop velocity, decelleration and ramp down time.

The maximum emergency stop deceleration of the axis must be configured in the Dynamic emergency stop configuration window. In the event of an error, the axis is brought to a stop at this deceleration rate. The velocity values that were configured in the General dynamic response screen are displayed in this information area.

Deceleration

The deceleration limit for emergency stop is set in the Emergency stop ramp-down time or Emergency stop deceleration fields.

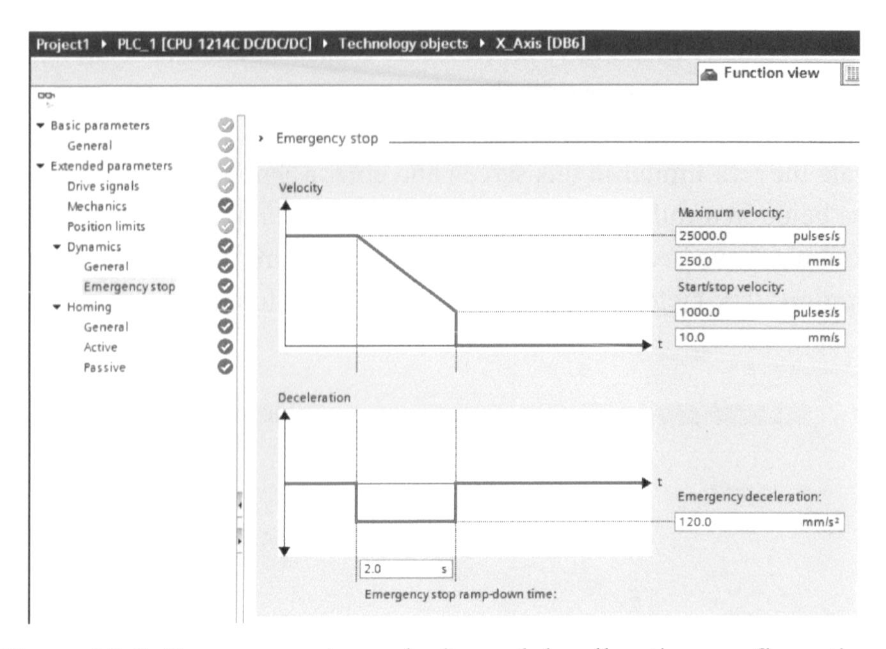

Figure 10-8. Emergency stop velocity and decelleration configuration.

Homing Configuration

Parameters for active and passive homing are set in in the Homing configuration screen (Figure 10-9). The homing mode of the active reference point approach is set up by means of the statement MC_Home Mode = 3; passive homing is set up with Mode = 2. The movement for passive homing must be triggered by the user (e.g., using an axis motion command).

Home Reference Input

Select the digital input for the home reference switch from the drop-down list box (Figure 10-9). The input used must support alarm functionality. The integrated CPU inputs and the inputs of a signal board can be used for the home reference point switch. Then enter the tagname or address of the input to be used as the reference (home) switch. The select the desired level to be used: high (normally-closed type switch) or low (normally-open type switch).

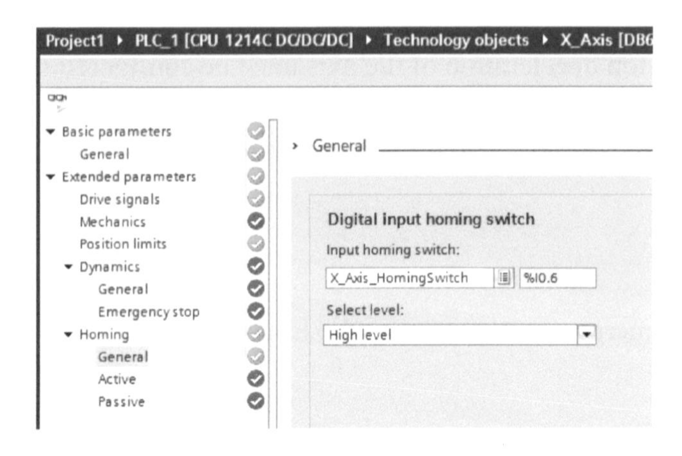

Figure 10-9. Configuration screen for specifying the input homing switch and the desired state for homing: high or low.

Approach Direction (Active Homing Only)

The approach direction for sensing the reference point (homing) switch is set here.

Reference Switch (Active and Passive homing)

Choose either the left or the right side of the reference point switch as homing position. Note that depending on the start position of the axis, the reference point approach can be different from the sequence displayed in the diagram in Figure 10-10.

There are two velocities involved in homing: approach velocity and referencing velocity. The approach velocity specifies the velocity at which the axis will move to find the reference (home) switch. The referencing velocity is the velocity at which the axis will move during referencing the home position (see Figure 10-9).

Startup Velocity (Active Homing Only)

The velocity of the axis motion during the reference point approach is specified here. Figure 10-9 also shows the configuration screen for homing and reference point direction. Activate the check box for Permit autoreverse at hardware limit switch to use the hardware limit switch as a reversing point for the reference point approach (homing). This is very useful if the axis is on the wrong side of the home switch when homing begins.

After the axis has reached the hardware limit switch during active homing, it is ramped down at the set deceleration rate and then reversed. The reference point switch is then sensed in reverse direction. If this function is not active and the axis reaches the hardware limit switch during active homing, the reference point approach is canceled with an error and the axis is stopped by means of emergency stop deceleration ramp.

Approach Velocity (Active Homing Only)

The velocity at which the axis approaches the reference point switch for homing is also configured here.

If it is desired to have a different position for the home reference point switch and home reference point position, in this field enter the respective reference point shift. In other words, if we would like to home the axis and then have that be 5.0 inches instead of 0.0, we can configure that here. The axis approaches the homing position at approach velocity.

The absolute reference point coordinate is specified at the position parameter in the MC_Home statement.

Figure 10-10 shows the parameter screen for configuring passive homing. One parameter is set to tell the motion Technology object which way the homing switch will be approached from: top or bottom. A tag name or address must also be entered into the Home position entry.

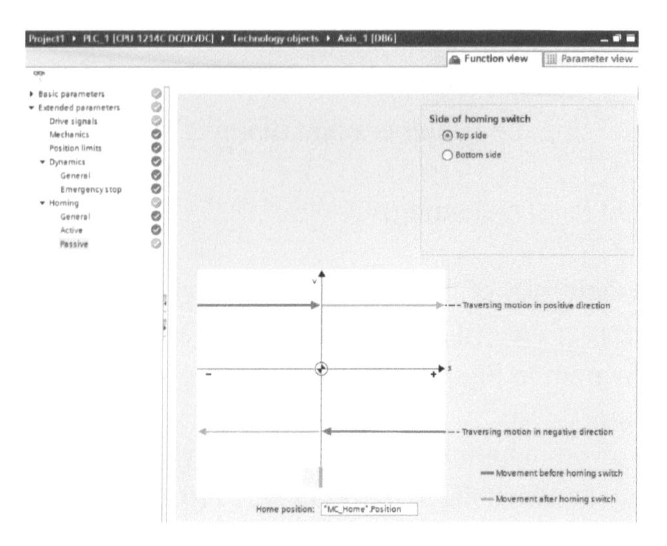

Figure 10-10. The configuration screen for passive homing.

This completes the configuration of the axis. If you have correctly configured all values there are three possible messages you will see.

The Configuration Contains Default Values and is Complete.

The configuration only contains default values. With those default values you can use the technology object without further changes.

The Configuration Contains User-Defined Values and is Complete

All values in the input fields of the configuration are valid and at least one default value was modified.

Incomplete or Incorrect Configuration

At least one input field or a drop-down list box has no value or an invalid value. The corresponding field, or the drop-down list box, is displayed on a red background. You can click on the roll-out error message to indicate the cause of error.

Commissioning Tool

Once you have an axis configured you can test the functioning of an axis by using the axis Commissioning tool. The commissioning tool is located under the axis being tested in the project tree. This allows the axis to be tested without having to develop a user program.

Diagnostics Tool

The axis Diagnostics tool is fond under the axis in the project tree. It can be sued to check the current status and error information of an axis and drive.

Motion Control Instructions

Now that the axis has been configured correctly it is possible to use motion control instructions. The Axis technology object is controlled by the use of motion control instructions. Motion control instructions will be covered next. Figure 10-11 shows some of the motion control instructions that are available. The usual sequence to enable and use an axis would be to use an MC_Power instruction to enable the axis for motion followed by an MC_Home instruction to establish a home position for the axis. Once this has been done the other axis commands can be used.

MC_Power	The MC_Power instruction enables or disables an axis.
MC_Reset	The MC_Reset instruction is used to acknowledge all Motion Control errors that require acknowledgment. Fatal errors are acknowledged by cycling power or by downloading the project data to the module again. If the cause of the errors has been eliminated, the values of the Error, ErrorID, and ErrorInfo parameters of the Motion instructions are reset upon acknowledgment.
MC_Home	The MC_Home instruction is used to establish a home position. Homing is required for absolute positioning of the axis. The following types of homing can be used: Active homing The homing procedure is executed automatically. Passive homing The homing procedure must be executed by the user. Direct homing absolute The home position is set in absolute terms. Direct homing relative The home position is set relative to the current position.
MC_Halt	The MC_Halt instruction stops all motions and brings the axis to a standstill. The command is completed when the axis has reached a standstill or it is canceled by a new motion instruction.
MC_MoveAbsolute	The MC_MoveAbsolute instruction starts the move of an axis to an absolute position. The command is terminated when the target position is reached.
MC_MoveRelative	The MC_MoveRelative instruction starts a move relative to the start position.
MC_MoveVelocity	The MC_MoveVelocity instruction causes the axis to move at the preset velocity.
MC_MoveJog	The MC_MoveJog instruction is used to move an axis with closed-loop speed control in jog mode.

Figure 10-11. Motion control instructions.

Enabling and Disabling an Axis

An axis must first be enabled to operate. The MC_Power instruction is used to enable or disable an axis (see Figure 10-12). To use this instruction the axis must have been configured correctly and there cannot be a pending error that would inhibit the enabling of the axis.

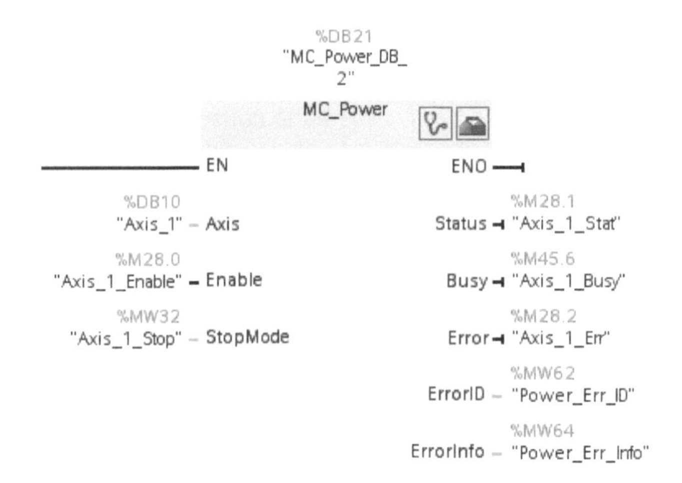

Figure 10-12. A MC_Power instruction.

The MC_Power instruction inputs and their function in are shown in Figure 10-13.

Inputs for a Power (Enable/Disable) Axis Instruction			
Parameter	Data Type	Default Value	Description
Axis	TO_Axis_1		Axis to be used.
Enable	BOOL	False	If true, motion control attempts to enable the axis. If false, all active jobs are aborted according to the configured StopMode and the axis is stopped.
StopMode	INT	0.0	If 0 then emergency stop. If a request to disable the axis is pending, the axis brakes at the configured emergency stop deceleration. The axis is disabled after reaching standstill. If 1 then Immediate stop. If a request to disable the axis is pending, this axis is disabled without deceleration. The pulse output is stopped immediately. *If 2 then* emergency stop with jerk control. If a request to disable the axis is pending, the axis brakes at the configured emergency stop deceleration. If the jerk control is activated, the configured jerk is taken into account. The axis is disabled after reaching standstill.

Figure 10-13. Inputs to a power axis instruction.

The MC_Power instruction outputs and their functions in a MC_Power instruction are shown in Figure 10-14.

Outputs from a Power (Enable/Disable) Axis Instruction			
Parameter	Data Type	Default Value	Description
Done	BOOL	False	If false, the axis is disabled. The axis does not execute motion control jobs and does not accept any new jobs (exception: MC_Reset command). The axis is not homed. Upon disabling, the status does not change to FALSE until the axis reaches a standstill. If true, the axis is enabled. The axis is ready to execute motion control jobs. Upon axis enabling, the status does not change to TRUE until the signal "Drive ready" is pending. If the "Drive ready" drive interface was not configured in the axis configuration, the status changes to TRUE immediately.
Busy	BOOL	False	The instruction is active.
Error	BOOL	False	An error occurred during execution of the command. The cause of the error can be found in parameters ErrorID and ErrorInfo.
ErrorID	WORD	16#0000	This is the Error ID for the error.
ErrorInfo	WORD	16#0000	This is the Error Information ID for the error.

Figure 10-14. Outputs from a power axis instruction.

MC_Reset

The MC_Reset instruction is used to acknowledge axis errors (see Figure 10-15). The MC_Reset instruction can be used to acknowledge the Operating error with axis stop error and the Configuration error. The errors that require acknowledgement can be found in the List of ErrorIDs and ErrorInfo.

To use the MC_Reset instruction, the axis must have been configured correctly.

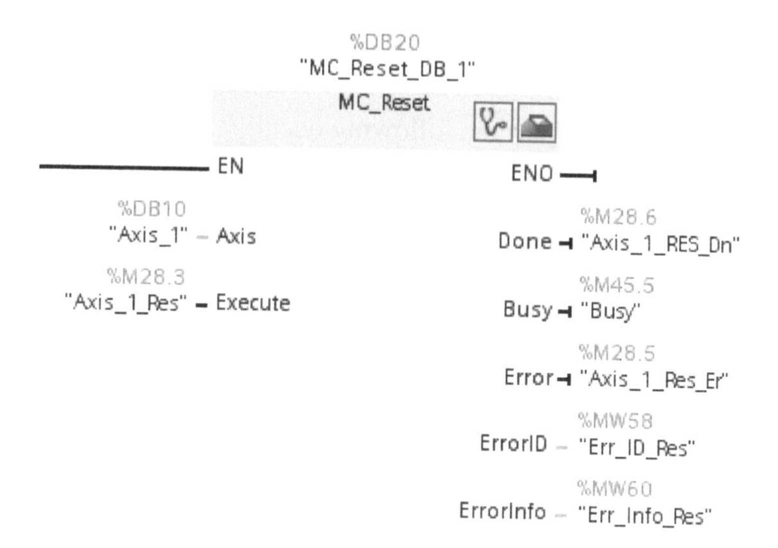

Figure 10-15. A MC_Reset instruction.

The MC_Reset instruction inputs and their function are shown in Figure 10-16.

Inputs to a Reset (Acknowledge Error) Instruction			
Parameter	Data Type	Default Value	Description
Axis	TO_Axis_1		Axis to be used.
Execute	BOOL	False	Start of the command with a positive edge transition.
Restart	INT	0.0	If true, download the axis configuration from the load memory to the work memory. The command can only be executed when the axis is disabled. If false, acknowledges pending errors.

Figure 10-16. Inputs to a reset instruction.

The MC_Reset instruction outputs and their function are shown in Figure 10-17.

Outputs from a Reset (Acknowledge Error) Instruction			
Parameter	Data Type	Default Value	Description
Done	BOOL	False	Error has been acknowledged.
Busy	BOOL	False	The instruction is being executed.
Error	BOOL	False	An error occurred during execution of the command. The cause of the error can be found in parameters ErrorID and ErrorInfo.
ErrorID	WORD	16#0000	This is the Error ID for the error.
ErrorInfo	WORD	16#0000	This is the Error Information ID for the error.

Figure 10-17. Outputs from a reset instruction.

MC_Home Instruction

Homing is performed to establish a home reference position so that the control can establish a reference point to track position. An axis normally homed one time at startup of the axis.

The homed state of the axis is lost if any of the following occur:

- A CPU memory reset
- A power transition of power off to power on.
- A CPU restart
- A reset of the axis enable at the Motion Control statement MC_Power

With the exception of a MC_MoveAbsolute instruction, all motion commands can be executed in a non-homed state. An MC_MoveAbsolute instruction requires that the axis be homed.

Axis homing is performed with a MC_Home instruction (see Figure 10-18). During homing, the reference point coordinate is set at a defined mechanical position of the axis.

Homing Modes

- Active Homing

In active homing mode, the MC_Home instruction is used to perform the required reference point approach. All other active motions are cancelled.

- Passive homing

In passive homing mode, the MC_Home instruction does not execute a reference point approach. Other active motions are not cancelled. Approaching the reference point switch must be executed by the user via motion control instructions or by mechanical movement.

- Direct Homing Absolute

The axis position is set irrespective of the reference switch position. Other active motions are not cancelled. The value of the Position parameter in the MC_Home instruction is activated immediately as a reference point and position value of the axis. The axis must be at a standstill in order to allow the precise assignment of the reference point to a mechanical position.

- Direct Homing Relative

The axis position is set irrespective of the reference switch position. Other active motions are not cancelled. The reference point and axis position will be:

New axis position = current axis position + value of the position parameter.

The value for the reference point and axis position is activated immediately. The axis must be at a standstill in order to allow the precise assignment of the reference point to a mechanical position.

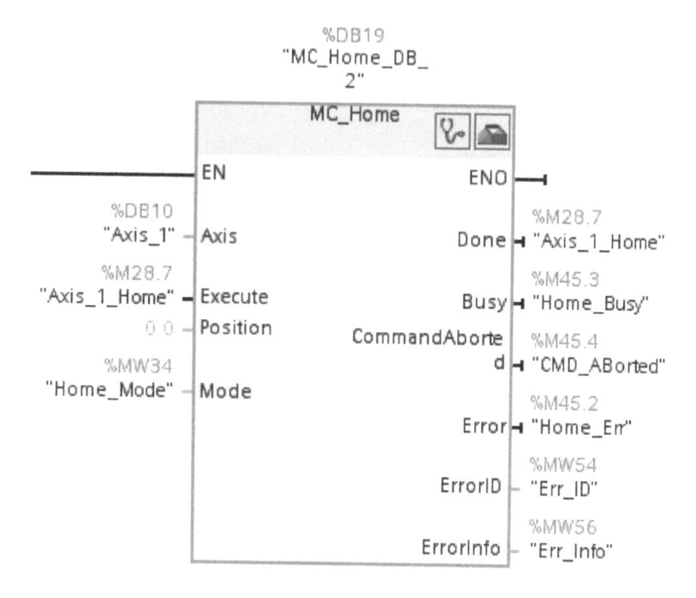

Figure 10-18. A MC_Home instruction.

To use this instruction the axis must have been been configured correctly and the axis must be enabled. If Mode = 0, 1 or 2, no MC_CommandTable command may be active.

Figure 10-19 shows the inputs and their function for a home instruction.

Inputs to a Home Instruction			
Parameter	Data Type	Default Value	Description
Axis	TO_Axis_1		Axis to be used.
Execute	BOOL	False	Start of execution with a positive edge transition.
Position	REAL	0.0	If Mode = 0, 2, and 3, Absolute position of axis after completion of the homing operation. If Mode = 1, Correction value for the current axis position. Limit values: $-1.0e^{12} \leq$ Position $\leq 1.0e^{12}$
Mode	INT	0	0 = Direct homing absolute. New axis position is the position value of parameter "Position". 1 = Direct homing relative. New axis position is the current axis position + position value of parameter "Position". 2 = Passive homing. Homing according to the axis configuration. Following homing, the value of parameter "Position" is set as the new axis position. 3 = Active homing. Reference point approach in accordance with the axis configuration. Following homing, the value of parameter "Position" is set as the new axis position.

Figure 10-19. Inputs to a home instruction.

Figure 10-20 shows the outputs and their function for a home instruction.

Outputs from a Home Instruction			
Parameter	Data Type	Default Value	Description
Done	BOOL	False	Command completed
Busy	BOOL	False	The instruction is being executed
CommandAborted	BOOL	False	During execution the command was aborted by another command.
Error	BOOL	False	An error occurred during execution of the command. The cause of the error can be found in parameters ErrorID and ErrorInfo.
ErrorID	WORD	16#0000	This is the Error ID for the error
ErrorInfo	WORD	16#0000	This is the Error Information ID for the error

Figure 10-20. Outputs from a home instruction.

MC_Halt Instruction

The MC_Halt instruction stops all movements and brings the axis to a standstill with the configured deceleration. The standstill position is not defined. An MC_Halt instruction is shown in Figure 10-21.

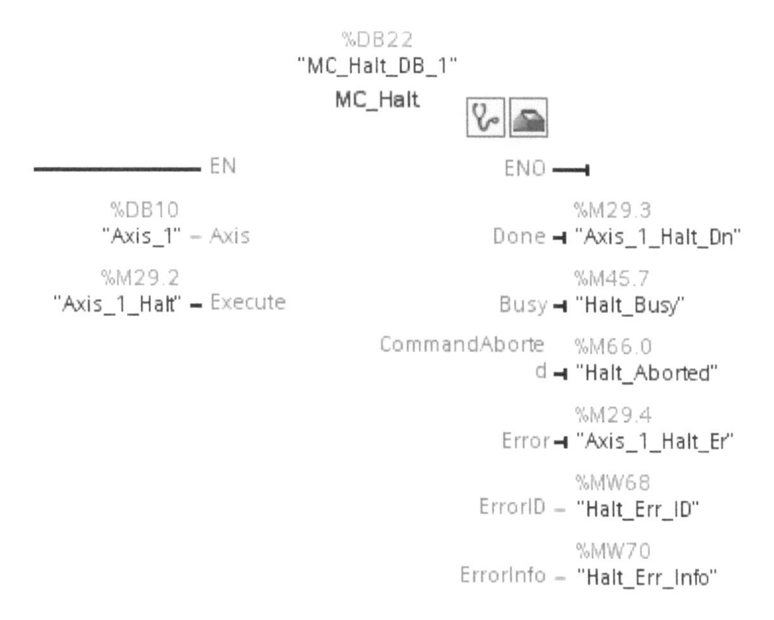

Figure 10-21. A MC_Halt instruction.

To use the MC_Halt instruction the axis must have been configured correctly and the axis must be enabled. The inputs to a MC_Halt instruction and their function are shown in Figure 10-22.

Inputs to a Halt Instruction			
Parameter	Data Type	Default Value	Description
Axis	TO_Axis_1		Axis to be used.
Execute	BOOL	False	Start of execution with a positive edge transition.

Figure 10-22. Inputs to a halt instruction.

The outputs from a MC_Halt instruction and their function are shown in Figure 10-23.

Outputs from a Halt Instruction			
Parameter	Data Type	Default Value	Description
Done	BOOL	False	Zero velocity reached.
Busy	BOOL	False	The instruction is being executed.
CommandAborted	BOOL	False	During execution the command was aborted by another command.
Error	BOOL	False	An error occurred during execution of the command. The cause of the error can be found in parameters ErrorID and ErrorInfo.
ErrorID	WORD	16#0000	This is the Error ID for the error.
ErrorInfo	WORD	16#0000	This is the Error Information ID for the error.

Figure 10-23. Outputs from a halt instruction.

MC_MoveAbsolute Instruction

The MC_MoveAbsolute instruction starts an axis positioning motion to move it to an absolute position. An MC_MoveAbsolute instruction is shown in Figure 10-24.

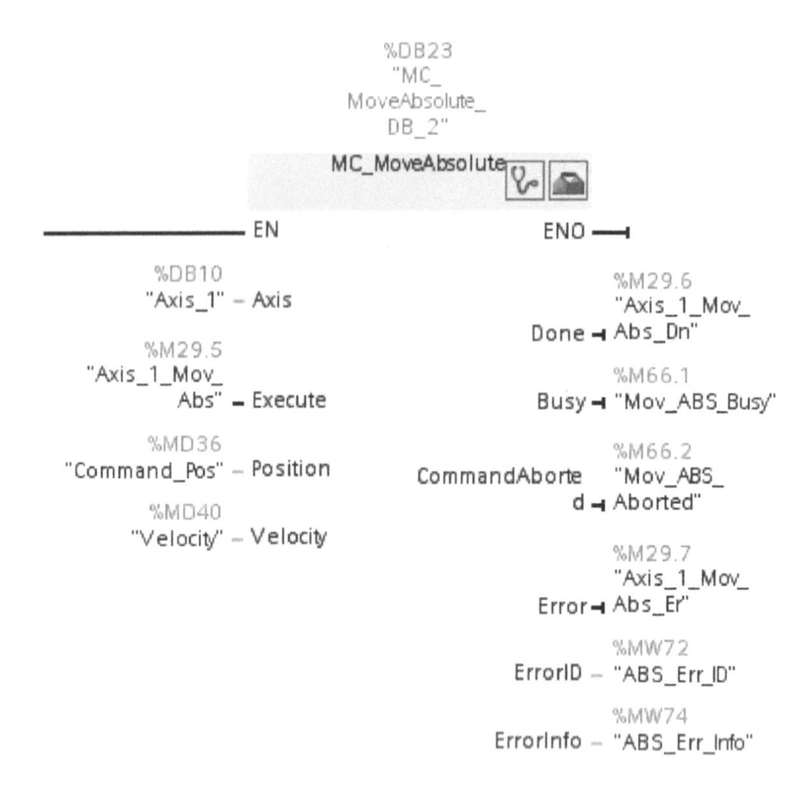

Figure 10-24. A MC_MoveAbsolute instruction.

To use the MC_MoveAbsolute instruction the axis must have been configured correctly, the axis must be enabled, and the axis must have been homed. The inputs to a MC_MoveAbsolute instruction and their function are shown in Figure 10-25.

Inputs to a Move Absolute Instruction			
Parameter	Data Type	Default Value	Description
Axis	TO_Axis_1		Axis to be used.
Execute	BOOL	False	Start of execution with a positive edge transition.
Position	REAL	0.0	Absolute target position. Limit values: $-1.0e^{12} \leq$ Position $\leq 1.0e^{12}$
Velocity	REAL	10.0	Velocity of axis This velocity is not always reached on account of the configured acceleration and deceleration and the target position to be approached. Limit values: Start/stop velocity ≤ Velocity ≤ maximum velocity.

Figure 10-25. Inputs to a move absolute instruction.

The outputs from a MC_MoveAbsolute instruction and their function are shown in Figure 10-26.

Outputs from a Move Absolute Instruction			
Parameter	Data Type	Default Value	Description
Done	BOOL	False	Absolute target reached.
Busy	BOOL	False	The instruction is being executed.
CommandAborted	BOOL	False	During execution the command was aborted by another command.
Error	BOOL	False	An error occurred during execution of the command. The cause of the error can be found in parameters ErrorID and ErrorInfo.
ErrorID	WORD	16#0000	This is the Error ID for the error.
ErrorInfo	WORD	16#0000	This is the Error Information ID for the error.

Figure 10-26. Outputs from a move absolute instruction.

MC_MoveRelative: Position Axes Relative Instruction

The MC_MoveRelative instruction starts a positioning motion relative to the start (current) position (see Figure 10-27). For example, a 3.0 move in the positive direction would move it three inches from its current (start) position. Another 3.0 relative move would move it 3 more inches in a positive direction. This is often called incremental positioning.

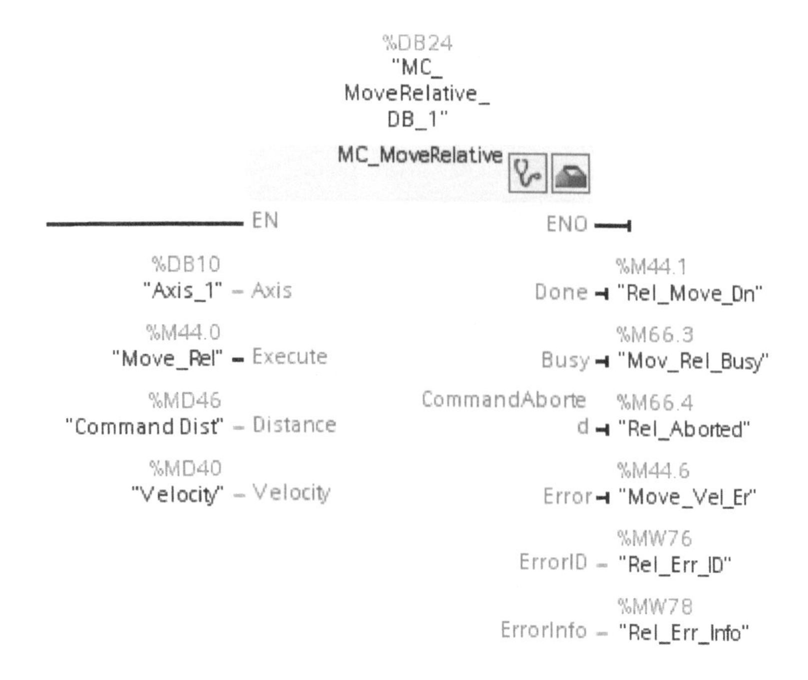

Figure 10-27. A MC_MoveRelative instruction.

To use the MC_MoveRelative instruction the axis must have been configured correctly and the axis must be enabled. Note that the axis does not have to be homed to use this instruction. The inputs to a MC_MoveRelative instruction and their function are shown in Figure 10-28.

Inputs to a Move Relative Instruction			
Parameter	Data Type	Default Value	Description
Axis	TO_Axis_1		Axis to be used.
Execute	BOOL	False	Start of execution with a positive edge transition.
Position	REAL	0.0	Travel distance for the positioning operation.
Velocity	REAL	10.0	Velocity of axis. This velocity is not always reached on account of the configured acceleration and deceleration and the target position to be approached. Limit values: Start/stop Start/stop velocity ≤ Velocity ≤ maximum velocity.

Figure 10-28. Inputs to a move relative instruction.

The outputs from a MC_MoveRelative instruction and their function are shown in Figure 10-29.

Outputs from a Move Relative Instruction			
Parameter	Data Type	Default Value	Description
Done	BOOL	False	Target position reached.
Busy	BOOL	False	The instruction is being executed.
CommandAborted	BOOL	False	During execution the command was aborted by another command.
Error	BOOL	False	An error occurred during execution of the command. The cause of the error can be found in parameters "ErrorID" and "ErrorInfo".
ErrorID	WORD	16#0000	This is the Error ID for the error.
ErrorInfo	WORD	16#0000	This is the Error Information ID for the error.

Figure 10-29. Outputs from a move relative instruction.

MC_MoveVelocity: Move Axes with Speed Preset Instruction

The MC_MoveVelocity instruction moves an axis constantly at the specified velocity. Figure 10-30 shows a MC_MoveVelocity instruction.

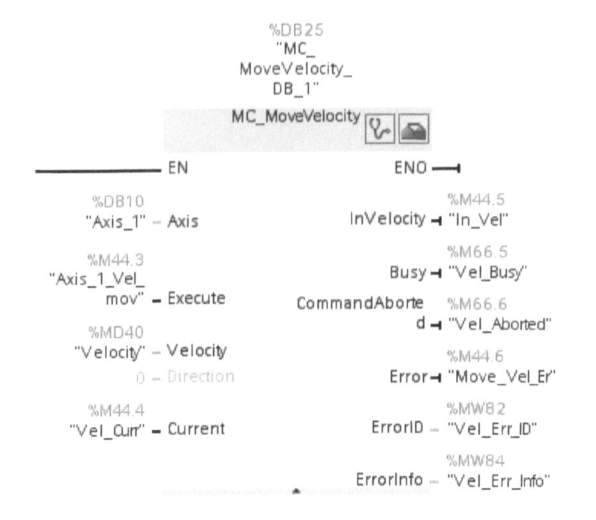

Figure 10-30. A MC_MoveVelocity instruction.

To use the MC_MoveVelocity instruction the axis must have been configured correctly and the axis must be enabled. Note that the axis does not have to be homed to use this instruction. The inputs to a MC_MoveVelocity instruction and their function are shown in Figure 10-31.

Inputs to a Move Velocity Instruction			
Parameter	Data Type	Default Value	Description
Axis	TO_Axis_1		Axis to be used.
Execute	BOOL	False	Start of execution with a positive edge transition.
Velocity	REAL	10.0	Velocity specification for axis motion. Limit values: Start/stop velocity ≤ Velocity ≤ maximum velocity. (Velocity = 0.0 is permitted).
Direction	INT	0	If 0 the direction of rotation corresponds to the sign of the value in parameter "Velocity". If 1 then positive direction of rotation. (The sign of the value in parameter "Velocity" is ignored). If 2 then negative direction of rotation. (The sign of the value in parameter "Velocity" is ignored).
Current	BOOL	False	If false then “maintain current velocity" is deactivated. The values of parameters "Velocity" and "Direction" are used. If true, then "Maintain current velocity" is activated. The values in parameters "Velocity" and "Direction" are not taken into account. When the axis resumes motion at the current velocity, the "InVelocity" parameter returns the value TRUE.

Figure 10-31. Inputs to a velocity instruction.

The outputs from a MC_MoveVelocity instruction and their function are shown in Figure 10-32.

Outputs from a Move Velocity Instruction			
Parameter	Data Type	Default Value	Description
InVelocity	BOOL	False	• If true then, "Current" = FALSE: The velocity specified in parameter "Velocity" was reached. • If true then, "Current" = TRUE: The axis travels at the current velocity at the start time.
Busy	BOOL	False	The instruction is being executed.
CommandAborted	BOOL	False	During execution the command was aborted by another command.
Error	BOOL	False	An error occurred during execution of the command. The cause of the error can be found in parameters ErrorID and ErrorInfo.
ErrorID	WORD	16#0000	This is the Error ID for the error.
ErrorInfo	WORD	16#0000	This is the Error Information ID for the error.

Figure 10-32. Outpus from a velocity instruction.

MC_MoveJog: Move axes in Jog Mode Instruction

A MC_MoveJog instruction moves the axis at the specified velocity in jog mode (see Figure 10-33). You use this motion control instruction, for example, for (jogging) moving an axis.

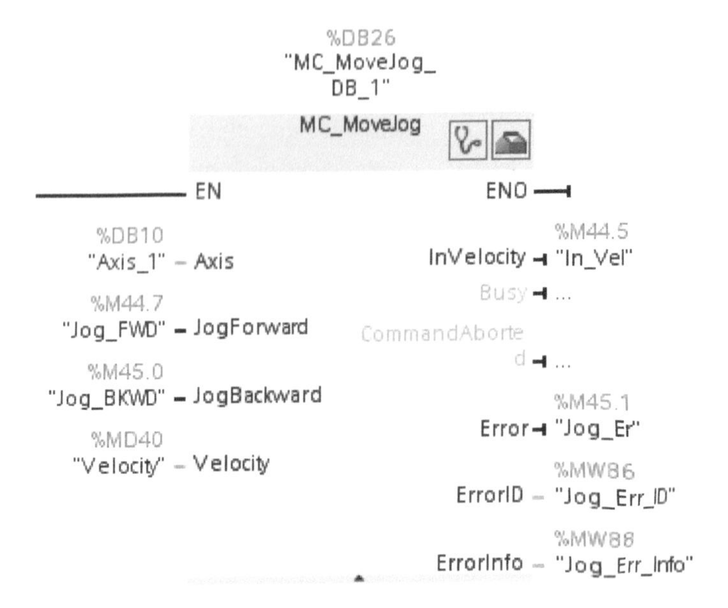

Figure 10-33. MC_MoveJog instruction.

The axis must have been configured correctly and the axis must be enabled to use this instruction. The inputs to a MC_MoveJog instruction and their function are shown in Figure 10-34.

Inputs to a Jog Instruction			
Parameter	Data Type	Default Value	Description
Axis	TO_Axis_1	TO_Axis_1	Axis to be used.
JogForward	BOOL	False	If true, axis moves in positive direction.
JogBackward	BOOL	False	If true, axis moves in negative direction.
Velocity	BOOL	10.0	Velocity to be used.

Figure 10-34. Inputs to a jog instruction.

The outputs from a MC_MoveJog instruction and their function are shown in Figure 10-35.

Outputs from a Jog Instruction			
Parameter	Data Type	Default Value	Description
InVelocity	BOOL	False	If True, the velocity specified by the Velocity parameter has been reached.
Busy	BOOL	False	If True, the instruction is being executed.
CommandAborted	BOOL	False	If during the execution of the instruction the command was aborted by another command this will be True.
Error	BOOL	False	An error occurred during execution of the command. The cause of the error can be found in parameters "ErrorID" and "ErrorInfo".
ErrorID	WORD	16#0000	This is the Error ID for the error.
ErrorInfo	WORD	16#0000	This is the Error Information ID for the error.

Figure 10-35. Outputs from a jog instruction.

Motion Status Diagnostics

The diagnostics function, Motion status, can be used to monitor the motion commands of the axis. Diagnostics functions displays are available in online mode in the Manual control mode and in Auto mode. The meanings of the displayed fields are shown in Figure 10-36 and 10-37.

Status	Description
Target position	The Target position indicates the current target position of positioning command (e.g., by means of the MC_MoveAbsolute or MC_MoveRelative statement). The Target position is only displayed if the axis is homed and a positioning command is active.
Current position	The Current position field indicates the current axis position. The field only indicates the current position of a homed axis.
Current velocity	The Current velocity field indicates the actual axis velocity.

Figure 10-36. Motion status fields in diagnostics.

Dynamic limits

Dynamic limit	Description
Velocity	The Velocity field indicates the configured maximum velocity of the axis.
Acceleration	The Acceleration field indicates the configured maximum acceleration of the axis.
Deceleration	The Deceleration field indicates the configured maximum deceleration of the axis.

Figure 10-37. Limits for velocity, acceleration, and deceleration in diagnostics.

Chapter Questions

1. What is the difference between an axis and a drive?

2. What is the function of the enable output?

3. Describe what hardware limits are and how they function.

4. Describe what software limits are and how they function.

5. What is the purpose of homing an axis?

6. What are the basic steps that must be completed before an axis can be controlled with motion instructions?

7. What is a MC_Power instruction used for?

8. Describe the 4 modes of homing available in a MC_Home instruction.

9. What is the difference between incremental and absolute moves?

10. What instruction can be used to acknowledge axis errors?

Chapter 11

Developing HMI Applications

Objectives

Upon completion of this chapter, the reader will be able to:

Add an HMI to a project.

Assign IP addresses.

Develop HMI applications.

Introduction

This chapter will examine the development of HMI applications for the Siemens Simatic Touch Panel HMI. An HMI is a Human Machine Interface. It is a user-configured application that is designed to give an operator a window-to-the-world of what is happening in an application. This chapter will cover how to add hardware, assign IP addresses, configure the Simatic Touch Panel, assign tags to objects, and how to connect to the hardware and load a project.

The HMI is a useful tool for operators in operating and troubleshooting an application. Data is exchanged between the HMI and the PLC via tags. Siemens HMIs can communicate with the PLC via PROFIBUS, Industrial Ethernet, and Siemens Multi-Point Interface (MPI). MPI is based on the EIA-485 (formerly RS-485) standard and works with a speed of 187.5 kBd.

HMIs are very configurable. HMIs can be simple or complex. They can range from a few bits of information to a fully integrated machine control platform. The main influences that affect HMI functionality are both the application itself as well as the user developing it. The main test for HMI functionality depends on the solution it provides. If enough meaningful information is presented to allow operators to keep machines running with minimal downtime, then the application is effective. A lot of time can be spent developing HMI applications. It is up the developer to have a good understanding of the scope of an application along with the responsibilities of those involved with running the equipment. Knowing these things will aide in developing appropriate controls. Unnecessary controls written into an application can become costly if a lot of time is used to write them. It can also be costly if the HMI application does not give operators and maintenance personnel the capabilities they need. Trying to find a balance between too much and not enough information and control is the ultimate goal.

Adding a New HMI Device

First you must add the HMI hardware we want to use in our project (see Figure 11-1). In the following examples we will be using a KTP600 color touch panel. From the portal view under the Devices & Networks tab, select Add new device.

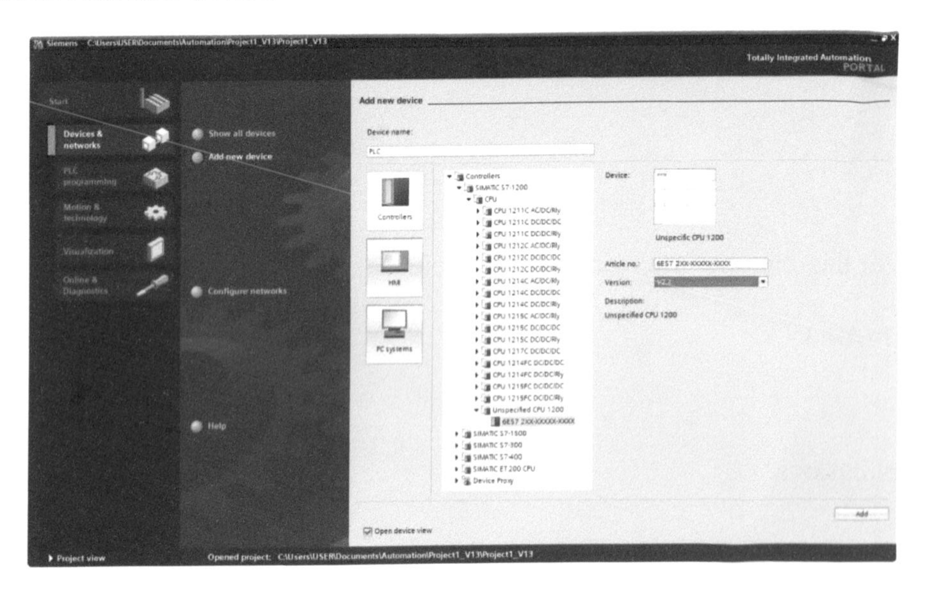

Figure 11-1. Adding a new device.

Select HMI from the list of available devices. After adding a device, it will show up under devices and networks. Once the hardware has been added, an HMI Device Wizard will appear if the checkbox "Start device wizard" (lower left) is checked. The device wizard will help you configure your HMI to set up the application to the correct model parameters. Features can be selected at this point but they are limited in configurability. The main one to focus on is PLC connections. In Figure 11-2, under PLC connections, you can select the PLC we want to connect to. For this example, PLC_1 is the one we want to select. For details on adding a PLC to your project refer to Appendix A.

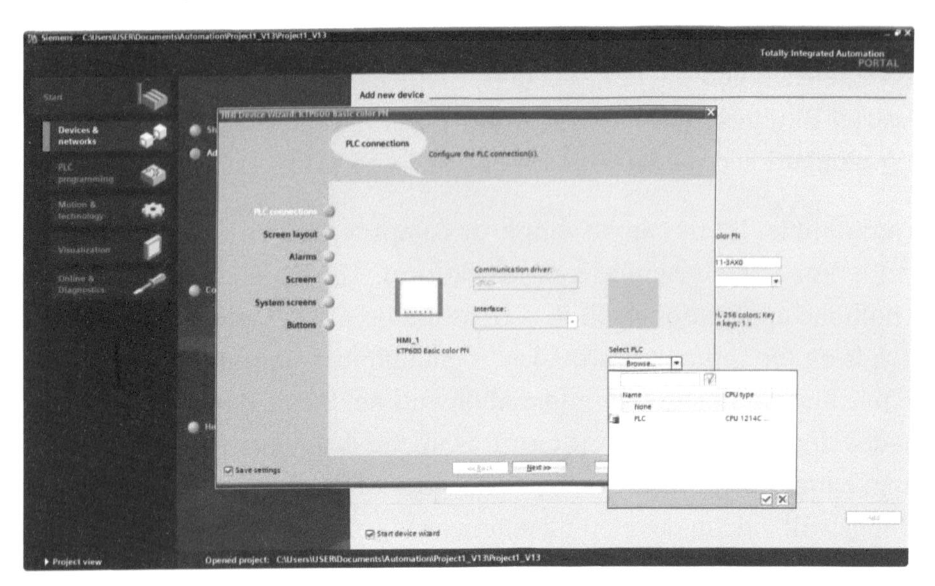

Figure 11-2. PLC connections, select PLC.

If a connection is not initially made between the HMI and PLC, it can be made later. In Project View under Devices & networks a connection can be manually made by drawing a line between the HMI and the PLC it will communicate with (see Figure 11-3).

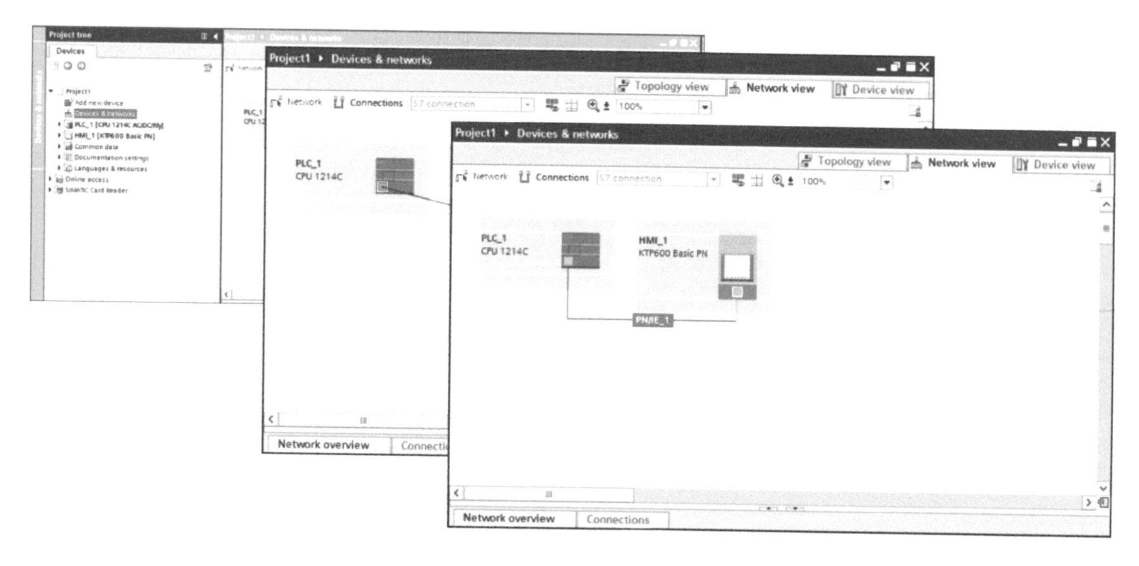

Figure 11-3. Manual partnering PLC and HMI connections.

Setting a network connection between the PLC and HMI makes a logical connection to the tag databases. This allows you to browse tags from the connected PLC within the HMI configuration. This allows you to assign object attributes to specific tags and their values.

After finishing the HMI Device Wizard, in project view, the initial HMI configuration screen will appear (see Figure 11-4). Basic object, advanced elements and additional controls can be found in the toolbox area. Additional features and configuration can be found under the HMI device itself within the Project tree.

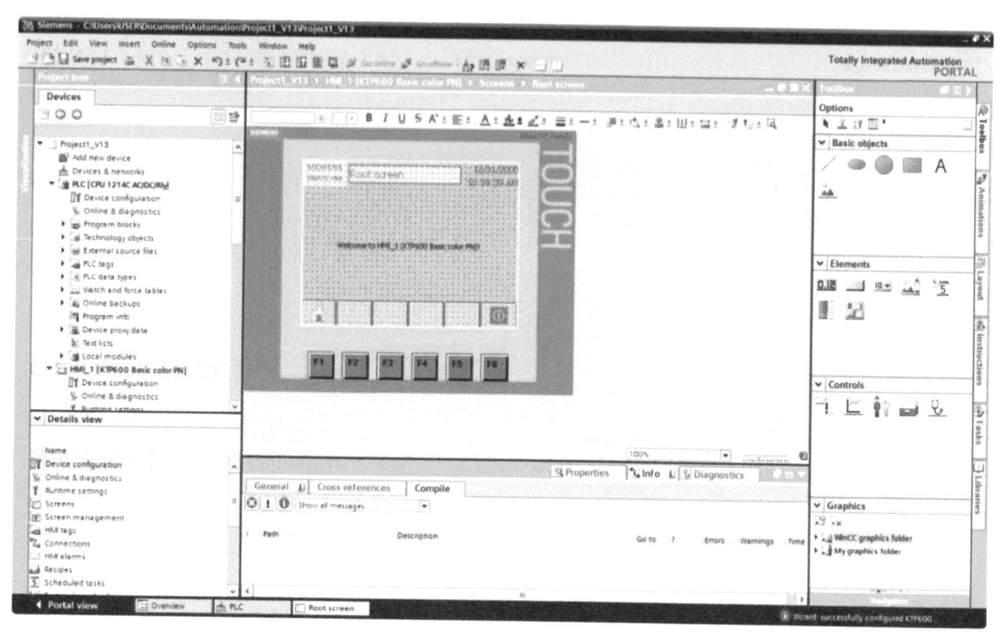

Figure 11-4. Initial HMI screen.

Configuring Project and Hardware Device IP Addresses

Working from within the project tree, open the properties of the HMI device. The IP address of the HMI can be configured under device properties. Some HMI hardware may already have an IP address assigned. In this example configuring the HMI project to match the HMI hardware device IP address is the only thing that needs to be done. If an IP address has not already been assigned to the hardware or you wish to change an IP address, this section will show you how to assign one.

To configure an IP address within a project, right click on the HMI device within the Project tree and select properties. Within properties, under the general tab, select the PROFINET Interface (X1) configuration. As shown in Figure 11-5, the checkbox, "Use IP protocol", and the radio button, "Set IP address", in the project are both selected along with an IP address and Subnet mask. Assign the appropriate IP address, confirm settings and select OK.

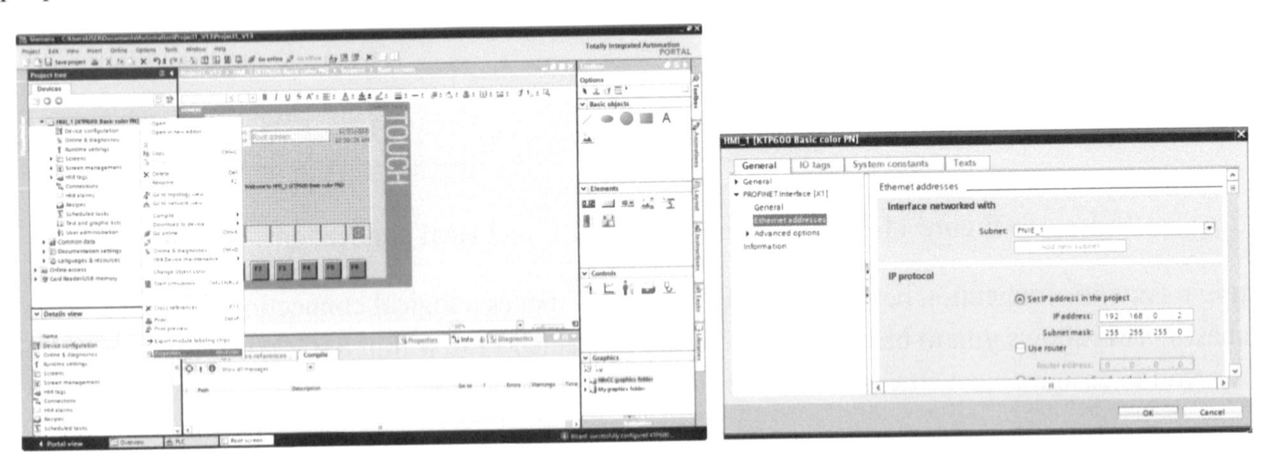

Figure 11-5. Configuring the projects HMI IP address.

In order to clearly see how a hardware IP address is assigned, close any open project and return to the Portal View as shown in Figure 11-6. By closing any open project, we can demonstrate how to assign an IP address to any accessible hardware device independent of a project. These same steps can be used while a project is open but it helps to show how these tools operate separate from a project. Once within the Portal View click on the Online & Diagnostics tab. Then, click on Accessible devices.

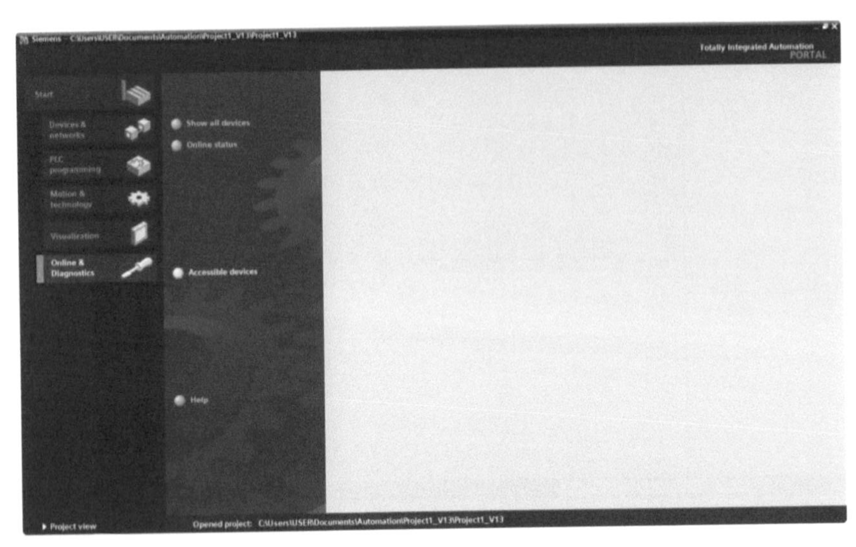

Figure 11-6 Online Diagnostics and Accessible devices.

After clicking on the Accessible devices button, an accessible devices window will appear (see Figure 11-7). From this widow select the appropriate network interface to connect to the accessible device and select the Refresh button. In this instance we are looking for the device type SIMATIC-HMI. Select the SIMATIC-HMI from the list and click the Show button.

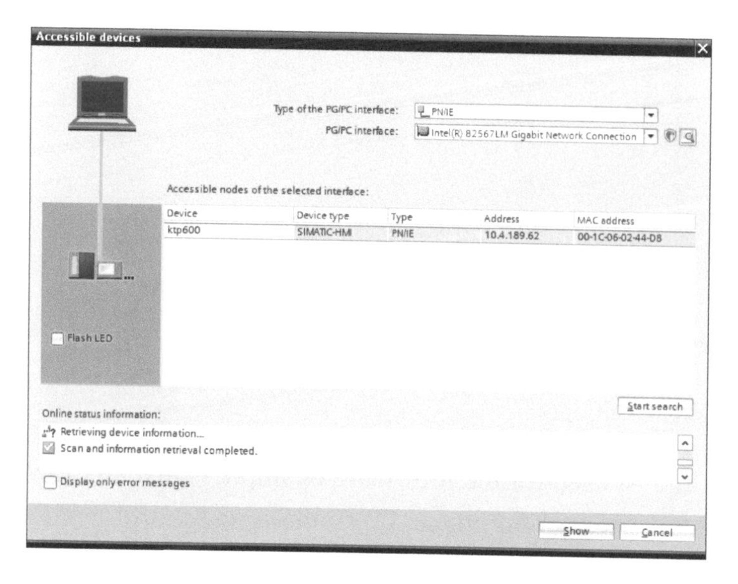

Figure 11-7. Finding Accessible devices.

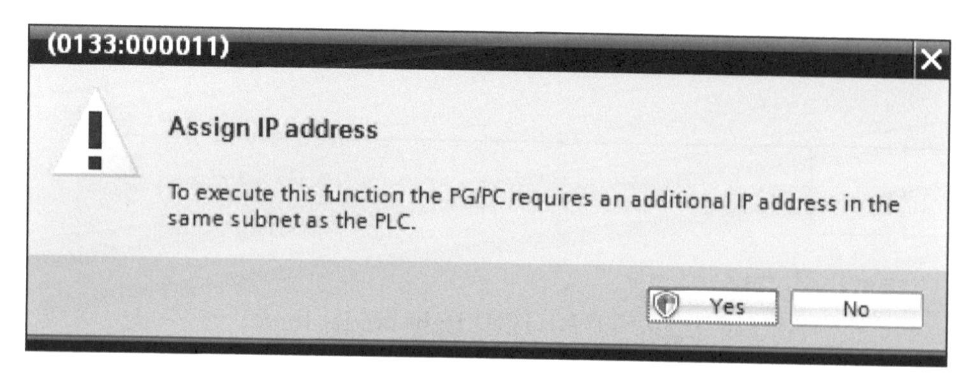

Figure 11-8. IP assignment Prompt.

A prompt may appear to add an addition IP address in the same subnet as the PLC. Since we are only assigning an IP address at this time, select No (see Figure 11-8). The reason for the assign IP address prompt, is that in order to communicate with any IP addressed hardware, your workstation must be on the same IP range and subnet.

As shown in Figure 11-9, we return to the Project view where we can see the generic Project tree. Under Online access, the accessible HMI device will appear. Double click on the Online & diagnostics for the device. An Online access window will appear. Under Functions, navigate to Assign IP address. Enter the appropriate IP address and subnet mask. Confirm the MAC address of the destination device. The MAC address is a unique hardware address that is assigned to the devices from the manufacture. It helps in identifying equipment so you do not inadvertently assign an IP to the wrong device. Once you have confirmed the MAC address matches the target device, click the Assign IP address button. The IP address will be assigned. Once the device matches the project configured IP address an HMI project can be downloaded.

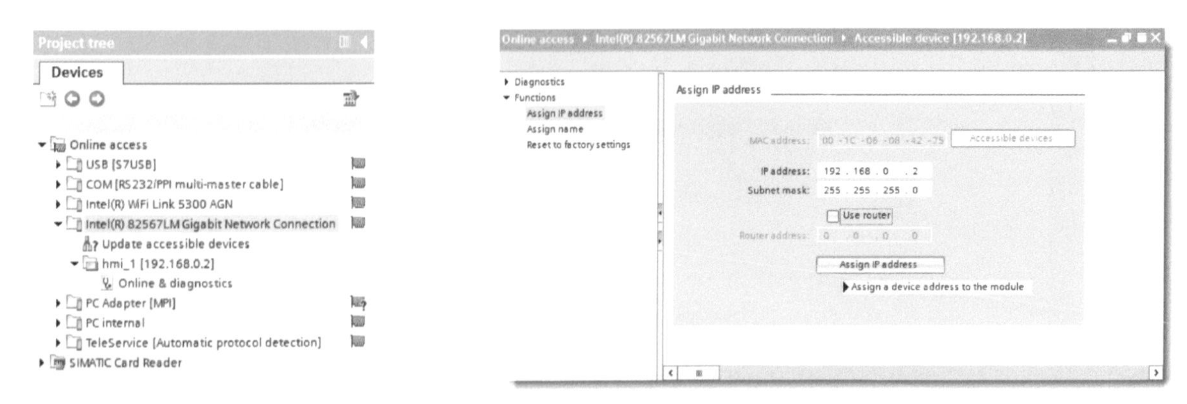

Figure 11-9. Assigning an IP to an accessible HMI device.

Working with Tags and Basic Objects

The ladder programming that is used for the following example is just an example to base our object configurations on. The ladder program in Figure 11-10 has a real-world input named Input_0, and a memory tag named, HMI_Button. We are also using a memory tag named HMI_Light as our output. To begin we will create the first object base on Input_0.

Figure 11-10. HMI light ladder logic.

To start we will add a basic object from the toolbox. Select the circle object from basic objects, then click and drag a circle anywhere on the HMI screen (see Figure 11-11).

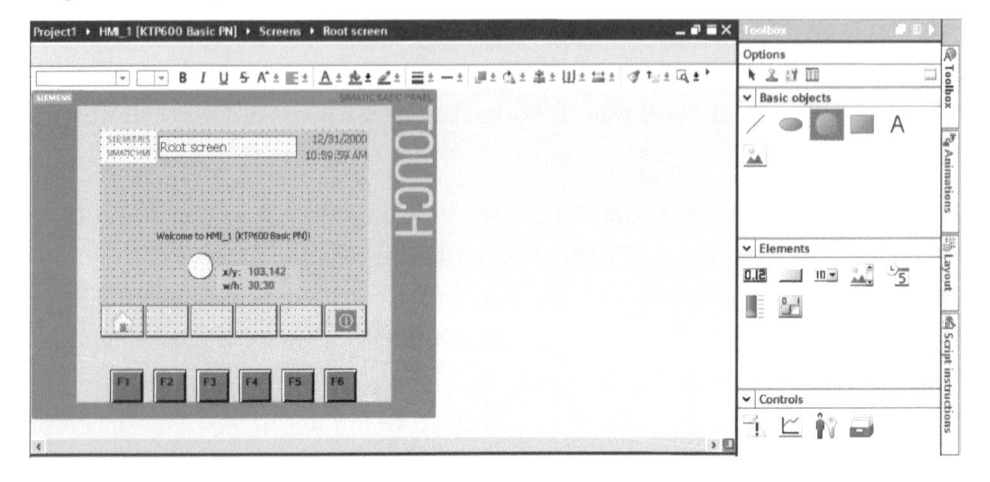

Figure 11-11. Adding a circle to the display.

Once we have added a circle, we will add animation to it. Animations alter the appearance of an object based on a tag value. Animations are a great way to give objects meaningful at-a-glance information to operators. With our circle selected, we can see some general information about the object listed under the properties tab. This object will be configured under the animations tab of the object (see Figure 11-12). Under animation, click on the arrow adjacent to the Display configuration then select Add new animation followed by selecting Appearance. Under Appearance select the tag name you want to associate with the animation. In this example it is Input_0. Input_0 is a configured tag within PLC_1. When selecting a tag from the drop-down menu the full path may look something like this:

PLC_1/ControllerTagsFolder/Default tag table/Input_0

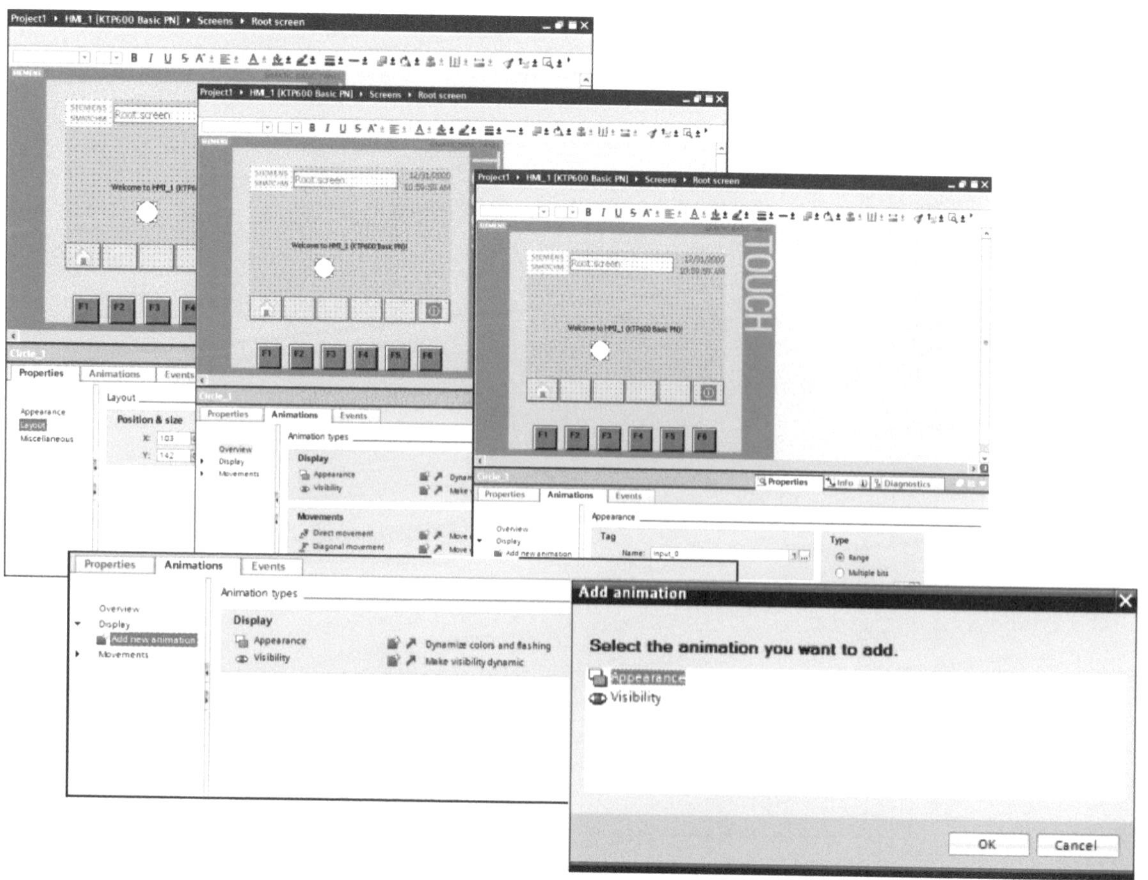

Figure 11-12. Animation configuration.

The syntax for this tag's path starts with the PLC we are looking at (PLC_1), the containing folder (ControllerTagsFolder), the tag table (Default tag table), and finally the tag (Input_0). If the tag has already been associated with the HMI the tag path may look something like this:

Default tag table/Input_0

The syntax for this tag's path starts with the HMI tag default table (Default tag table), and the tag within the table (Input_0). Under the HMI default tag table, additional columns show connections and PLC names for individual tags. If a tag does not have a connection to a PLC, under the connection's column, it will be labeled as <Internal tag>. Internal tags are used as variables that only interact with the HMI and do not directly communicate with the PLC. The reason for the two different tag paths is because once a PLC tag has been assigned to an attribute within the HMI, it will then be added to the HMI's default tag table. This allows users to select from a tag table of tags already assigned to HMI attributes.

There are a range of values associated with the appearance of the circle object under the Appearance tab. To add a range value, click "add new". In this instance we show two states for Input_0. When Input_0 is equal to zero (0), the circle will have a white background with a grey border. When Input_0 is equal to one (1) the circle will have a white background with a green border. Colors can be changed by selecting the pull-down menu for each background and foreground color on each specified range.

The same steps to animate Input_0 are followed to animate HMI_Light (see Figure 11-13).

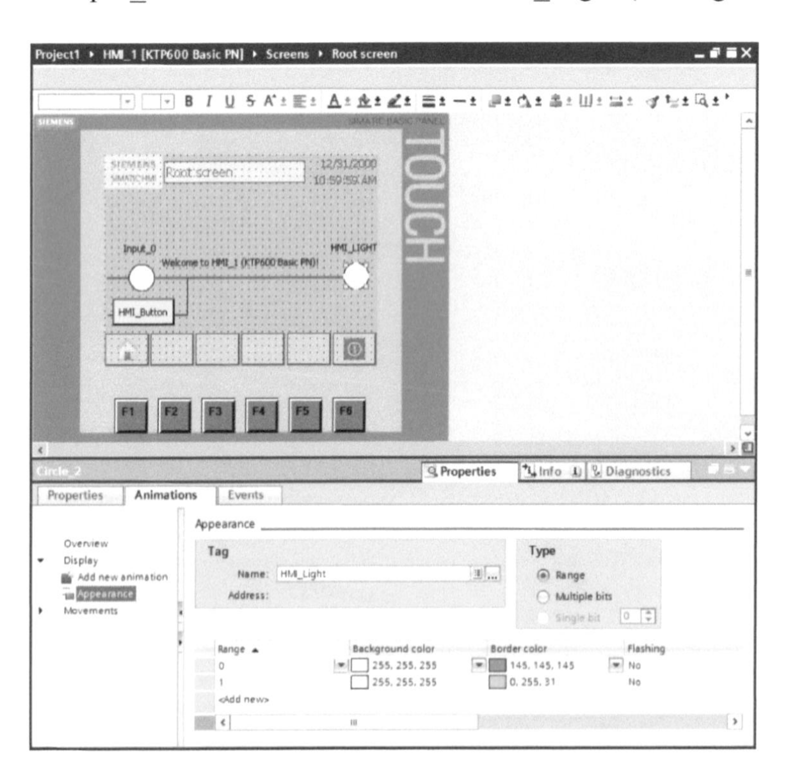

Figure 11-13. Animation for HMI_Light.

Button Object

The HMI objects in this example were intentionally arranged to best illustrate the LAD program. Objects can be arranged in any configuration that represents the application.

Our HMI button is assigned to a memory address that is read/write. This allows the HMI to change the value of the tag from the HMI. Looking at Figure 11-14, under the events tab for our button we assign a Press action. To assign a Press action select "Add function". Under "Add function", expand "Edit bits" followed by selecting the "InvertBit" function.

Once the function is added, in the tag column for Tag (input/output), assign HMI_Button. The InvertBit action takes the current state of a memory bit and changes it to the opposite state. This provides the functionality of a maintained push-button. To start with we will assume that HMI_Button is in the off or zero state. By pressing the button once the state of HMI_Button will be change from a zero (0) to a one (1). The resulting logic sets HMI_Light to one (1). Pressing the button again will invert signal of the button from one (1) back to zero (0) and the HMI_Light turns off.

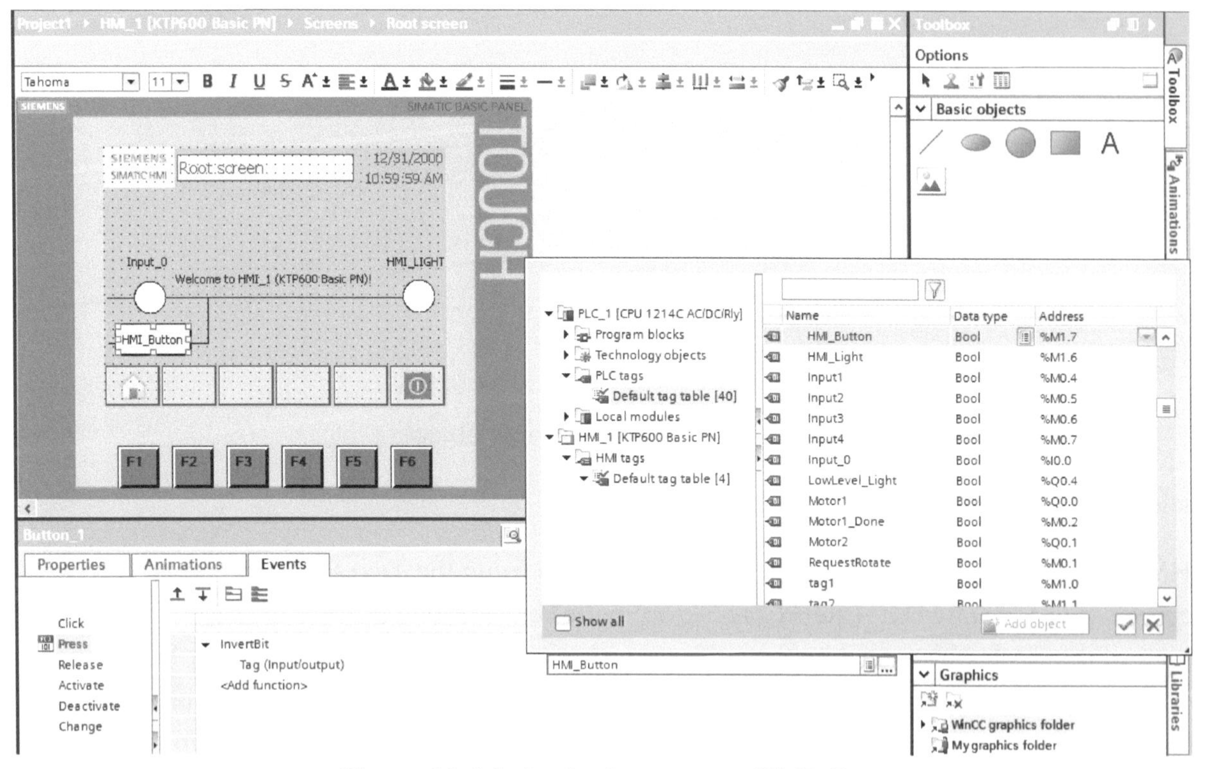

Figure 11-14. Assigning a tag to HMI_Button.

Graphic View Object

In the last example we showed how a basic object like a circle can be used to represent the state of a tag value. We also used simple lines to represent those in the ladder program. The Line, Ellipse, Circle, and Rectangle objects all draw and configure similarly. The variety of these objects will allow users to draw graphics that represent the process. Users will find that using a combination of these objects along with animations will enable the developer to accurately represent any application.

It is useful to be able to add and alter an object within a screen but, sometimes it is useful to use an existing graphic that represents the process. A Graphic View object can be used to insert a picture. One benefit of using the Graphic View object is the ability to take a picture of real equipment then overlay animations and text to better document and represent the process. The Graphic View object can also be used for icons as well as company logos. The one drawback to this is that the graphics used will have to be edited outside of the Step 7 programming environment. Draw a Graphic View object on your screen, then by right clicking on the object, select Add graphic. Browse for the graphic you want to use and select open (see Figure 11-15).

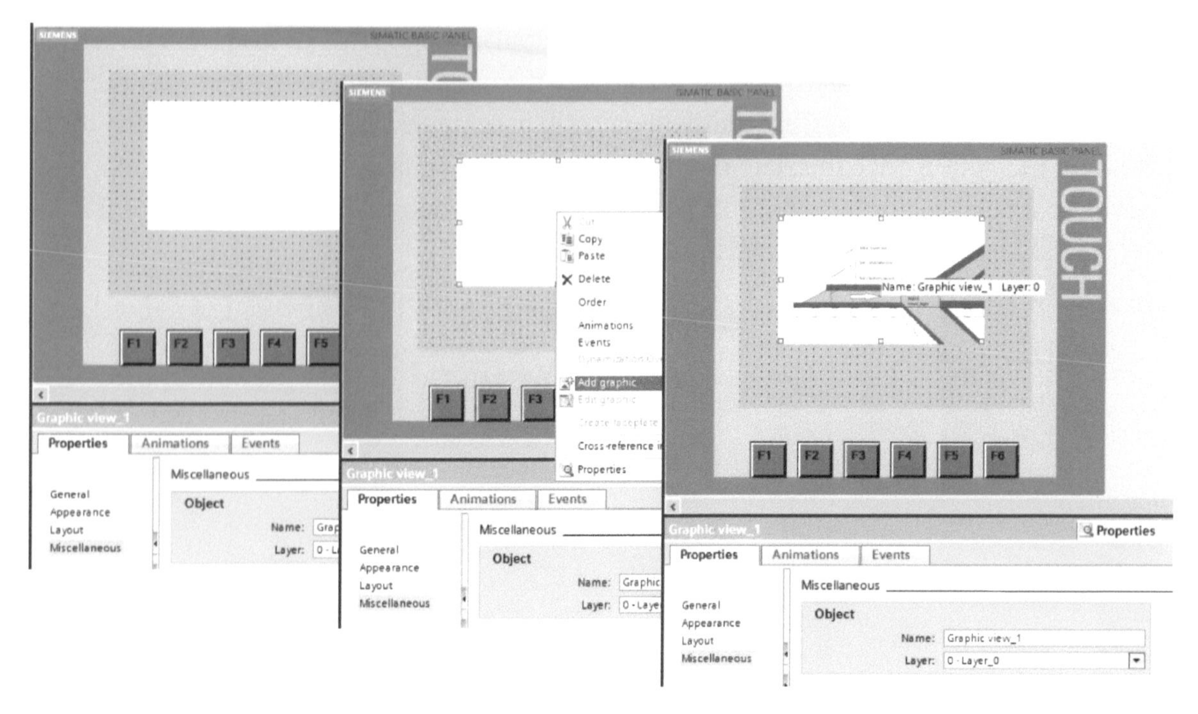

Figure 11-15. Graphic view object.

Text Field Object

In our HMI Light example, Figure 11-13, text was inserted using the Text Field object. We had text on the screen for Input_0 and HMI_LIGHT. These are both Text Fields that were inserted to provide a description of the circle objects below them. Text can be inserted by selecting the Text Field object, then clicking and dragging anywhere on a screen. Once the object has been drawn, general properties can be changed to give the text different sizes, colors, and fonts. Animations can also be assigned (see Figure 11-16).

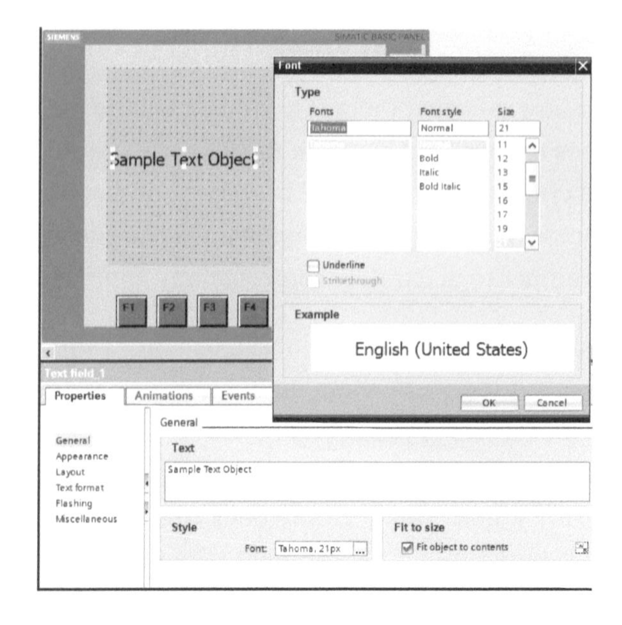

Figure 11-16. Text field object.

I/O Field Object

There is often a need for an operator to enter values to control a process. One way to enter values is to use the I/O Field object. This object has many built-in features. In Figure 11-17, under the properties tab, under General, we assign the tag whose value we want to change. Notice under Type, the mode is Input/Output. This means the operator has the ability to change the value to the HMI. Format can help when displaying at runtime. Number of decimal places and field length can also be changed.

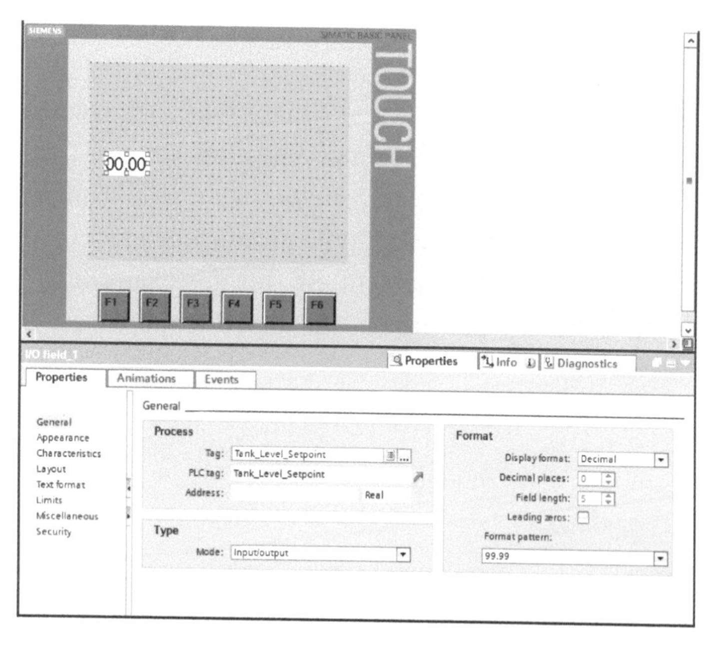

Figure 11-17. I/O field object.

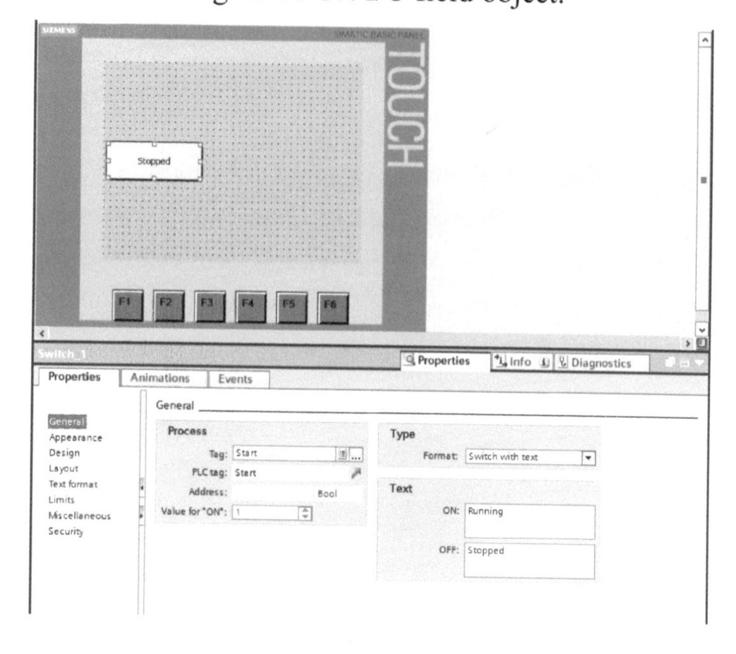

Figure 11-18. Switch object.

Switch Object

The Switch object is used when changing the value of a tag from zero to one (see Figure 11-18). The feedback method for a switch lies within the text for ON and OFF. In this instance when the switch is toggled from OFF to ON the text will change from Stopped to Running. When the Switch is toggled again, the state of the switch will return to zero and the text will read Stopped. The state of the Start tag can be used in logic to enable or disable logic.

Bar Object

Sometimes we need to know more than if something is true or false. We use analog values to give us more information about what is happening in a process. When raw PLC analog data comes in, it comes in as a value from 0-32767. If this is from a level transmitter, we would want to know what this range of values represents. It could be a distance, percent, or some other value representative of what it is being measured. An easy measurement to relate to is percent. So, for this example, 0-32767 will be equal to 0-100%. That means when the transmitter is putting out full signal and the PLC is reading in 32767, we will have 100% for our level reading.

Figure 11-19. Tank level ladder logic.

In this example we are taking raw data from analog channel AI0, memory address %IW64, which has been associated with the tag Tank_Level. Notice the min/max values on the NORM_X instruction in Figure 11-19. This instruction takes the range of 0-32767 and converts it to an equivalent 0.0 - 1.0 scale. Next, we take this adjusted value and use the SCALE_X instruction to convert from 0.0 - 1.0 to 0 - 100. The end result is a value that is representative of 0-100% (see Figure 11-20).

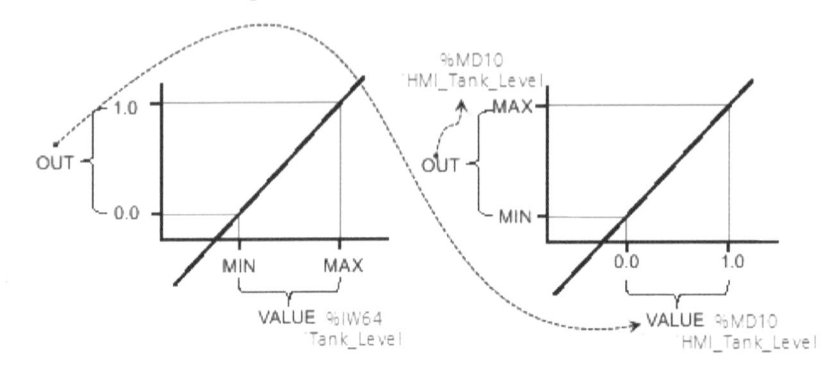

Figure 11-20. Data flow through NORM_X and SCALE_X instructions.

Programmers Tip: In this example the result from the NORM_X instruction's output is passed to HMI_Tank_Level. Since the final value of HMI_Tank_Level is written over by the SCALE_X instruction, HMI_Tank_Level can be used as a placeholder for values between instructions. This technique is often used to minimize the number of tags needed when using a series of instructions to get a single result. This only works if the final tag value is initially over-written at the start of each execution.

Now that we have a tag value to work with, a Bar object can be configured (see Figure 11-21). The Bar object can be configured in many ways to present information about the process. Limits and colors can be assigned to let operators know if a process is within normal ranges. Under Animations, ranges can be established for notification. Different background and foreground colors can be assigned based on the values of an assigned tag. One example might be a yellow foreground if the value of HMI_Tank_Level is between 80-90 and a red flashing foreground if the value is between 91-100. The first range would serve as a warning to the operator to monitor the process. The second range would indicate to an operator to make some decision to put the process back in operating range. Other things that can be changed include number of divisions and numeric intervals. A Bar object should be scaled so that operators know what the process is doing.

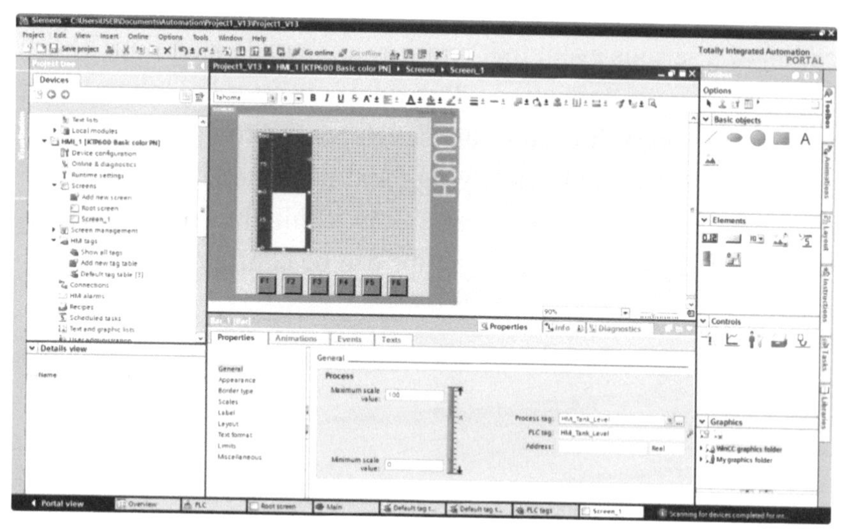

Figure 11-21. Bar object

Trend View Object

Having analog data is useful when monitoring the current value within a process. We can also use the analog data to provide historical data with trending. Trending allows us to look at a changing value over time. This is very useful when tuning or troubleshooting a process. Trends consist of an X and Y axis. Where X is the changing value and Y is time.

To configure a Trend View object, select Trend View object from the toolbar area. Click and drag the object on the desired screen. Under the properties tab, assign a tag value you wish to monitor. In Figure 11-22, our source is HMI_Tank_Level. Other advanced features can be found under the properties tab. Number of graph divisions, colors, and text are just a few of the attributes that can be assigned to the trend view properties.

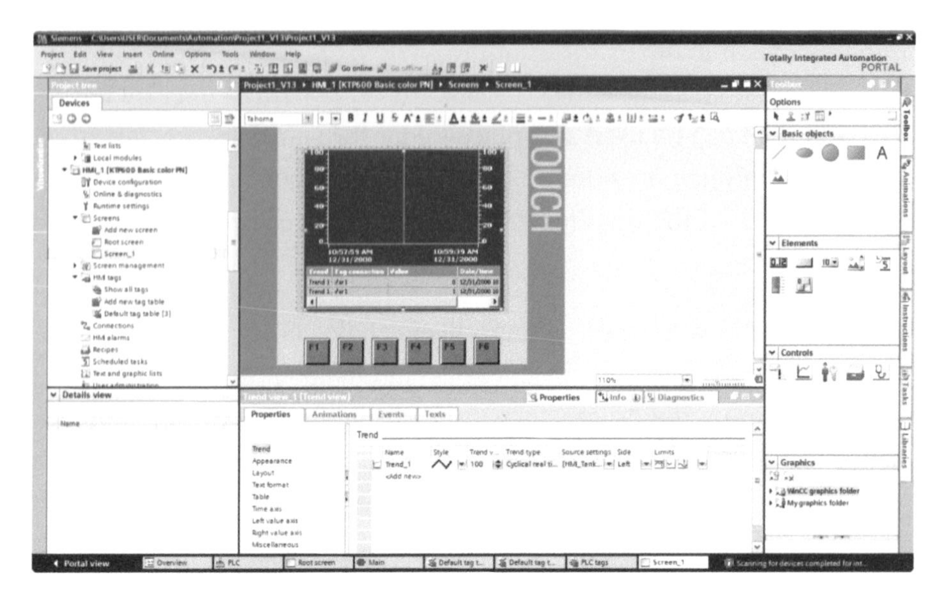

Figure 11-22. Trend View Object.

Alarming

Alarming is a user configured set of messages that notifies an operator of a condition they need to be made aware of. Alarming can be something as minor as a reminder to check something. Alarming can also serve as notification of a serious event that requires immediate attention, like nuclear reactor that is overheating. Alarming goes beyond animations. Instead of operator having to look at a screen they can be prompted to conditions that exist. It is up to the developer to provide alarming that will protect both the machine and the people operating it by notifying them when an alarm condition exists. The HMI alarm configuration is under HMI in the project tree. It is within the HMI alarms configuration. Discrete alarms, analog alarms, alarm classes and alarm groups can be configured. Figure 11-23 shows the configuration of a discrete alarm.

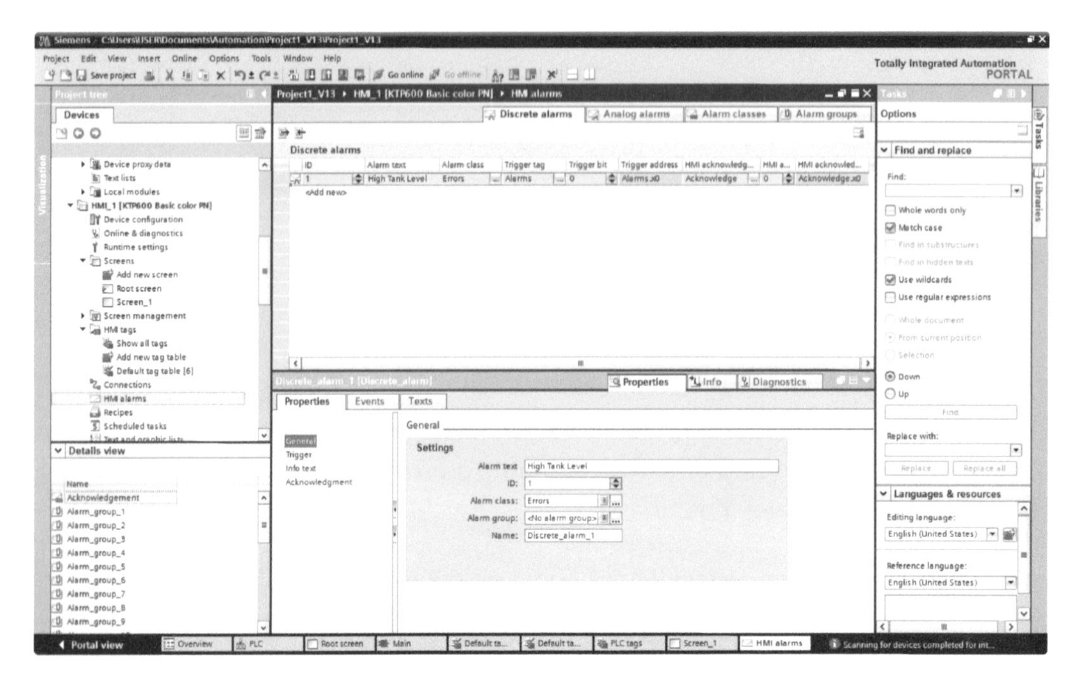

Figure 11-23. Discrete alarms.

Within the PLC an alarm tag called "Alarms" is configured. The Alarms tag is a whole word address that uses the bit values to trigger an alarm with an associated message. In our example bit X0 of Alarms is the trigger for High Tank Level. When bit X0 of the Alarms tag is set to 1 the message "High Tank Level" will appear in the alarm list.

Acknowledgement of discrete alarms.

Notice the HMI acknowledgment tag column. In Figure 11-24 under properties, acknowledgment can be assigned to HMI and PLC tags. HMI and PLC tag acknowledgment work differently. HMI acknowledgment is feedback status that an acknowledgment was made on a particular alarm. PLC acknowledgment is used to trigger an acknowledgment on a particular alarm. In addition to PLC and HMI acknowledgement there is also a built in HMI system acknowledgement. PLC acknowledgment and HMI system acknowledgment both acknowledge alarms. The HMI system acknowledgment can be only be used to acknowledge alarms.

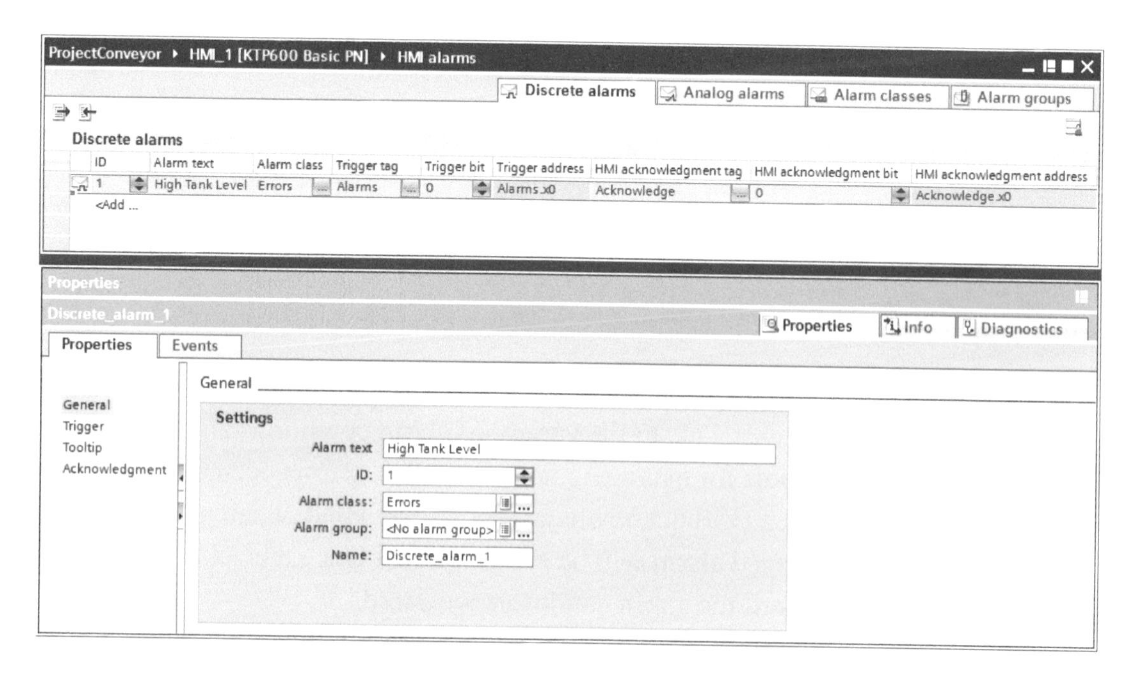

Figure 11-24. HMI Acknowledgement.

One benefit to using PLC tag acknowledgment is the ability to program conditions that need to be satisfied before an acknowledgment can occur. If conditions need to be met before an alarm can be acknowledged, PLC tag acknowledgment can provide application specific solutions. Neither tag is required in the acknowledgment properties of an alarm. The HMI system acknowledgment can be used for alarm acknowledgment by itself, but each acknowledgment source provides unique solutions.

It is because of the assigned alarm class in our example that acknowledgment is required. Alarm classes that require an acknowledgment can have acknowledgment tags assigned. Alarm classes define how alarm conditions are managed. In our example the alarm class is named Errors. Within the HMI alarms configuration window under Alarm classes, the definition for each alarm class is displayed. In Figure 11-25 we show three alarm classes.

The Errors class requires an acknowledgement whereas the System and Warnings classes do not. As shown, colors can be assigned to the alarm messages to differentiate the alarm classes. This can be helpful when operators have to determine the severity of an alarm and whether it needs to be acknowledged. For the Errors alarm class an "Alarm with Simple Acknowledgment" is assigned to the acknowledgement model. This means that when an Error class alarm occurs, the alarm will stay in the alarm list until it is out of alarm and acknowledged.

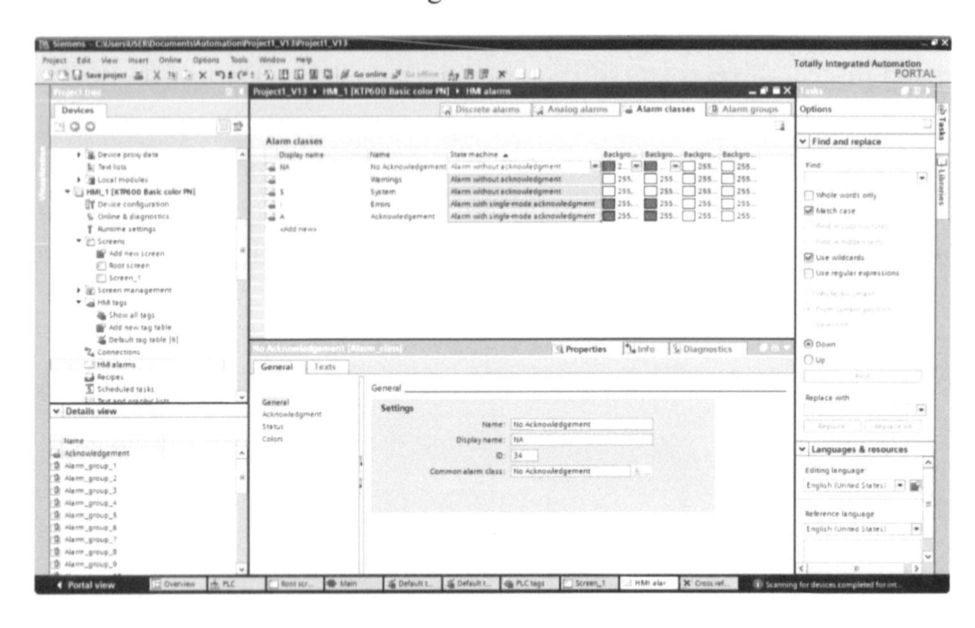

Figure 11-25 Alarm classes.

Setting status on an acknowledgment tag from the HMI is one way for the PLC to sense when an operator has acknowledged an alarm. Due to the variety of alarm conditions and severities it is up to the user to develop appropriate controls for managing how and when alarms occur as well as how they are acknowledged and cleared. With a HMI acknowledgment tag assigned, when an alarm is acknowledged the bit assigned to the acknowledged alarm will be set to one (1). Once the acknowledgement bit for an alarm is set, it will remain true until the alarm condition is cleared.

In addition to alarm classes, alarm groups can be assigned. Only alarms that require acknowledgment can be assigned to alarm groups. Alarm groups are a way to tie related alarms together. If a fault occurs and several alarms are triggered, those alarms could be associated to one alarm group related to a specific fault. Having an alarm group assigned to areas on a machine is another way to group alarm conditions. A facility may also have more than one of the same type of machine. Having an alarm group for each machine could help when monitoring and acknowledging alarms. Alarm groups must be tailored to the application and solution they will provide. It is up to the user to examine the application and provide a solution that allows operators to easily diagnose and fix the system. Alarm groups are just one more tool that provides a layer of organization in fault handling. The example of alarm groups in Figure 11-26 shows possible alarm groups that could exist. Alarm group "Fault Condition X" could be for our several related alarms. Machine Area Y could be for a specific area on a machine. Lastly Machine # would be an alarm group for an entire machine. These three examples of alarm groups demonstrate the range alarms can be grouped into, from very specific to more general.

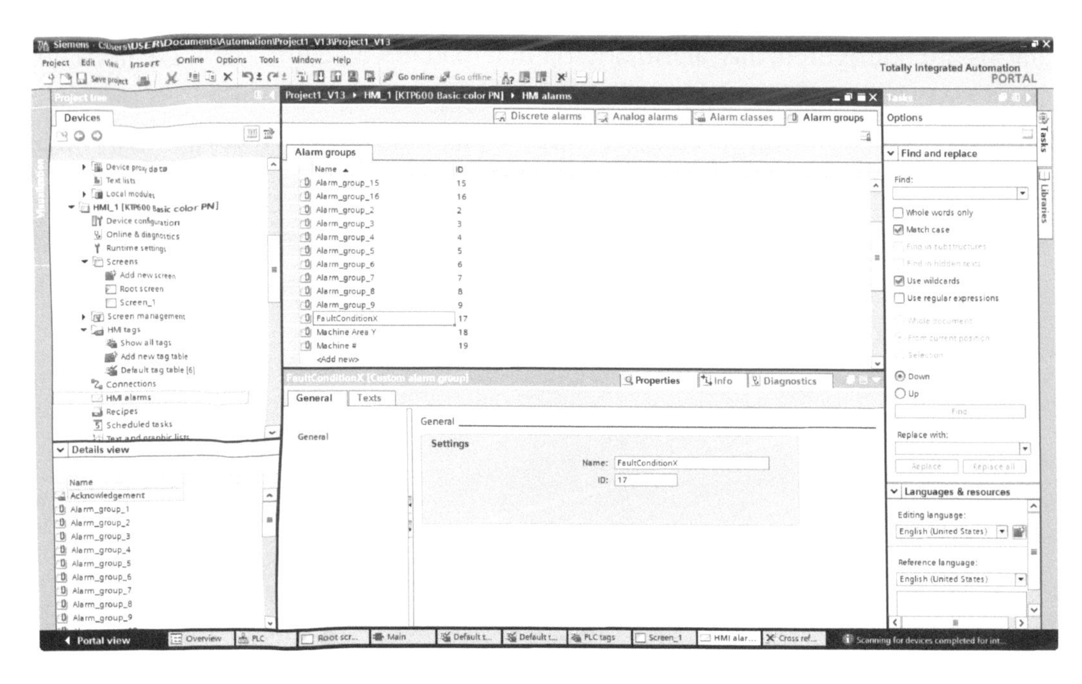

Figure 11-26. Alarm groups.

Analog alarms

Analog alarms are managed the same way discrete alarms are. Alarm classes and alarm groups can both be assigned. The main difference between discrete and analog alarms is a limit test that looks at high limit or low limit violations. In Figure 11-27, an alarm for Tank High Temperature is configured. When the value of Tank_Temperature is greater than or equal to 100 the message for Tank High Temperature will be displayed. Again, because this alarm is assigned to the Errors alarm class an acknowledgement is required. Analog alarms monitor the changing value of a tag and compare it to a limit and limit mode to trigger an alarm.

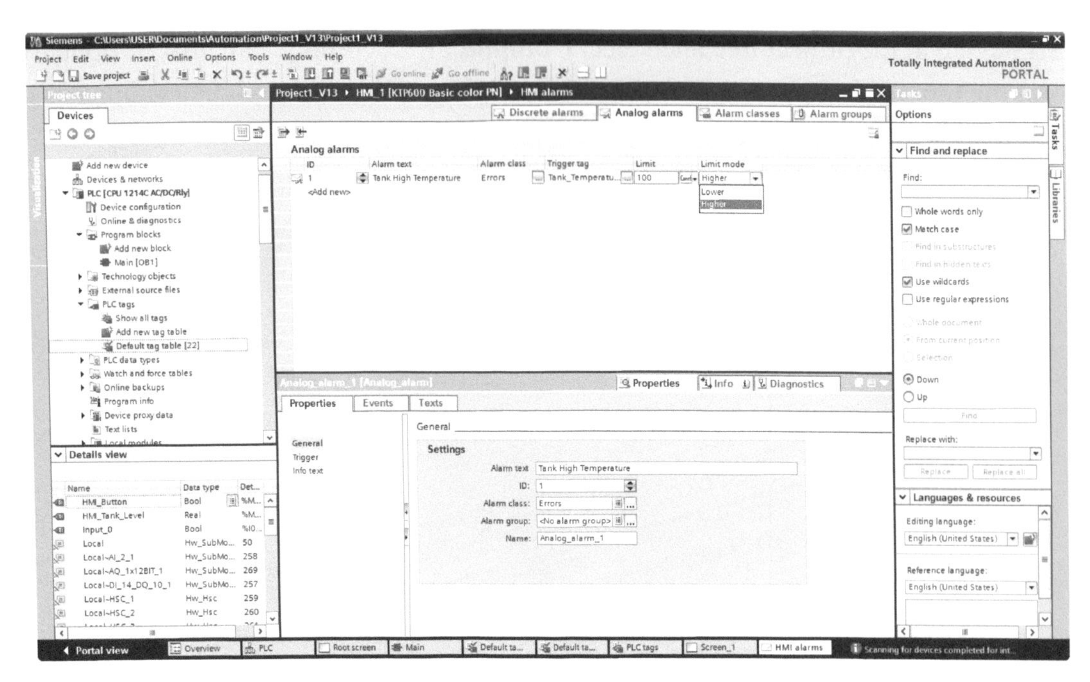

Figure 11-27. Analog alarms.

After HMI alarms are configured, any alarm criteria that matches an alarm view object's configuration, will appear in an HMI alarm view object. The alarm view object can be tailored to show specific alarm types and statuses. Alarm view objects can be configured to show different columns of information for alarms (see Figure 11-28).

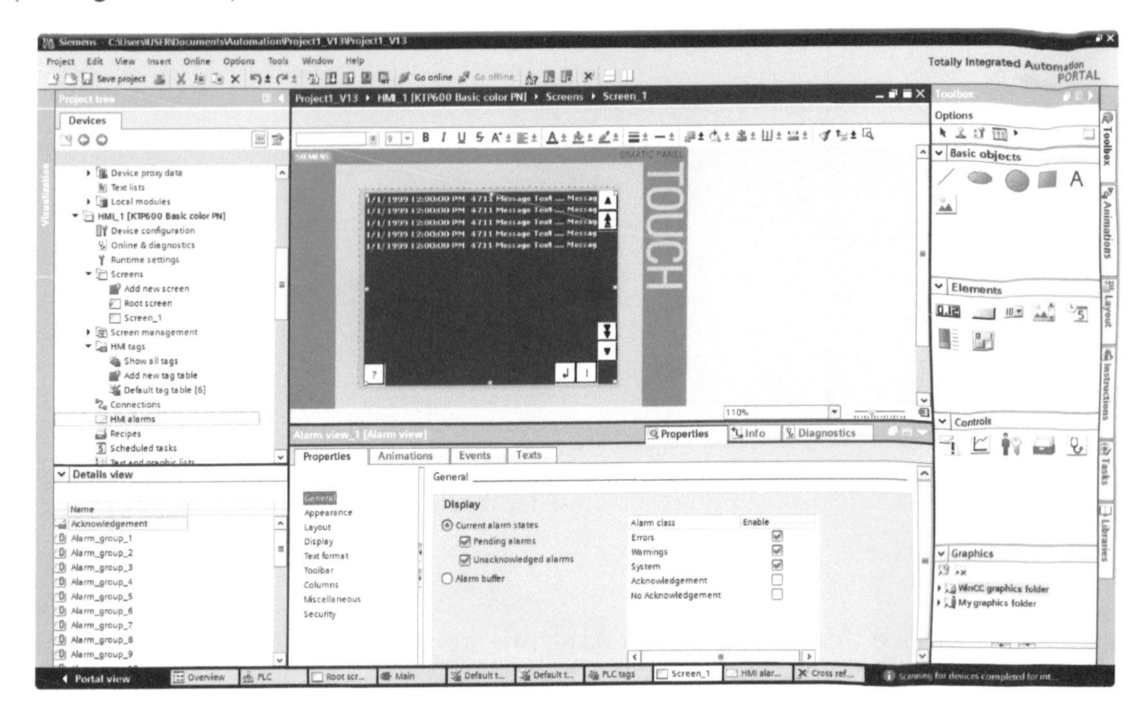

Figure 11-28. Alarm display object.

Compiling and Downloading

Once an HMI configuration is ready to download, the next step is to compile. A compile is required before a download can take place. To compile before downloading the HMI configuration right-click on the root HMI device in the project tree and select Compile all. A compile can also be done by selecting Compile from the top window area as shown in Figure 11-29. Once the applciation finishes compiling without error, it is ready to download.

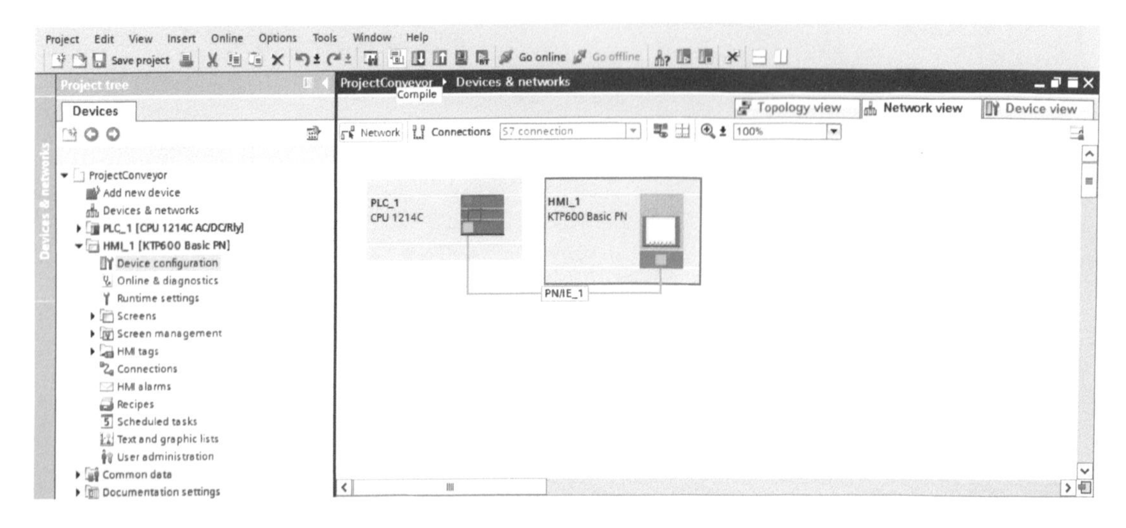

Figure 11-29. Compile HMI project.

Downloading can be done while online or offline. The first example will deal with offline downloading. We will start out by making sure the device we want to download to is an accessible device. To do this return to the Portal View, and under Online & Diagnostics click on Accessible devices. This is the same step as shown in Figure 11-6 where we assigned the IP address to the hardware. Once you have scanned for accessible devices and the SIMANTIC-HMI shows up under devices, select it from your devices and click the Show button. At this point you may be prompted again to assign an IP address. As Figure 11-30 describes, you will have to add at least one addition IP address in the same subnet as the PLC or in this example the HMI. By selecting yes to add an IP, your computer will be assigned an IP address that will allow your application to be downloaded to the HMI hardware.

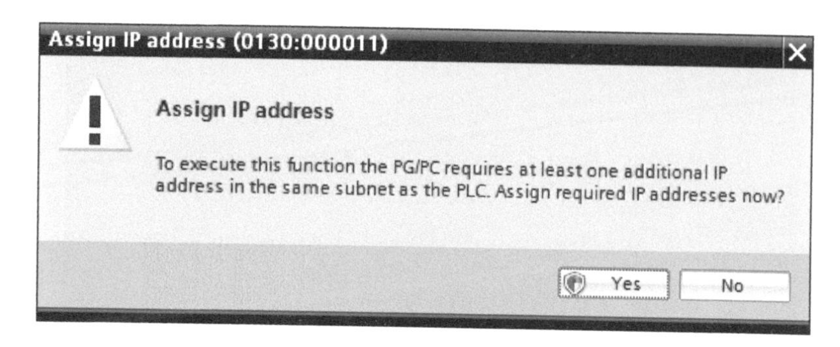

Figure 11-30. Assigning an IP address.

Once the IP address has been added, a prompt will appear to notify you that an IP address was added, what the IP address is, and which network interface it was added to (see Figure 11-31). Once you have added an IP address on the same range as the HMI device, return to Project view. In Project View we will continue with downloading to the device.

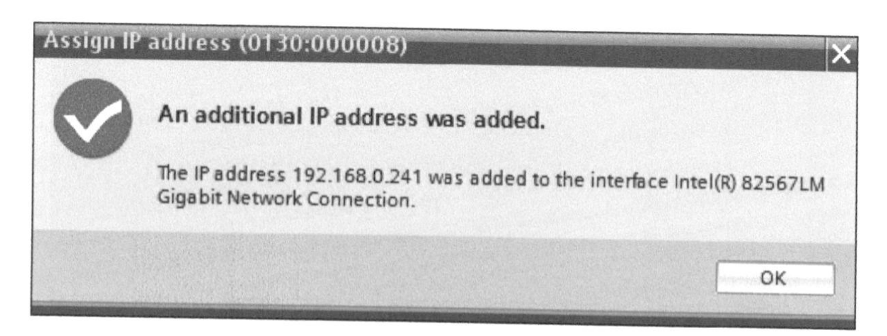

Figure 11-31 Added IP address.

Downloading Offline

To download your project offline, select "Download to device" from the shortcut menu of the HMI device under the Project tree. The same "Download to device" command can be issued from the top header menu as shown in figure 11-32. If using the "Extended download to device" window (Figure 11-33), make sure settings for loading match the transfer settings of the device. Within the Control Panel of the HMI hardware, under Transfer Setting, "Remote Control" must be selected if you wish to download without putting the HMI device to Transfer mode. In Transfer mode a prompt will be displayed with the text "Connecting to Host…".

This is only necessary if "Remote Control" is not selected. Extended download can be found under the menu command under "Online > Extended download to device". Extended download gives you more options when selecting network devices and devices to download to.

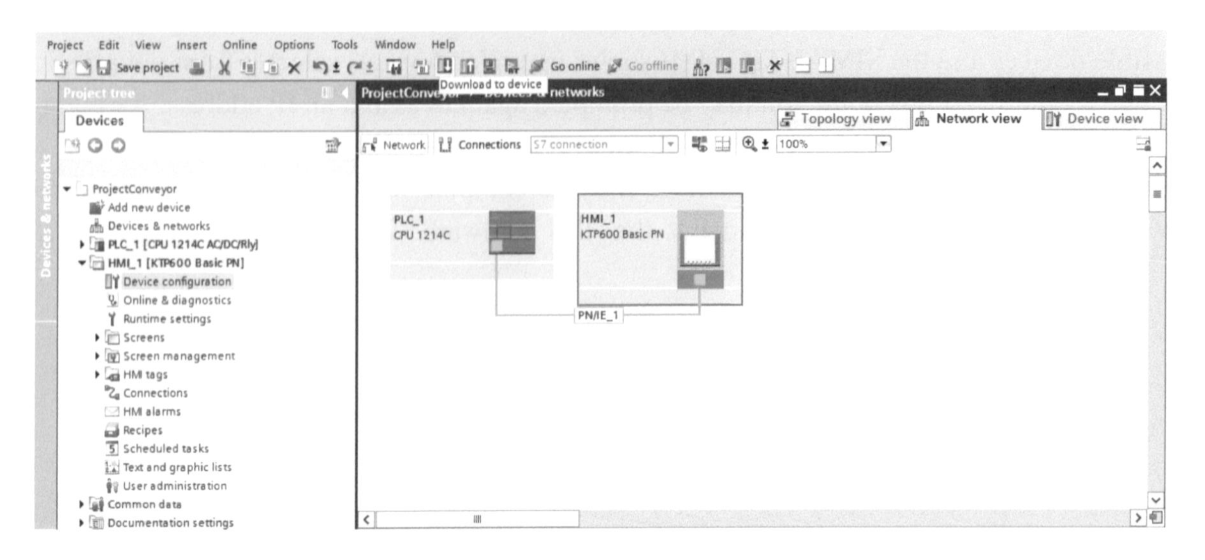

Figure 11-32. Download to device from header menu.

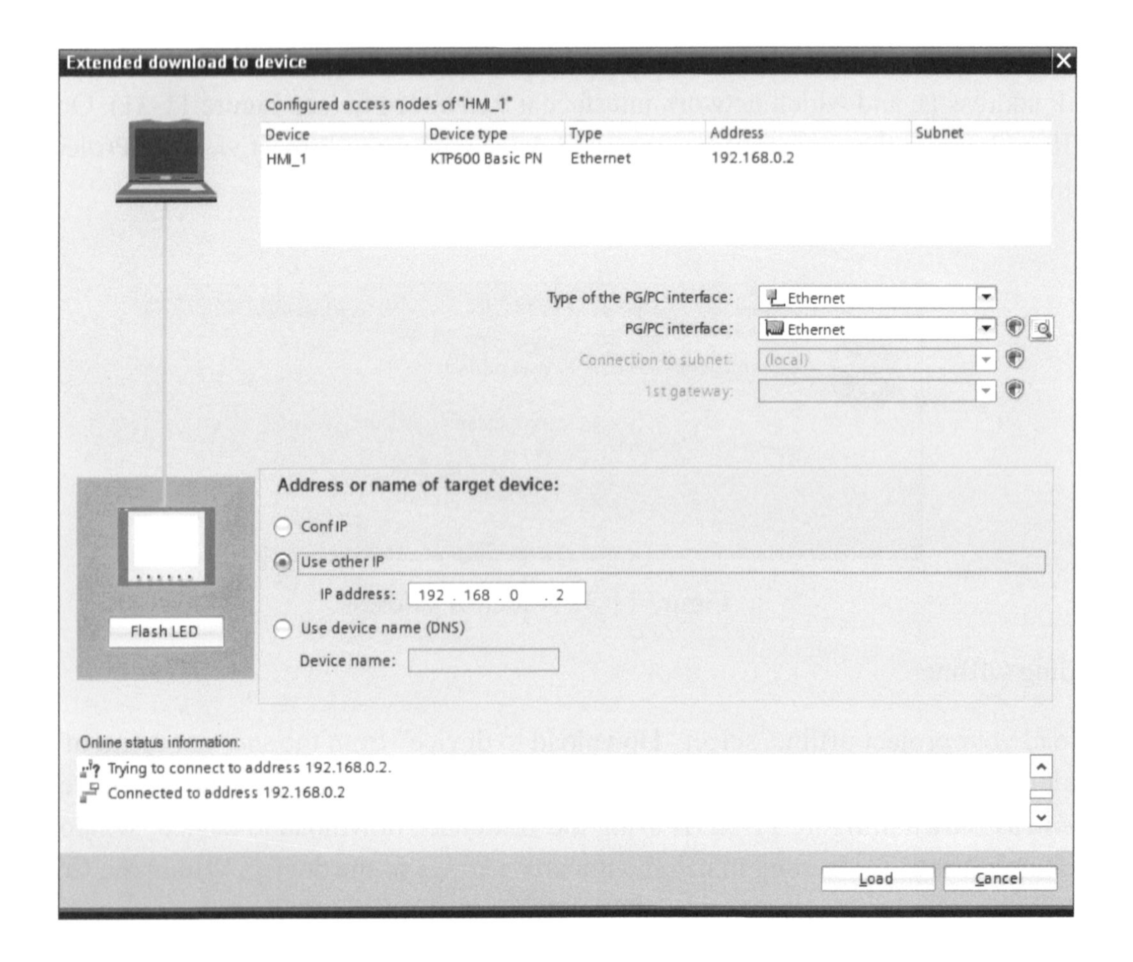

Figure 11-33. Extended download to device.

Once you select Download to device from the Project tree menu or the header menu, a brief Compile window will appear (see Figure 11-34) followed by a Load Preview window (see Figure 11-35).

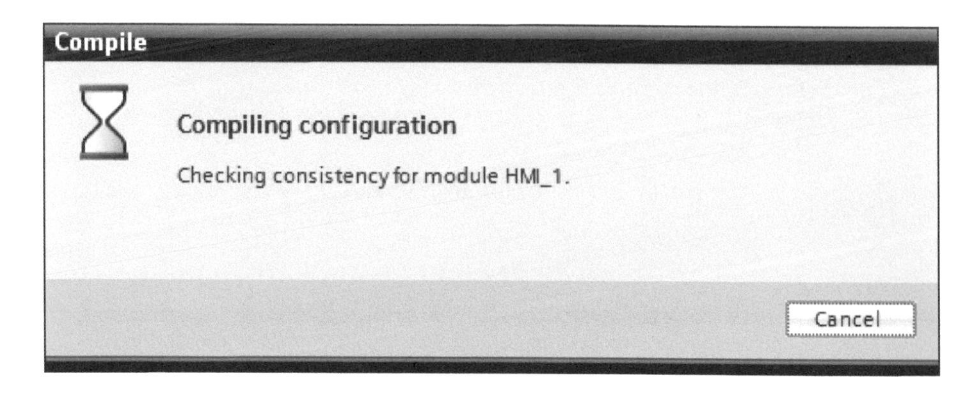

Figure 11-34. Compiling for download.

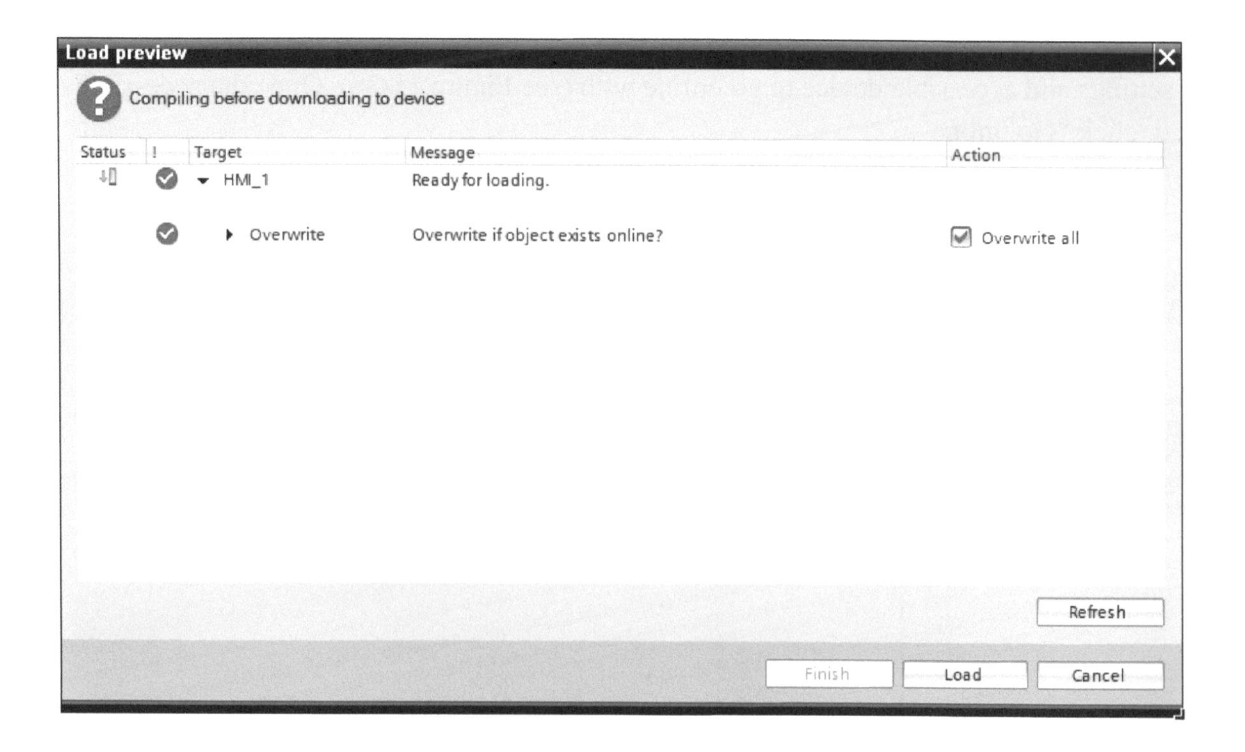

Figure 11-35. Load preview.

Within the Load preview window, click in Overwrite all and click Load.

After selecting "Load" the prompt shown in Figure 11-36 will appear showing the status of the download. Once the download is finished the HMI device will restart. After the HMI reboots, your application is ready to be started.

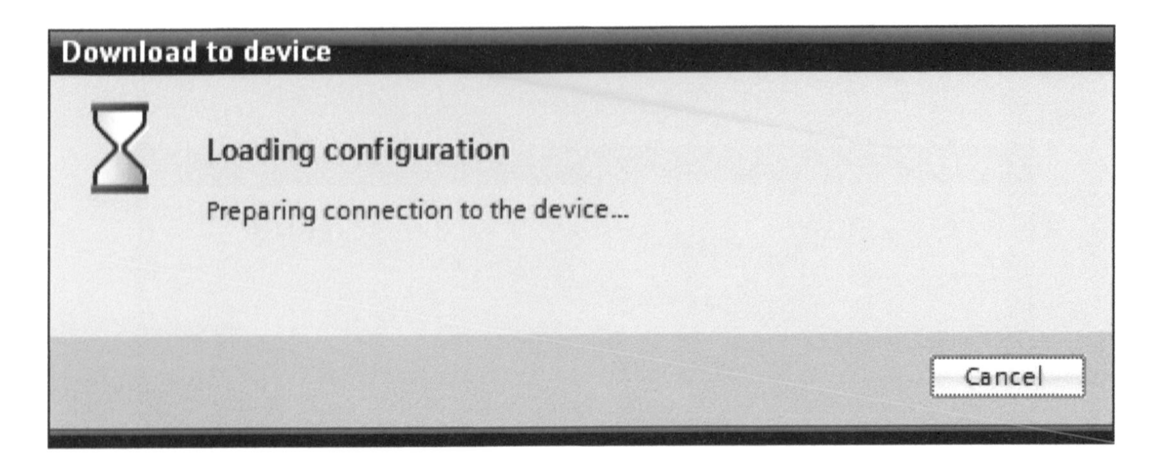

Figure 11-36. Download to device.

Downloading Online

To go online with the HMI device, select “Go Online” from the shortcut menu of the HMI device under the Project tree. The same “Go Online” command can be issued from the top header menu as shown in figure 11-37. After selecting to go online a “Go online” window will appear. Select the appropriate network settings and accessible device to go online with (see Figure 11-38). Once the accessible device is selected, click “Go online”.

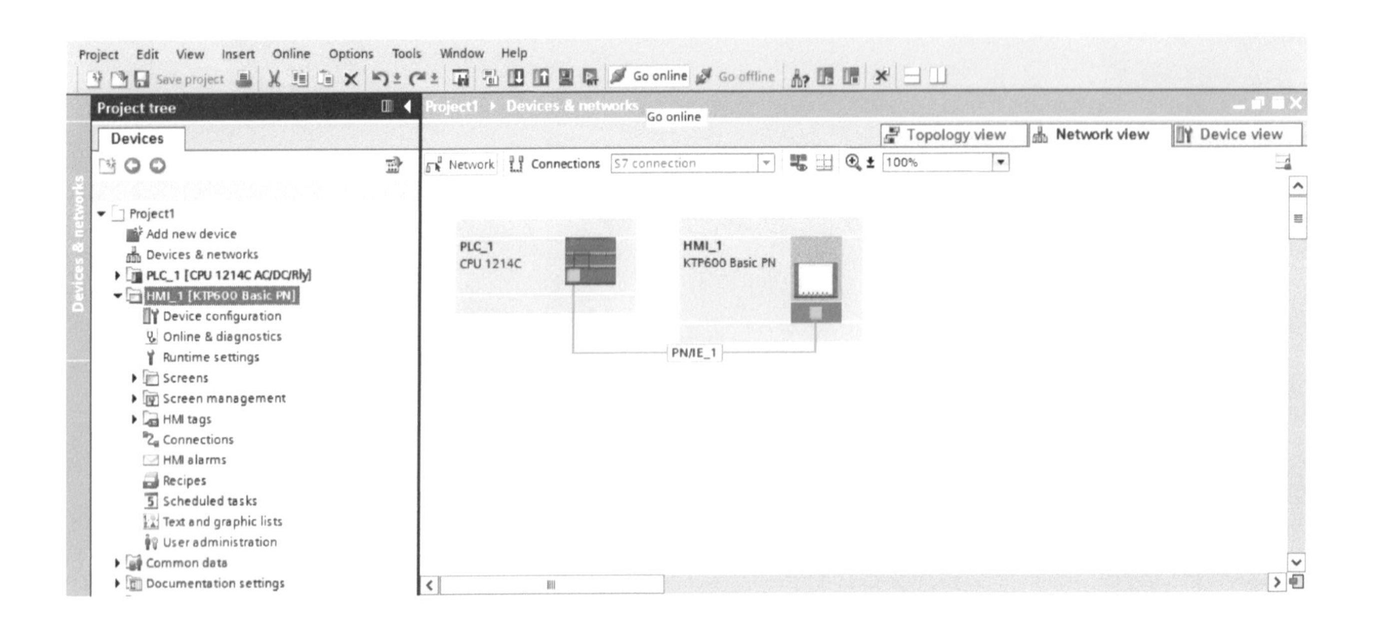

Figure 11-37. Going online.

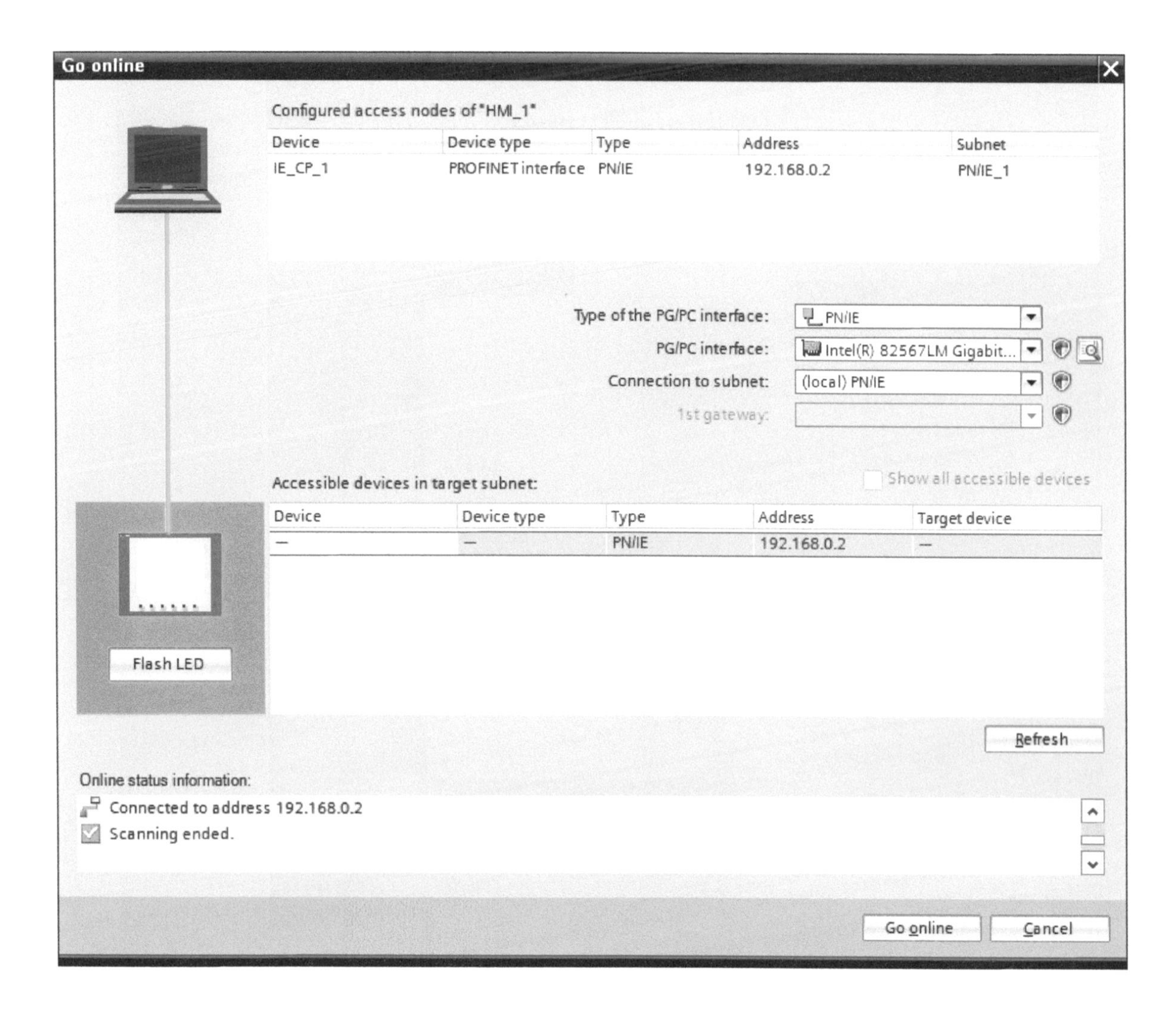

Figure 11-38 Go online window settings.

Once online your Project tree will display an orange-colored header and the "Go offline" icon in the header menu will no longer be grayed-out. Also, you will notice status boxes in your project tree. In Figure 11-39 the status boxes show green boxes with check marks through them indicating the online status.

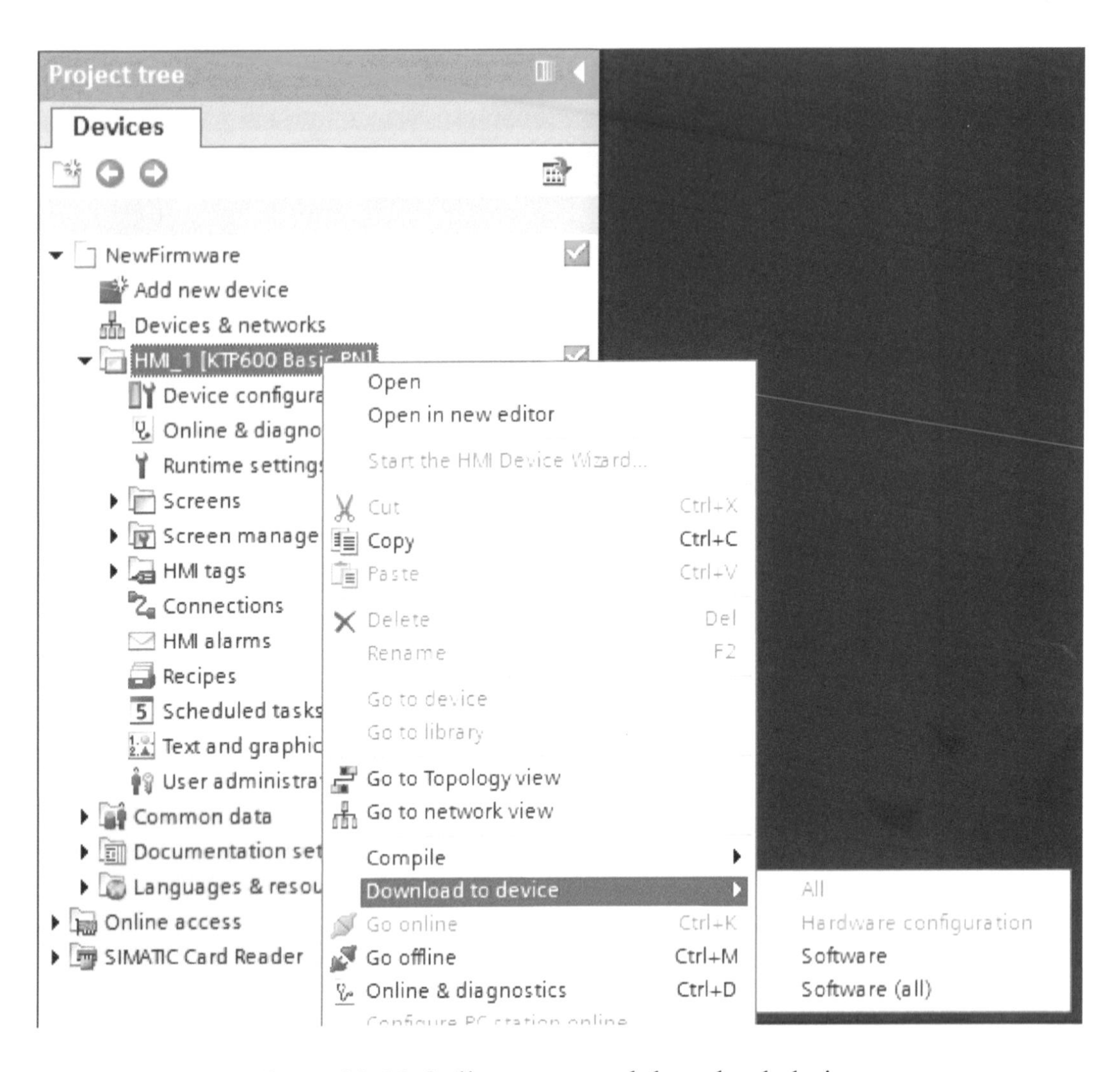

Figure 11-39 Online status and download choice.

While online the only way to download is from the shortcut menu within the Project tree under the HMI device as shown in Figure 11-39. From here the same compile and load preview windows will appear as in offline mode. Be sure to check "Overwrite all" in the Load preview window and Load your project. The downloading prompt will appear and HMI device will restart.

Chapter Questions

1. True or false? You can only assign an IP address while a project is open?

2. Data is exchanged between the PLC and the HMI via _______.

    a. Screens

    b. Flags

    c. Tags

    d. Addresses

3. HMIs can communicate with the PLC by using ______.

    a. PROFIBUS

    b. Industrial Ethernet

    c. MPI

    d. All of the above

4. True or false? HMIs are only useful to operators running the equipment?

5. True or false? There are only a couple of ways to configure an HMI?

6. True or false? It is very difficult to browse for tags in the PLC while configuring the HMI?

7. True or false? Alarms are only used for serious notifications?

8. True or false? Alarms can monitor increasing/decreasing values to trigger alarm conditions. These are known as analog alarms?

9. True or false? Any alarm can be assigned to an alarm group?

10. True or false? Downloading can only be done offline?

11. True or false? The HMI device must always be in "transfer" mode to download?

Appendix A

Creating a Siemens Step 7 Project

Note: You can download a trial version of Siemens Step 7 at the following link - www.siemens.com/sce/trial

This appendix will lead you through one method to create a S7-1200 project in Step 7 TIA Portal. The software is very powerful, versatile and intuitive. As you work with the software more you will learn other ways to do the same tasks.

Open TIA Software and you should see a screen similar to the one shown in Figure A-1. TIA offers two views of your application. The one shown in Figure A-1 is the Portal view. The Portal view is a step-by-step way to create, configure, and program an application. It is an intuitive, easy way to get started. Look at the lower-left corner of Figure A-1. In the lower-left corner you could choose the Project View. If you click on Project View you will get the screen shown in Figure A-2. Also note that on the lower-left corner of the Project View screen (Figure A-2) there is a choice to go to the Portal View. You can switch back and forth and use whichever one may be easier for you to use. On the left of the Project View screen in Figure A-2 you can see the Project Tree. The Project tree shows all of the resources in our project. We can access each of the resources in our project to configure them or program by double clicking on them. This appendix will concentrate on the Portal View to create and test a simple project. TIA software is very intuitive and you will discover many ways to do things in Portal and Project view.

Figure A-1. TIA Portal opening screen.

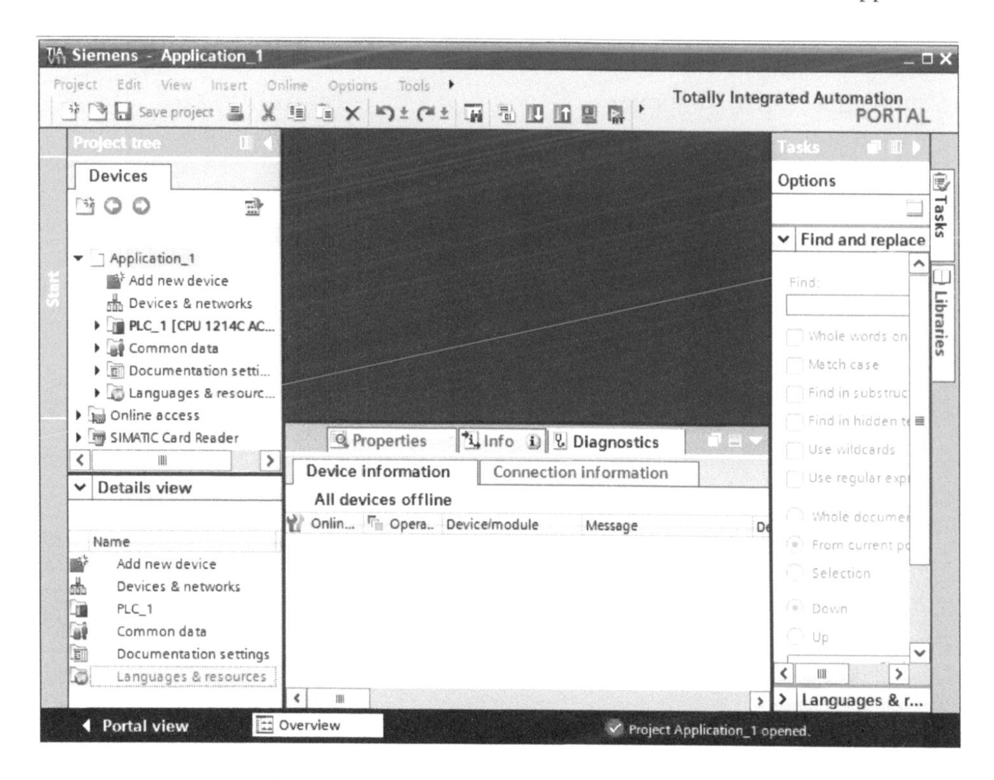

Figure A-2. TIA Portal opening screen.

Look at Figure A-1. Note that Create new project has been highlighted, if it is not highlighted on your screen, click on Create new project to highlight it. Enter a name for the project and click on the create button.

Next, the screen shown in Figure A-3 should appear. Note that the name of the project is Application_1 as we entered in the first screen. Note that the first task is to Configure a device. This will be done to add and configure the correct model of S7-1200 PLC. Click on the Configure a device choice.

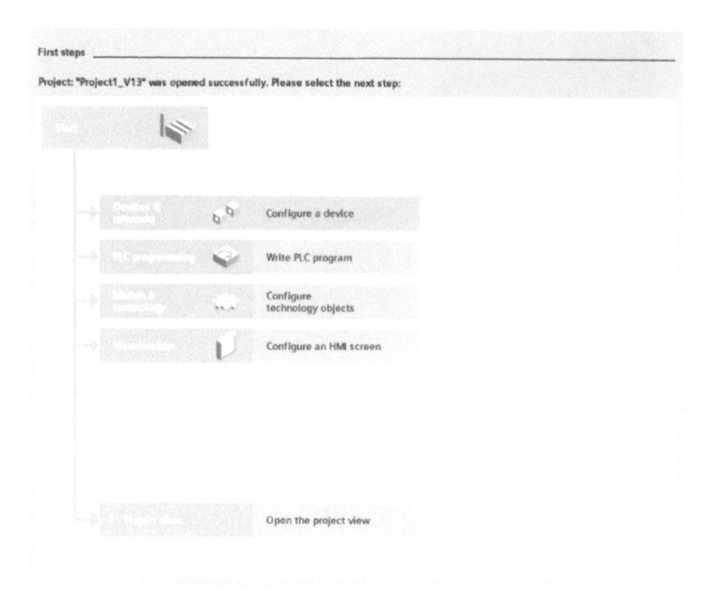

Figure A-3. Partial screen capture of screen to Configure a device.

Next the screen shown in Figure A-4 will appear. Click on the Add new device button. After a short wait the screen shown in Figure A-5 will appear.

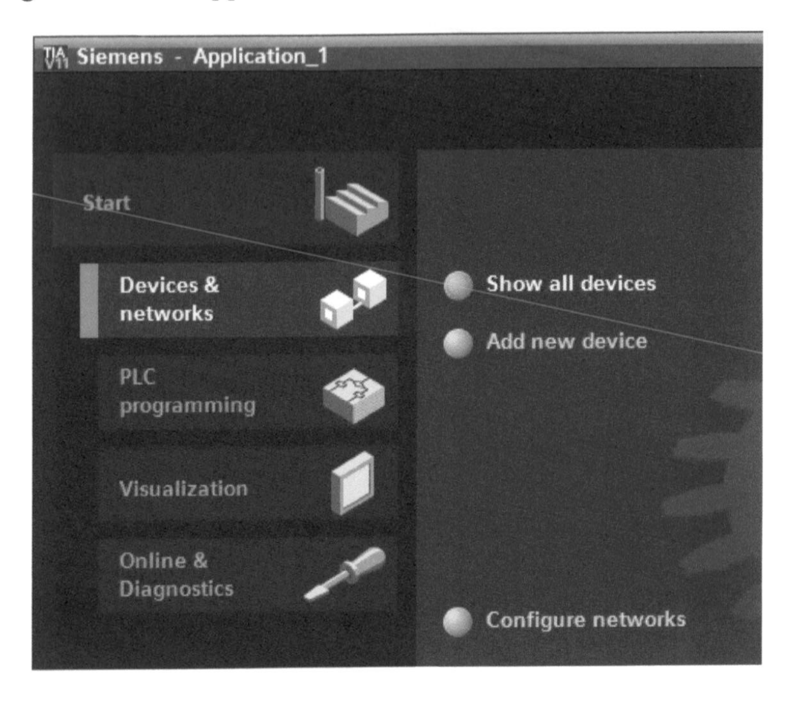

Figure A-4. Partial screen capture showing the Add new device choice.

Note the list of available S7-1200 PLCs in Figure A-5. You could choose the model of PLC you have from the list. For this example, we will let the software do the work for us. Click on the down arrow left of the Unspecified CPU 1200 choice and then highlight the 6ES7 2XX_XXXX_XXXX choice and then click on the Add button on the lower right. The screen shown in Figure A-6 will appear.

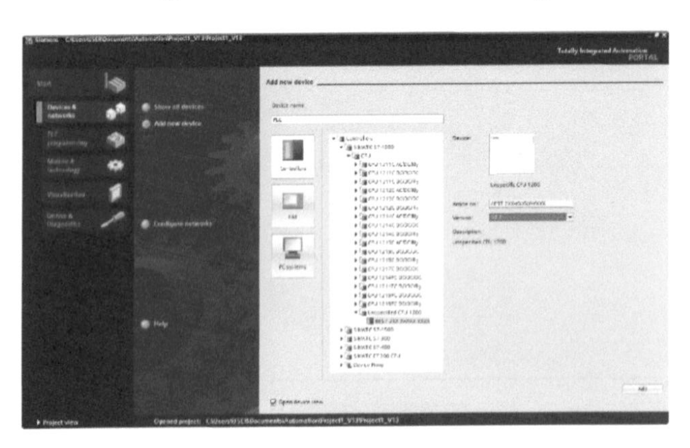

Figure A-5. Add device screen.

Figure A-6 shows that a generic CPU has been added to the project. Click on the Detect choice. The software will then begin the process of identifying the CPU that is connected to the computer and the screen shown in Figure A-6 will appear.

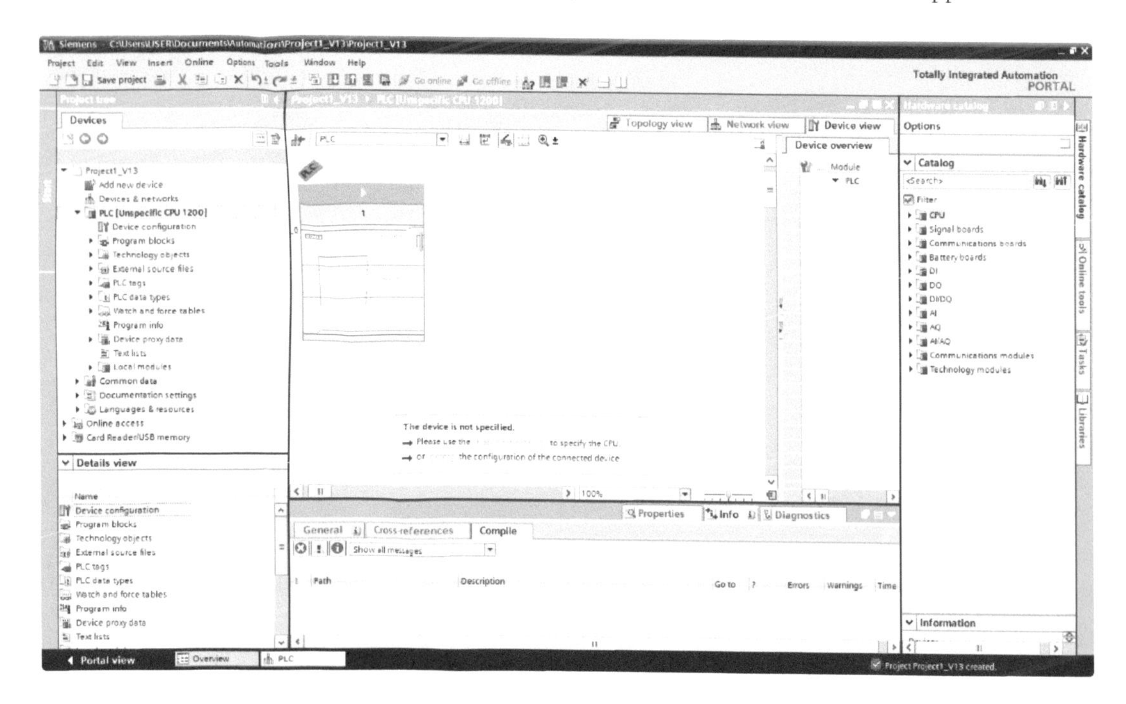

Figure A-6. Detect Hardware.

Figure A-7 shows that the user must choose the type of PG/PC interface must be chosen. This is the type of interface that will be used to communicate from the computer to the CPU. Click the down arrow to the right of the Type of PG/PC interface and choose PN/IE as shown in Figure A-8. In this example we are using an Ethernet connection to a SIMATIC CSM 1277 Ethernet switch that the PLC CPU is also plugged into.

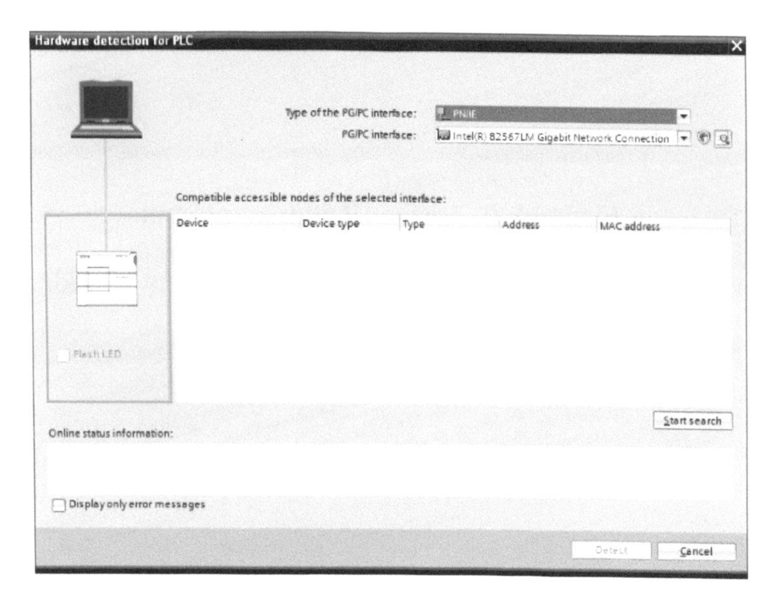

Figure A-7. Hardware detection screen.

The software will then connect to the S7-1200 CPU and the PLC will be added to the Accessible devices in target subnet list (See Figure A-8).

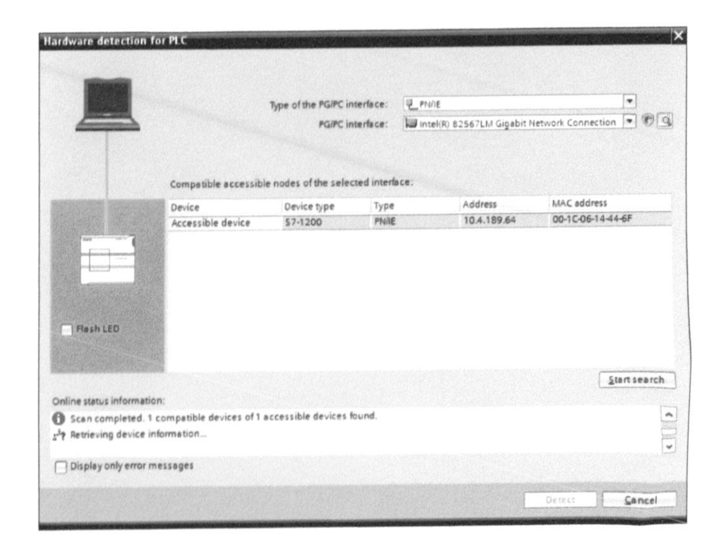

Figure A-8. Partial screen capture showing that the Hardware detection has found a S7-1200 CPU and has added it to the Accessible devices list.

Next you will be prompted to Assign an IP address to the PLC. Click on the Yes button as shown in Figure A-9.

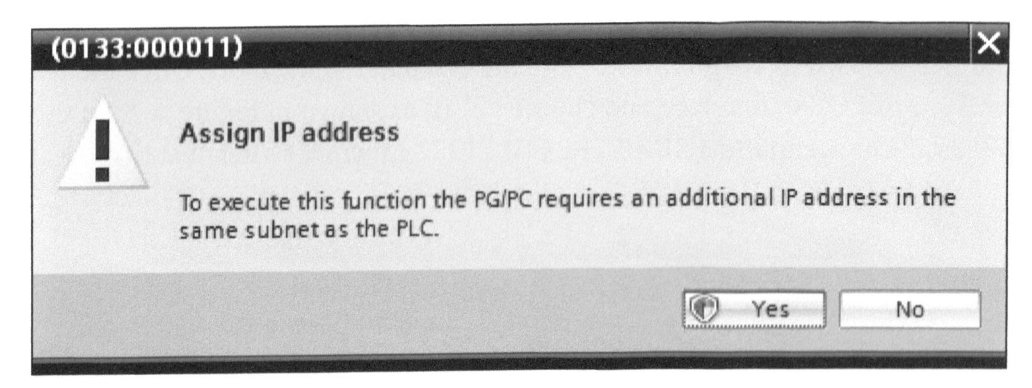

Figure A-9. Assign IP address prompt.

The message shown in Figure A-10 will appear to let you know that an IP address was added.

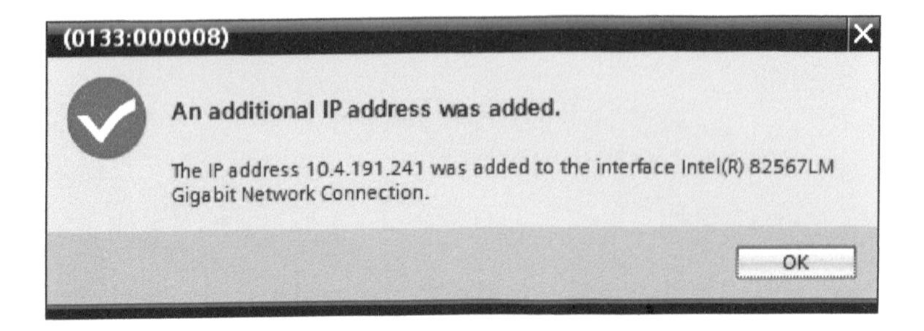

Figure A-10. Prompt showing that an IP address was added.

The software will then communicate with the PLC CPU and establish the exact specifications of the PLC CPU and the screen shown in Figure A-11 will appear. Note that the software now knows exactly what I/O the CPU has available, including a signal board (SB) module in this example.

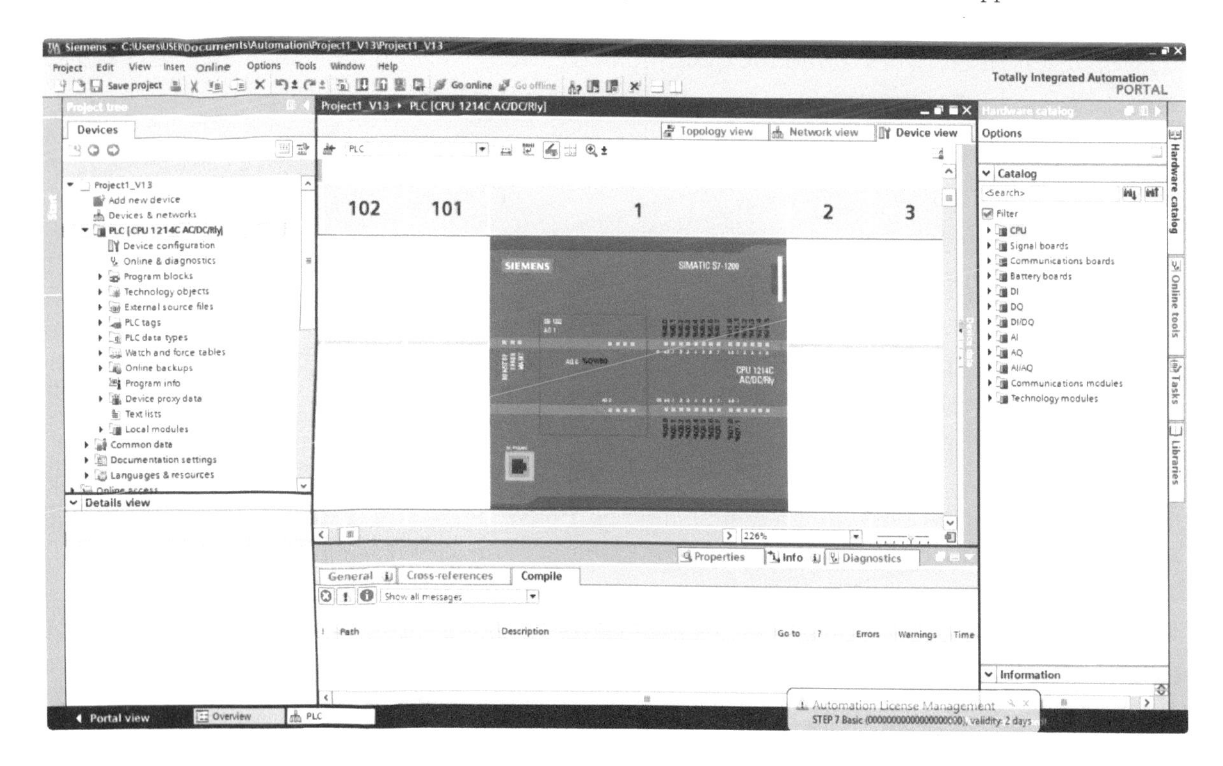

Figure A-11. Screen capture showing the CPU and the Project tree is on the left.

You are now ready to program.

Look at the left side of Figure A-12. You can drill down in the menu under the CPU PLC_1 in this example) to the program blocks. If you double click on the Block named Main [OB1] the programming screen will appear with Network 1. Note that there is no logic yet.

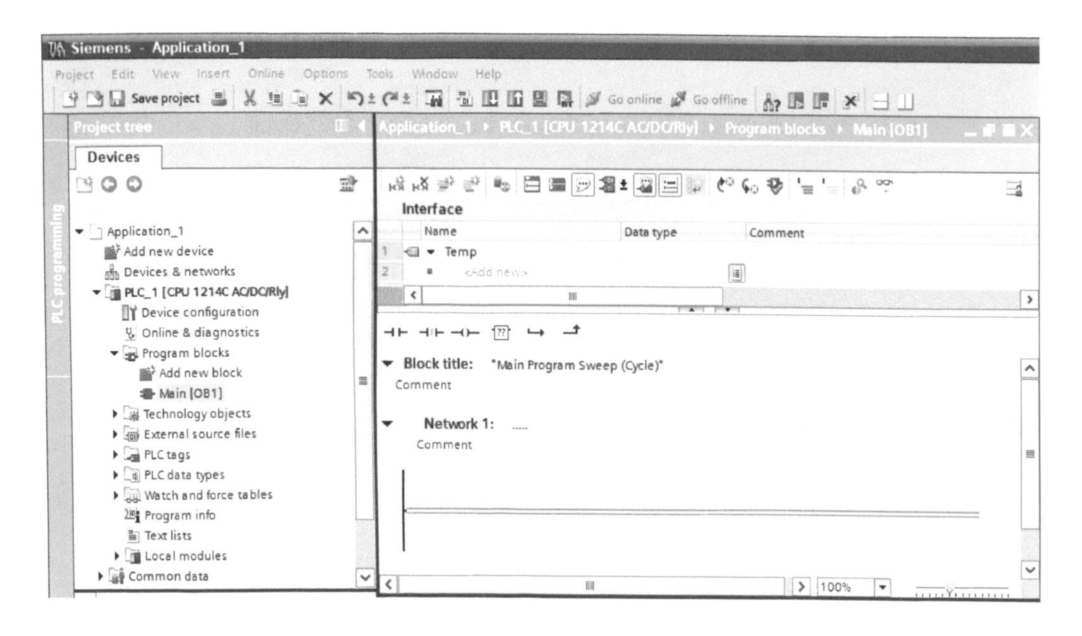

Figure A-12. Programming screen showing Network 1.

A normally open contact and a coil were pulled down from directly above Network 1 in Figure A-13.

Note the large X left of Network 1 and the question marks above the contact and coil. This is because they have not been addressed yet.

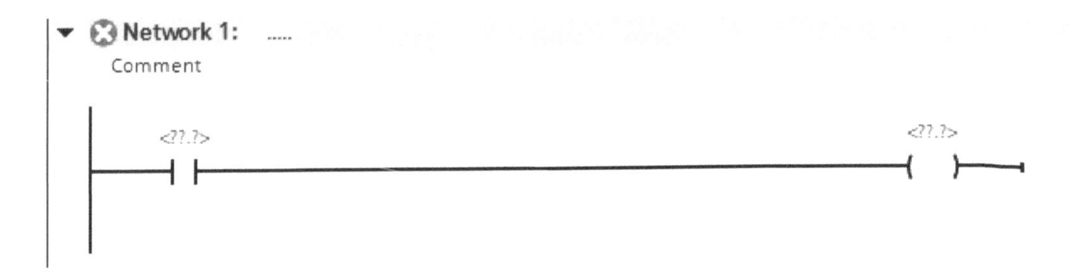

Figure A-13. Figure showing that a contact and coil have been added to Network 1. They have not been addressed yet.

Addressing was completed in Figure A-14. The normally open contact was assigned a real-world input addressed I0.0. The tag was given the default name: Tag_1. The coil was assigned a real-world output address of Q0.0 and given the default name: Tag_2. Note that the X left of the Network disappeared once the logic was addressed correctly.

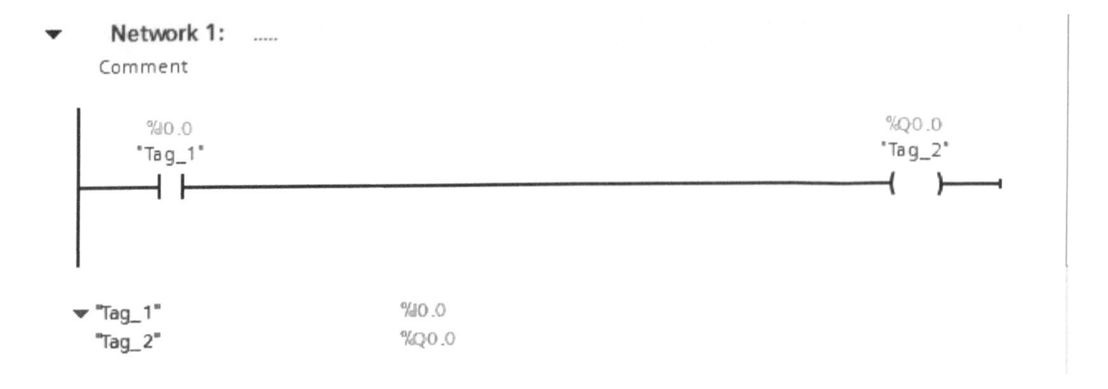

Figure A-14.

Next right-click on PLC_1 on the left of the screen in the Project tree, and choose Compile/Software. Your program will be compiled so it is ready to download to the CPU. Your logic must be compiled before you can download it.

Next you can right-click on PLC_1 and choose Download to device/software.

Next the screen shown in Figure A-15 will appear. You must click on the Load button to begin the load process.

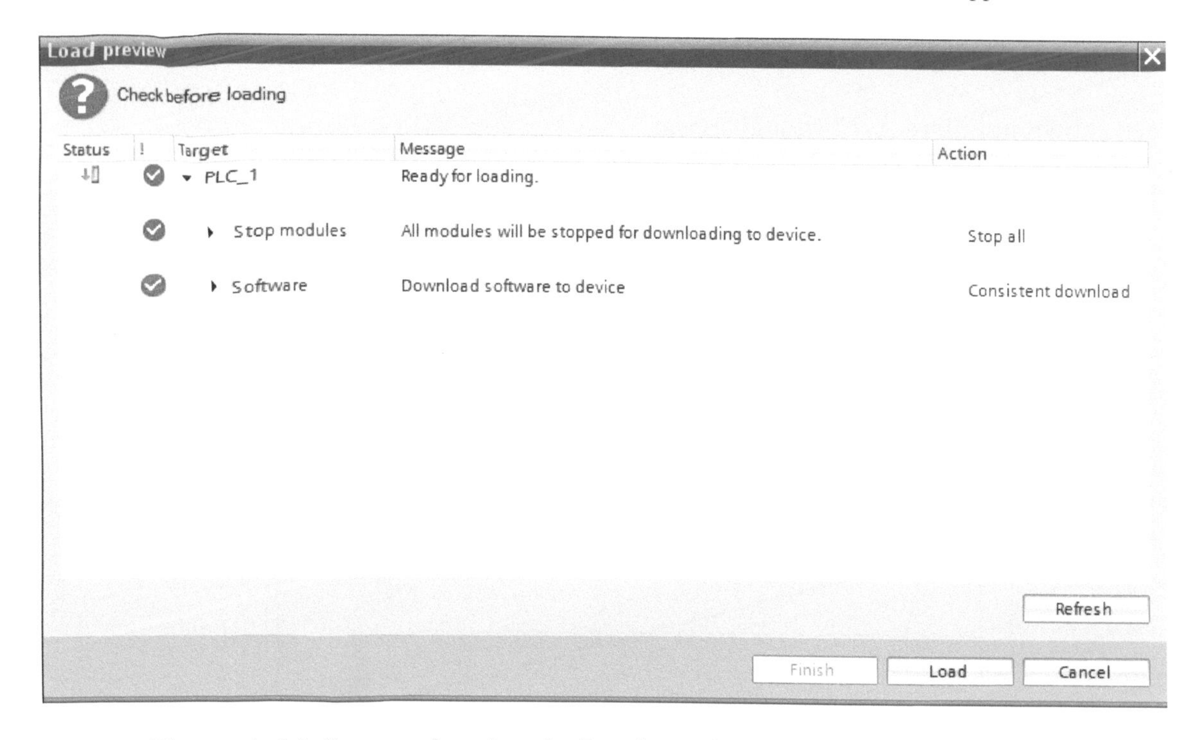

Figure A-15. Screen showing the Load preview screen and Load button.

Next click on the Finish button and the program will be downloaded to the CPU (see Figure A-16).

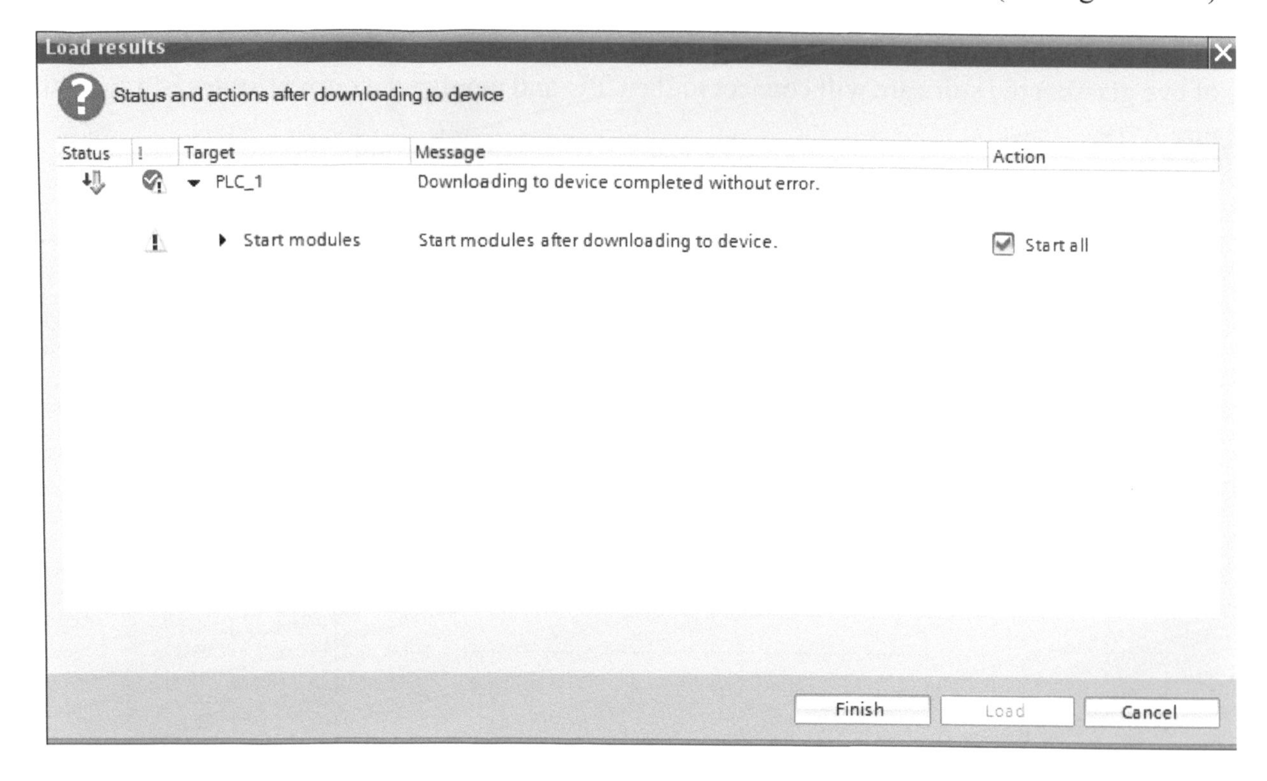

Figure A-16. Load results screen and Finish button.

The screen shown in Figure A-17 should appear.

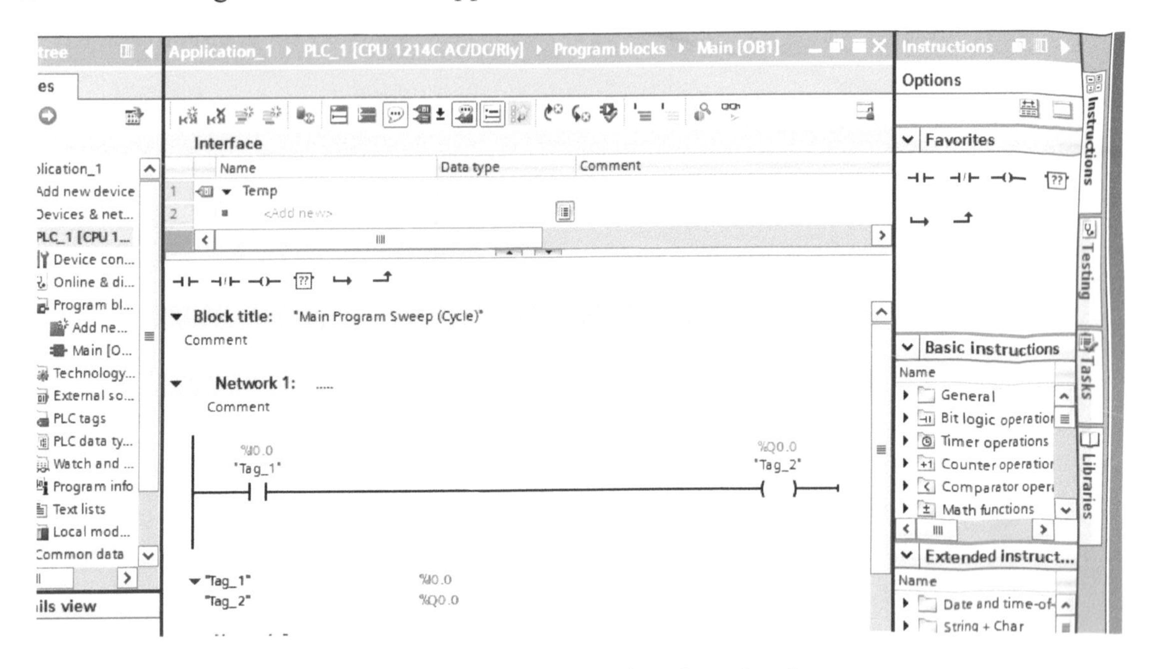

Figure A-17. Screen after download.

If you click on the Monitoring On/Off icon in the top menu as shown in Figure A-18 (the icon looks like a set of eye glasses) the software will connect to the CPU and display the current states of I/O as shown in Figure A-19.

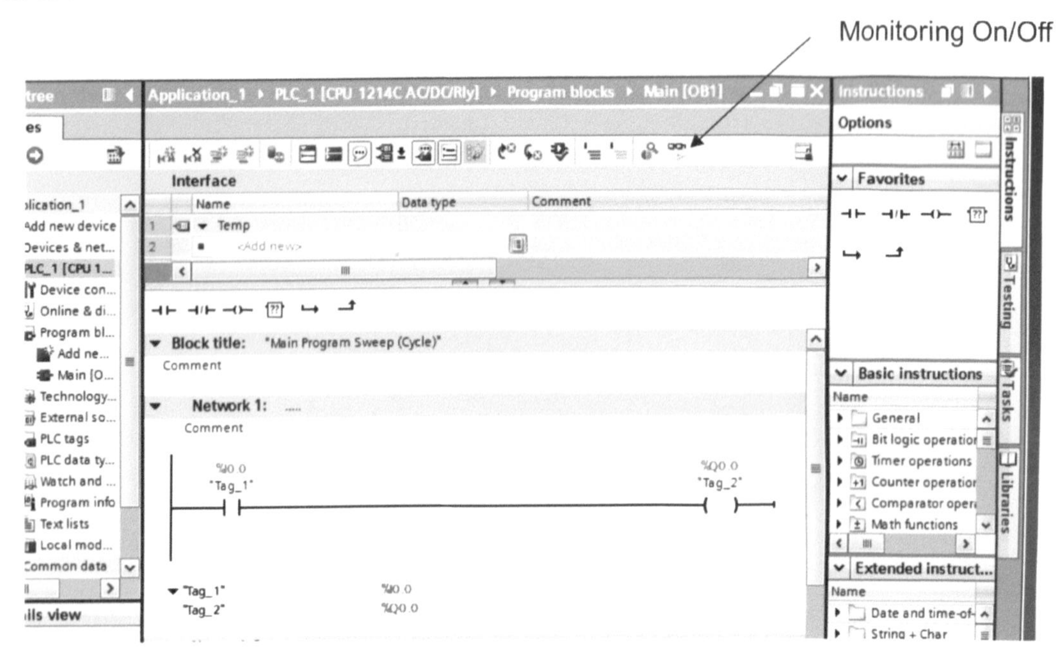

Figure A-18. Monitoring On/Off icon.

Figure A-19 shows that the CPU is now connected and monitoring is on. Note that the input (I0.0) has not been turned on yet. Note also that the CPU is in Run mode. You may use the Run/Stop buttons to change the state of the CPU.

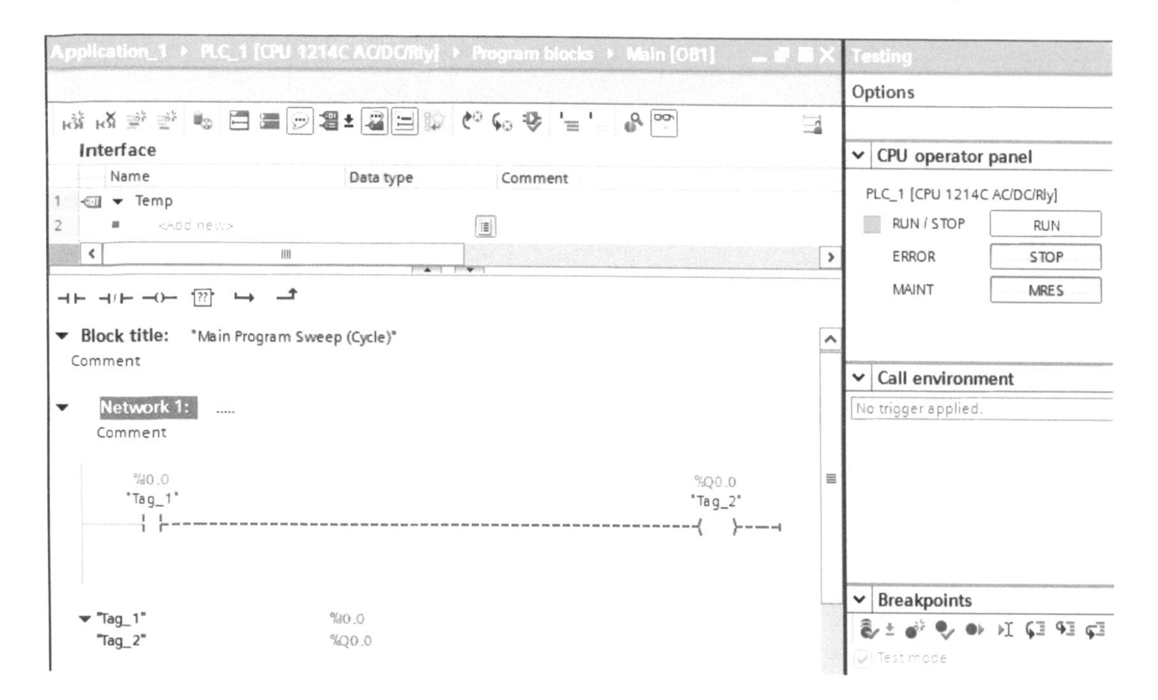

Figure A-19. Monitoring on.

If input I0.0 is turned on the rung should energize the output coil 9 (Q0.0). Figure A-20 shows that the input contact (I0.0) is now true and the run logic is true which has energized the output coil (Q0.0).

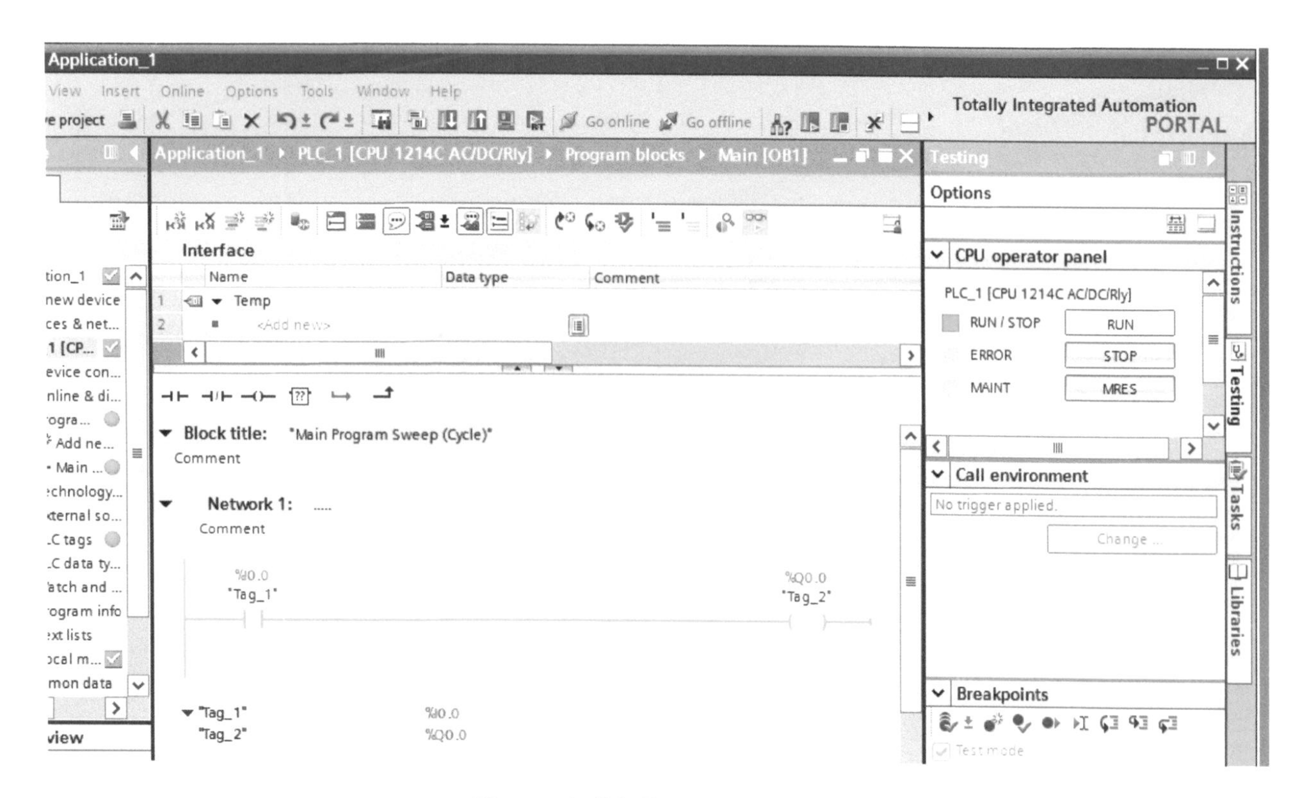

Figure A-20. Rung energized.

A mentioned earlier TIA portal is very versatile software and there are many ways to perform tasks in it. The more you work with the software the more ways you will discover to do things.

Appendix B

Answers to Even Numbered Chapter Questions

Chapter 1 Questions

2 - Name the main components that are found in a basic PLC system.

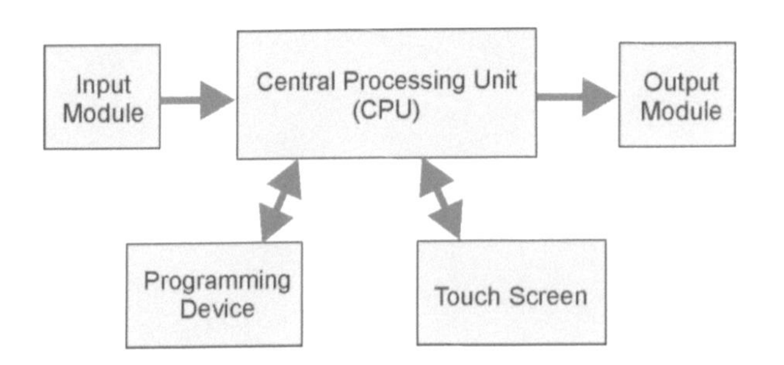

4 - What is the difference between discrete and analog?

A discrete device only has two states an analog device has a range of values. Analog modules are commonly available to output 0 – 10 VDC, -10 VDC to + 10 VDC, and 4-20 mA. These are useful for controlling analog output devices. A motor drive is one example. A drive that is capable of clockwise and counter-clockwise rotation at various velocities might require a -10 VDC to + 10 VDC signal from a PLC to control direction and velocity. Analog input modules are also available.

6 - Describe ladder logic.

Ladder logic is the most commonly used PLC programming language. The vertical lines in ladder logic are sometimes called power rails. The rails represent power. If we connect the left rail to the right rail, power flows. The horizontal lines represent rungs of logic. The symbols on the left of rungs (contacts) represent input states or conditions. The symbol on the right of the rung (coil) represents an output. If the conditions on a rung are true, the output is turned on.

8 - What is S7 an acronym for?

Step 7.

10 - What typical output types are available in analog modules?

4-20mA, -10 to + 10 VDC, and 0 to 10 VDC.

Chapter 2 Questions

2 - What is an OB and what types are available.

Organizational blocks represent the interface between the operating system and the user program. The CPU's operating system calls the organizational block when specific events occur such as hardware or cycle events. The main program is in organizational block 1. An Organizational Block (OB) is a code block that is used to structure or organize a program. It is possible to just have one OB that contains all of the user's program logic. The main organizational block is called a program cycle OB. You could put your whole program in the program cycle OB.

In addition to the user's program cycle OB, there are other types of organizational blocks that can be used (see the figure below). These other types of OBs are used for specific purposes.

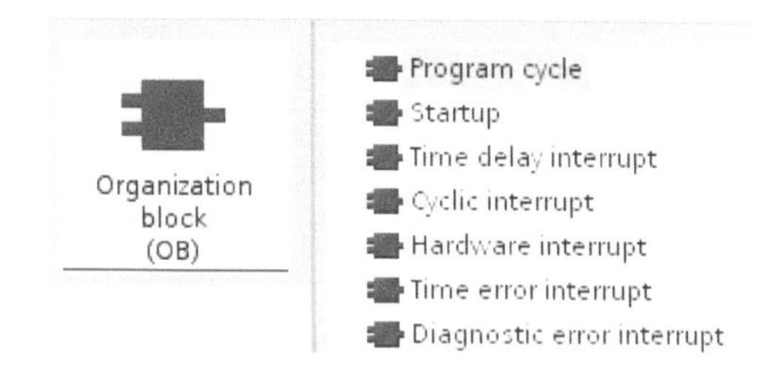

4 - What is an FC and what are they used for?

An FC is a subroutine that is called from another code block. An FC can be called from an OB, FB, or a FC. A function (FC) is like a subroutine. An FC is a code block that typically performs a specific operation on a set of input values. The FC stores the results of this operation in memory locations. FCs are typically used to perform the following types of tasks: standard and reusable operations, such as calculations and functional tasks, such as for individual controls using bit logic operations.

An FC can also be called several times at different points in a program. This reuse simplifies the programming of tasks that are used multiple times. Unlike an FB, an FC does not have an associated instance DB. The program block that calls the FC passes parameters to the FC. The output values from an FC must be written to a memory address or to a global DB.

6 - What is a DB?

There are two types of data blocks: global and instance. All program blocks in the user program can access the data in a global DB. An instance DB is used to store data for a specific function block (FB).

A data block (DB) can be used for fast access to data stored within the program itself. DBs can also be defined as being read-only. The data stored in a DB is not deleted when the data block is closed or the execution of the associated code block comes to an end.

Instance data blocks are assigned directly to a function block (FB). The structure of an instance data block is determined by the declaration of the function block. An instance data block (DB) contains the block parameters and static tags that are declared there. The user can define instance-specific values in instance data blocks, such as initial values for the declared tags.

8 - Convert the following numbers to binary, decimal, or hex as required.

Binary	Decimal	Hex
0011	3	3
1101	13	C
10001	17	11
011011	27	1B
100101	37	25
100000000	256	100
10001001	137	89
1001001	73	49
00100111	39	27
10110111	183	B7
10101111	175	AF
11000110	198	C6

10 - Explain the concept of byte addressing for an analog module.

The figure below shows addressing for a S7-1200 PLC. There is an analog combination module installed in slot 6. It has 4 analog inputs and 2 analog outputs on it. Its analog input addresses will be IW160, IW162, IW164, and IW166. The analog outputs addresses will be QW160 and QW162. The W in the analog addresses means word. Each analog input or output uses 2 bytes (one word).

Device overview

Module	Slot	I address	Q addre...	Type	Order no.	Firmware
PLC_1	1			CPU 1214C AC/DC/Rly	6ES7 214-1BE30-0XB0	V1.0
DI14/DO10	1.1	0..1	0..1	DI14/DO10		
AI2	1.2	64..67		AI2		
AO1 x 12bit	1.3		80..81	AO1 signal board	6ES7 232-4HA30-0XB0	V1.0
HSC_1	1.16			High speed counters (HSC)		
HSC_2	1.17			High speed counters (HSC)		
HSC_3	1.18			High speed counters (HSC)		
HSC_4	1.19			High speed counters (HSC)		
HSC_5	1.20			High speed counters (HSC)		
HSC_6	1.21			High speed counters (HSC)		
Pulse_1	1.32			Pulse generator (PTO/PWM)		
Pulse_2	1.33			Pulse generator (PTO/PWM)		
PROFINET in...	X1			PROFINET interface		
DI8 x 24VDC_1	2	8		SM 1221 DI8 x 24VDC	6ES7 221-1BF30-0XB0	V1.0
DI16 x 24VDC...	3	12..13		SM 1221 DI16 x 24VDC	6ES7 221-1BH30-0XB0	V1.0
DO16 x 24VD...	4		16..17	SM 1222 DO16 x 24VDC	6ES7 222-1BH30-0XB0	V1.0
DO8 x 24VDC...	5		20	SM 1222 DO8 x 24VDC	6ES7 222-1BF30-0XB0	V1.0
AI4 x 13bits /...	6	160..167	160..163	SM 1234 AI4/AO2	6ES7 234-4HE30-0XB0	V1.0
	7					
	8					
	9					

12 - Thoroughly describe each part of the following tag addresses.

I0.6 – Input, byte 0, bit 6

Q2.3 - Output, byte 2, bit 3.

I12.6 - Input, byte 12, bit 6.

QW112 – Output word 112.

IW96 – Input word 96.

MD48 – Memory double word 48.

I8.5:P – Input, Byte 8, bit 5, I/O memory area.

Q2.6:P – Output, byte 2, bit 6, I/O memory area.

IW0 Input word 0.

QW2 Output word 2.

Temp[5] Array named Temp, element 5.

Chapter 3 Questions

2 - What are transitional contacts used for?

An edge instruction records the change in a signal state. A positive (rising) edge is present if a signal changes state from 0 to 1. A negative (falling) edge is present if a signal changes state from 1 to 0. The program records the change of signal state. During the processing the CPU compares the actual result of the logic operation with the saved result of the previous logic evaluation. A signal edge is present if the two signal states (current and saved state) are different.

5 - Explain the term normally-closed.

A normally-closed contact is also called an examine-if-open contact. This type of contact can be confusing at first. A normally-closed contact will pass power until the real-world condition associated with it is energized. A normally-closed contact in a ladder diagram is only energized if the real-world input associated with it is false.

6 - Explain the terms true and false as they apply to contacts in ladder logic.

True would be equivalent to a 1 state or ON. False is equivalent to a 0 or OFF state.

8 - Design a ladder that has parallel input (OR logic). Use I0.2 and I0.7 for the contacts and Q0.6 for the output.

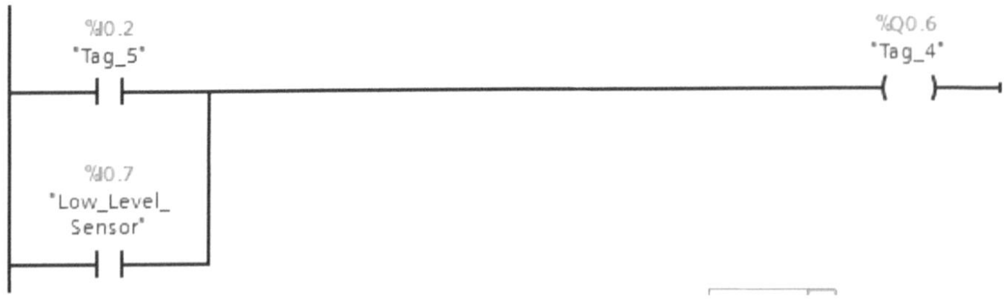

10 - Design a three-input ladder that uses AND logic and OR logic. The input logic should be I0.1 OR I0.3, AND NOT I0.5. Use Q1.2 for the output coil.

12 - What is a RET instruction used for?

The return instruction can be used to prematurely stop the execution of a block.

14 - Out_1 would be ON.

16 - Out_1 would be ON.

18 - Out_1 would be ON.

20. Write ladder logic for the application below. Your logic should have a start/stop circuit to start the application. Your logic should assure that the tank does not run empty nor overflow. Use the I/O names from the table for your logic.

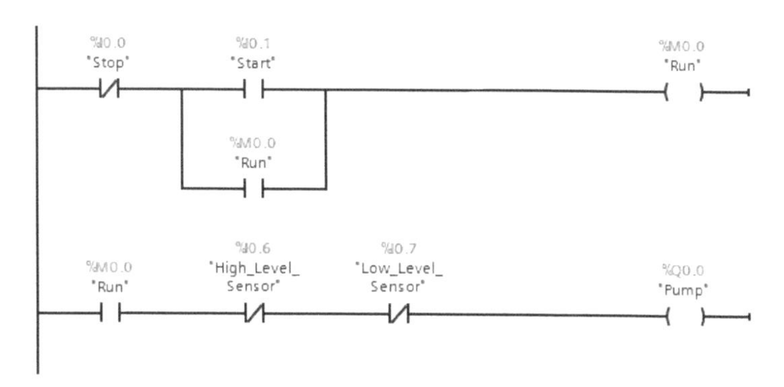

Chapter 4 Questions

2 - What is a TOF timer?

The timer off-delay (TOF) instruction can be used to turn an output coil on or off after the rung has been false for a desired time.

4 - What is the PT used for in a timer?

Timers have a PT time that must be set by the programmer. The PT time can be thought of as the number of time increments the timer must count before changing the state of the output.

6 - Describe two methods of resetting the elapsed time of a TON timer to 0.

To reset a timer's elapsed time to 0 a reset timer (RT) instruction is used. The figure below shows a RT instruction. An RT instruction uses the timer's name.

%DB2
"Timer_1"
[RT]

Timer's elapsed time can also be reset by de-energizing the IN input of the timer. The figure below shows an example in which the timer done bit (Part_Timer.Q) can be used to reset the timer ET to zero every time the timer elapsed time reaches the PT. This would turn Timer_1.Q on for one scan. This can be used to make things happen at regular intervals. For example, this could be used to execute an instruction or some logic every 5 seconds.

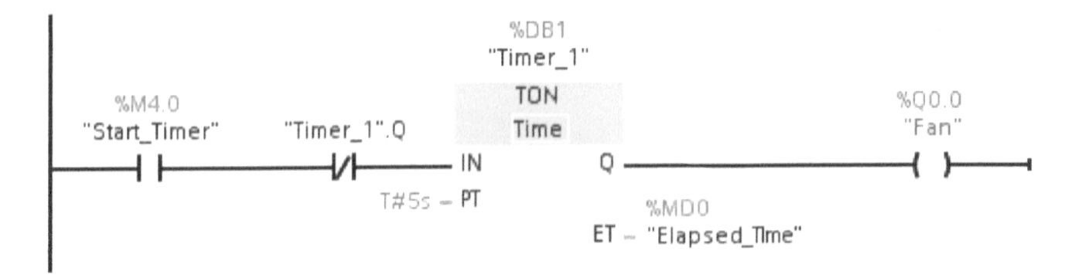

8 - Give an example of how the Q bit for a TOF timer could be used.

The figure below shows the use of a TOF timer. When the Cycle_Start contact becomes true, the timer Q bit becomes true. Note that Cycle_Time.Q is on, turning the output (Fan) on immediately. The timer's Q bit will stay on forever if the Cycle_Start input remains true. When Cycle_Start becomes false, it starts the timer timing cycle. When the rung goes false the time begins accumulating the elapsed time. When the elapsed time equals the preset, the Q bit will become false.

10 - In what way are counters and timers very similar?

Timers essentially count time increments.

12 - What is a CTD instruction?

Count-down instruction

14 - How can the accumulated count be reset in a counter?

The reset input on a counter or a RES instruction can be used.

16 - Write ladder logic for the following application.

This is a simple heat treat machine application. The operator places a part in a fixture then pushed the start switch. An inductive heating coil heats the part rapidly to 1500 degrees Fahrenheit. When the temperature reaches 1500 a discrete sensor's output becomes true. The coil turns off and a valve is opened which sprays water on the part to complete the heat treatment (quench). The operator then removes the part and the sequence can begin again. Note there must be a part present or the sequence should not start.

I/O	Type	Description
Part_Presence_Sensor	Discrete	Sensor used to sense a part in the fixture.
Temp_Sensor	Discrete	For simplicity assume this sensor's output becomes true when the temperature reaches 1500 degrees Fahrenheit.
Start_Switch	Discrete	Momentary normally-open switch.
Heating_Coil	Discrete	Discrete output that turns coil on.
Quench_Valve	Discrete	Discrete output that turns quench valve on.

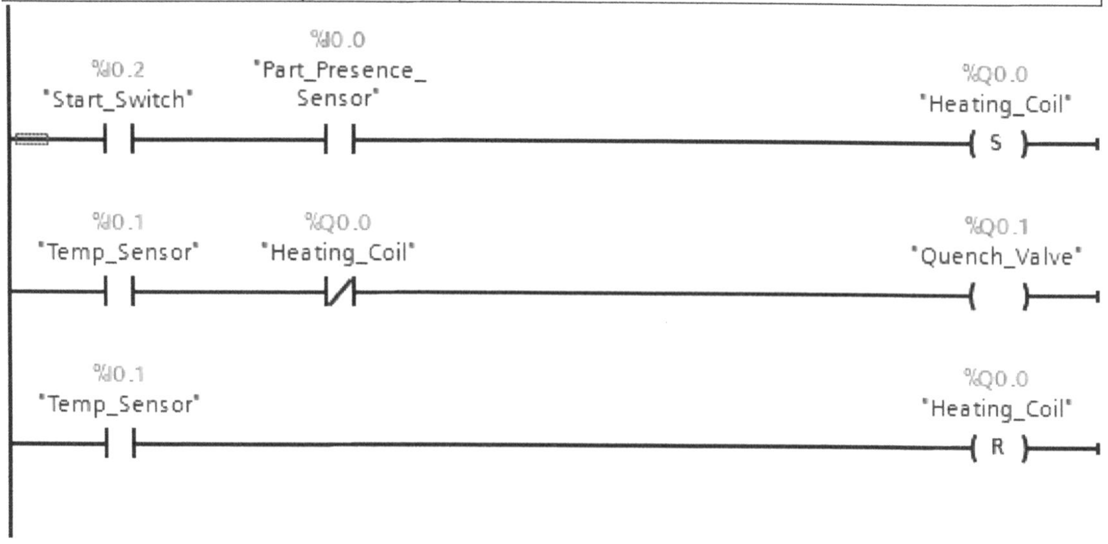

Chapter 5 Questions

2 - What is the difference between a signal board and a signal module?

Signal boards are plugged into the front of a S7-1200 PLC. Signal modules are individual I/O modules that are attached to right side of the CPU in a chassis or though the serial backplane bus.

4 - Look up the manual for a Siemens SM 1222 DQ 8 x 24 VDC (S7-1200) signal module or a SM 322; DO 16 x DC 24 V (S7-300) signal module on the Siemens website and draw a wiring diagram for 6 outputs.

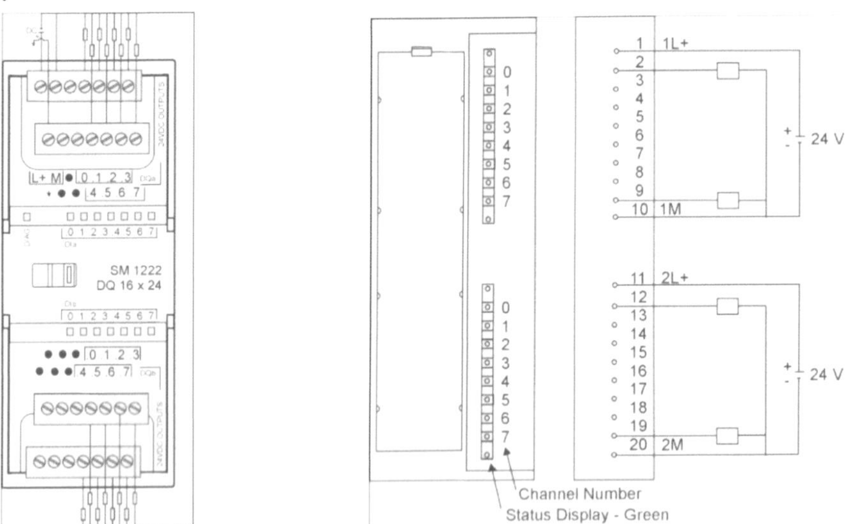

6 - Explain the term resolution as it applies to analog modules.

Resolution has to do with how closely something can be measured. Imagine a ruler. If the only graduations on the ruler were inches, the resolution would be 1 in. If the graduations were every 1/8 in., the resolution would be a 1/8 in. The closest we could measure any object would be l/8 inch. The CPU in a PLC only works with digital information. The analog-to-digital (A/D) card changes the analog source into discrete steps. The higher the resolution, the finer the measurement. Another way to think of resolution is in terms of a pie. If you have people over for thanksgiving the pie will be divided based on the number of people. The pie represents what we are measuring (maybe 0-10 volts) the number of people represents the size of each piece of pie. So, the higher the number of people (bits of resolution) the smaller each piece is.

Resolution is the smallest amount of change that a module can detect. Analog modules are available in different resolutions. Output modules are typically available in 12 to 16-bit resolution. A 16-bit module would have 65,536 counts. This can be calculated by raising 2 to the number of bits the module has. For example, a 16-bit module would be 2^{16} or 65,536.

8 - A 16-bit analog current input module (see Figure 5-34) is used to measure the level in a tank. The tank can hold between 0 and 8 feet of fluid. The sensor outputs 0 mA at 1 foot and 6 mA at 8 feet. What is the resolution in inches in this application?

.003 inches (84 inches / 27648)

10. What is scaling?

Scaling is used to change a quantity from one notation to another. Scaling is only available with the floating-point data format in CL modules. When a channel is scaled, two points along the module's operating range are chosen and low and high values are applied to the points.

You can use function call FC105 SCALE (scale values) and function call FC106 UNSCALE (unscale values) blocks to read and output analog values in STEP 7. These function calls are available in the STEP 7 standard library, in the TI-S7-Converting Blocks subfolder.

12. An S7-1200 SM 1234 AI 4 x 13 bit / AQ 2 x 14 bit (or you may use an S7-300 SM 332; AO 4 x 12 signal module for this question) will be used to output a 0-10 VDC output to control a valve. Look up the module to find its resolution. Calculate the counts that would be used to output 0 VDC and +10 VDC. Calculate the resolution in volts/count.

Answer is dependent on which module is chosen.

Chapter 6 Questions

2 - Why might a programmer use an instruction that would change a number to a different number system?

The numbers in the PLC are normally in binary. We might want to convert a binary number to a decimal or hex number for an application.

4 - Explain the following logic.

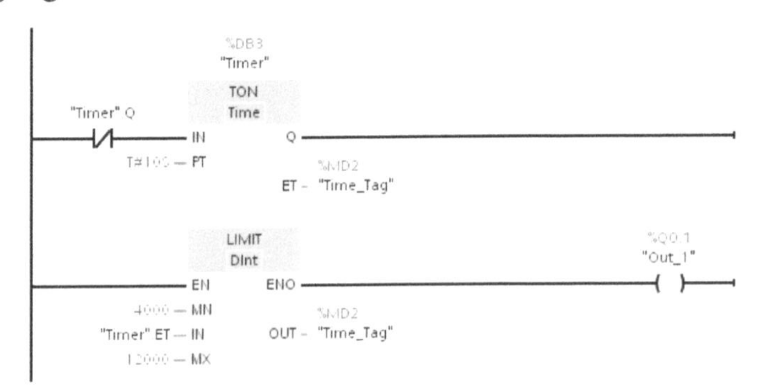

A ten second timer is being automatically reset by its Q-bit in the first rung. In the second rung a LIMIT instruction is used to turn on a tag named "Time_Tag" between 4 seconds and 12 seconds. Note that the timer only times up to 10 seconds, so in this example the output will only be on between 4 seconds and 10 seconds.

6. Write a rung of logic that checks to see if one value is equal to or greater than a second value. Turn on an output if true.

%MD14
"Num_5"
>=
DInt
%MD10
"Num_2"
%Q0.6
"Tag_1"

8 - Write a rung of logic to check if a value is less than 75 or greater than 100 or equal to 85. Turn on an output if the statement is true.

%MD14
"Num_5"
<
DInt
75
%Q0.6
"Tag_1"
%MD14
"Num_5"
>
DInt
100
%MD14
"Num_5"
==
DInt
85

10 - Write a ladder diagram program to accomplish the following. A tank level must be maintained between two levels. An ultrasonic sensor is used to measure the height of the fluid in the tank. The output from the ultrasonic sensor is 0 to 10 volts. This directly relates to a tank level of 0 to 5 feet. It is desired that the level be maintained between 4.0 and 4.2 feet. Output 1 is the inflow valve. The sensor output is an analog input to an analog input module. Utilize math comparison instructions to write the logic. Calculate the correct analog counts for the instruction. Use the tagnames shown in the table.

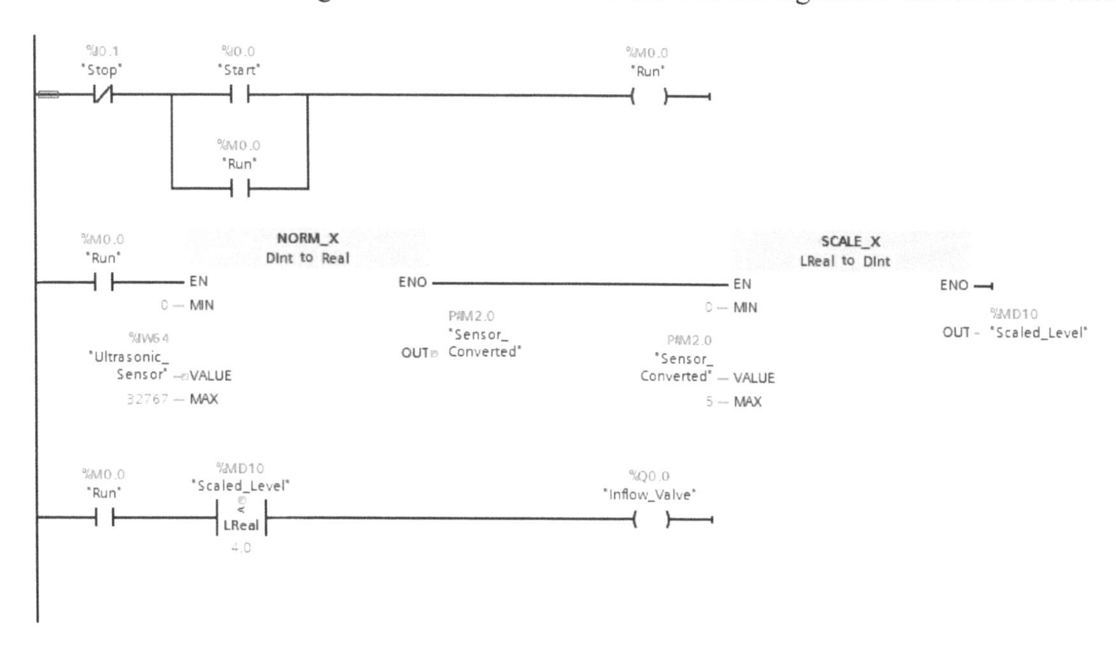

12. Write a ladder logic routine for the following application. Utilize comparison instructions.

This is a simple heat treat machine application. The operator places a part in a fixture then pushes the start switch. An inductive heating coil heats the part rapidly to 1500 degrees Fahrenheit. When the temperature reaches 1500 the coil turns off and a valve is opened which sprays water on the part to complete the heat treatment (quench). The operator then removes the part and the sequence can begin again. Note there must be a part present or the sequence should not start.

I/O	Type	Description
Part_Present_Sensor	Discrete	Sensor used to sense a part in the fixture.
Temp_Sensor	Analog	Assume this sensor outputs 0-2000 degrees.
Start_Switch	Discrete	Momentary normally-open switch.
Heating_Coil	Discrete	Discrete output that turns coil on.
Quench_Valve	Discrete	Discrete output that turns quench valve on.

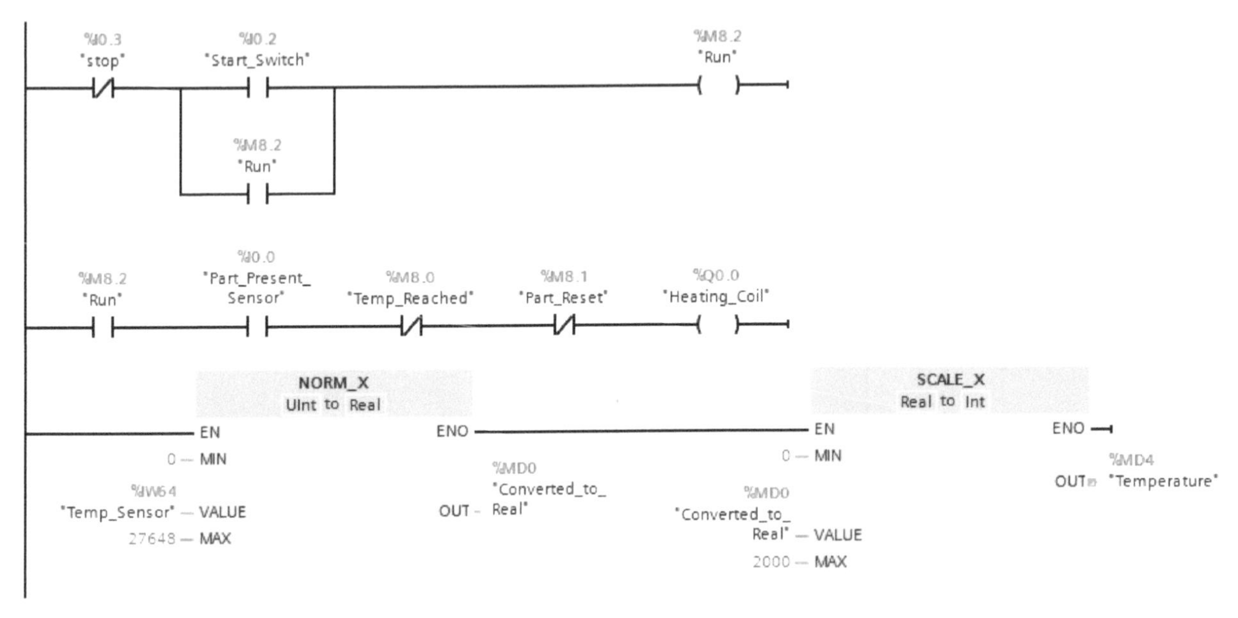
%I0.3
"stop"
%I0.2
"Start_Switch"
%M8.2
"Run"
%M8.2
"Run"
%M8.2
"Run"
%I0.0
"Part_Present_
Sensor"
%M8.0
"Temp_Reached"
%M8.1
"Part_Reset"
%Q0.0
"Heating_Coil"
NORM_X
UInt to Real
EN
ENO
0 — MIN
%IW64
"Temp_Sensor" — VALUE
27648 — MAX
OUT
%MD0
"Converted_to_
Real"
SCALE_X
Real to Int
EN
ENO
0 — MIN
%MD0
"Converted_to_
Real" — VALUE
2000 — MAX
OUT
%MD4
"Temperature"

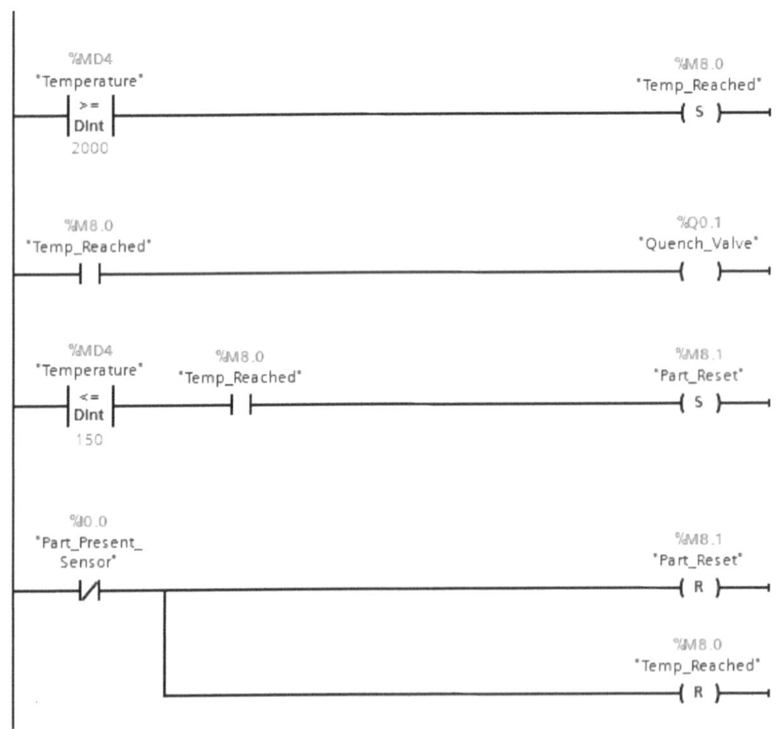
%MD4
"Temperature"
>=
DInt
2000
%M8.0
"Temp_Reached"
S
%M8.0
"Temp_Reached"
%Q0.1
"Quench_Valve"
%MD4
"Temperature"
<=
DInt
150
%M8.0
"Temp_Reached"
%M8.1
"Part_Reset"
S
%I0.0
"Part_Present_
Sensor"
%M8.1
"Part_Reset"
R
%M8.0
"Temp_Reached"
R

Chapter 7 Questions

2 - What instruction could be used to move an integer in memory to an output module?

A MOVE value instruction can be used to transfer the value of the tag or constant at the IN input to the tag or memory address at the OUT1 output. The transfer is always made in ascending address order.

4 - Explain an RS instruction.

The reset/set flip-flop instruction can be used to reset or set based on the signal state of the inputs at R and S1 (see Figure 7-15). If the signal state at input R is 1 and is 0 at input S1, the specified operand is reset to 0. If the signal state at input R is 0 and is 1 at input S1, the specified operand is set to 1.

Input S1 takes priority over input R. If the signal state is 1 at inputs R and S1, the signal state of the specified operand is set to 1.

If the signal state at the two inputs R and S1 is 0, the instruction is not executed. The state of the output (Q) then remains unchanged.

Bit memory M 0.0 (Output_Bit) and output Q 0.0 (Output_5) are reset when the following conditions are fulfilled: Inp_1 is true, Inp_2 is false.

Tags Output_Bit and Output_5 are set when the following conditions are fulfilled: Input Inp_1 is false and Inp_2 is true, both inputs Inp_1 and Inp_2 are true.

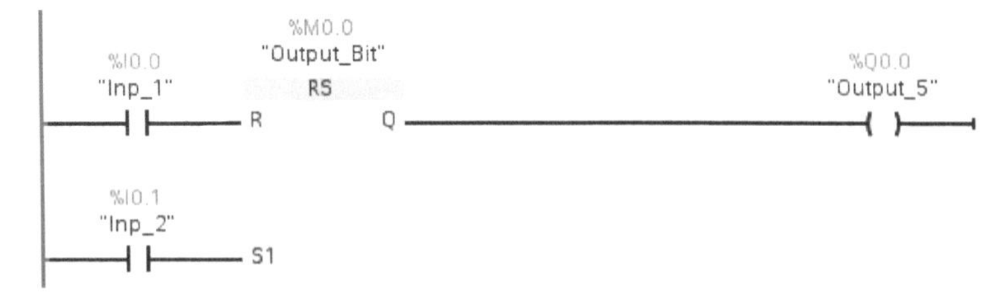

6 - What does a MUL instruction do?

The multiplex instruction is used to copy the content of a selected input to the OUT output. The number of selectable inputs of the MUX box can be expanded. The inputs are automatically numbered in the box. Numbering starts at IN0 and is incremented continuously with each new input. You can use the K parameter to determine the input whose content should be copied to the OUT output. If the value of the K parameter is greater than the number of available inputs, the content of the ELSE parameter is copied to the OUT output and the enable output ENO is assigned signal state 0.

8 - How can process values be saved?

Data logging instructions can be used to save process values to data logs. Data logs can be saved in the internal load memory or on a SD (Secure Digital) memory card (see Figure 7-29). The data logs are saved in Comma Separated Value (CSV) format.

There are data logging instructions to create or open a data log, to write an entry and to close the data log file. The programmer chooses which values are logged when the data buffer is created. The data buffer is used as a memory location for new data log entries. New values must be written to the buffer before the DataLogWrite instruction is executed. During the execution of the DataLogWrite instruction, data is written from the buffer into a data log record.

Chapter 8 Questions

2 - What is the difference in the exact same logic written in either LAD or FBD?

The only difference is the visual representation of the logic. The same input condition result in the same logical output results.

4 - Is it best to put all logic in one Network location?

No, it is best to group related logic that corresponds to machine devices in separated networks. This makes it easier to find and edit logic.

Chapter 9 Questions

2 - What does PID stand for?

Proportional, integral, and derivative.

4 - What are the two types of outputs available in a PID compact instruction?

Output_PER and Output_PWM. An Output_PER output uses an analog output as the manipulated variable output. An Output_PWM output uses a digital switching output and controls it by means of a pulse width modulation. The result at the PWM output is variable turn-on and turn-off times.

6 - Describe what the integral portion of PID control corrects.

The I gain is the integral gain. The integral gain is used to correct for small errors that persist over time. The proportional gain cannot correct for very small errors. The integral gain is used to correct for these small errors over time.

8 - How can the initial gains settings for PID be established?

The PID technology object can calculate the proportional, integral, and derivative gains automatically by self-tuning during the initial start. The parameters can be optimized further by means of a self-tuning in the operating point.

Chapter 10 Questions

2 - What is the function of the enable output?

A drive must get an input from a controller (PLC) to enable the drive to operate. The enable output from the PLC will be used for this. This is where you specify which output will be used. After the drive receives the enable input it will turn on an output that is used to tell the PLC that the drive is ready to be operated.

4 - Describe what software limits are and how they function.

If the software limits are to be used the user sets a low and a high number (position) that will act as software limts. Software limits are set so that the drive will run into the software limits before it would hit the hardware limits. If we would command the drive to move outside of the software limits it will not make the move.

6 - What are the basic steps that must be completed before an axis can be controlled with motion instructions?

The usual sequence to enabling and using an axis would be to use an MC_Power instruction to enable the axis for motion followed by an MC_Home instruction to establish a home position for the axis. Once this has been done the other axis commands can be used.

8 - Describe the 4 modes of homing available in a MC_Home instruction.

Active homing: the homing procedure is executed automatically.

Passive homing: the homing procedure must be executed by the user.

Direct homing absolute: the home position is set in absolute terms.

Direct homing relative: the home position is set relative to the current position.

10 - What instruction can be used to acknowledge axis errors?

The MC_Reset instruction is used to acknowledge axis errors.

Chapter 11 Questions

2 - HMIs are only useful to operators running the equipment?

False, HMIs are useful to both operators and maintenance personal. Operators use specific information from HMIs to run machinery. Maintenance personal use HMI information such as fault diagnostics and other machine status information to troubleshoot and maintain machine operations.

4 - It is very difficult to browse for tags in the PLC while configuring the HMI?
False, under the HMI device within the project, any associated PLC's tags can be browsed to from the HMI configuration.

6 - Alarms can monitor increasing/decreasing values to trigger alarm conditions. These are known as analog alarms?

True, in many instances monitoring changing analog values can provide early detection to conditions before they become a real problem.

8 - Downloading can only be done offline?

False, downloading can be done while in online or offline mode.

ABOUT THE AUTHORS

David Deeg has worked in a variety of industrial settings. David worked as a Controls Designer and Field Service Technician for Trident Automation. During his time with Trident, he traveled the country working on ethanol plant distributed control systems (DCS), boiler house user interfaces and controls, corn grain handling systems, and iron ore ships on the Great Lakes. After working for Trident Automation, David worked at Neenah Foundry. Neenah Foundry makes the majority of man-hole covers, iron grates surrounding trees on urban sidewalks. In addition to the municipal castings, Neenah produces many of the power transmission housings found in large tractor trailer semis on the road today. While at Neenah Foundry David implemented new control techniques not previously used there. David developed control and operator information systems that made it easier for operators and maintenance personnel to troubleshoot and adapt to changing production needs. After working on almost every major automation system at Neenah, David took a teaching position in the Automated Manufacturing Systems (AMS) program at Fox Valley Technical College. David enjoys the challenging work industrial controls and programming provides. As an instructor he gets to share his industrial experience in preparing students with the tools they will need to be successful.

Jon Stenerson is the author of several books on PLCs and industrial automation. He taught in the Automated Manufacturing Systems Technology program at Fox Valley Technical College. There he taught regular program students and students from industry. He has also taught and consulted in business and industry.

Lightning Source UK Ltd.
Milton Keynes UK
UKHW050044210522
403325UK00002B/98
* 9 7 8 1 5 1 5 0 3 6 5 7 9 *